Lecture Notes in Computer Scie

FoLLI Publications on Logic, Language and Information

Subline of Lectures Notes in Computer Science

Lecture Notes in Computer Science 7395

Commenced Publication in 1973

Founding and Former Series Editors:

Gerhard Goos, Juris Hartmanis, and Jan van Leeuwen

Editorial Board

FoLLI Publications on Logic, Language and Information

Subline of Lectures Notes in Computer Science

Subline Editors-in-Chief

Subline Area Editors

Philippe de Groote   Mark-Jan Nederhof (Eds.)

Formal Grammar

15th and 16th International Conferences
FG 2010, Copenhagen, Denmark, August 2010
FG 2011, Ljubljana, Slovenia, August 2011
Revised Selected Papers

Volume Editors

Philippe de Groote
INRIA Nancy - Grand Est
615 rue du Jardin Botanique
54600 Villers-ls-Nancy, France
E-mail: degroote@loria.fr

Mark-Jan Nederhof
University of St. Andrews
School of Computer Science
North Haugh, St. Andrews
Fife, KY16 9SX, Scotland
E-mail: mjn@cs.st-andrews.ac.uk

ISSN 0302-9743 e-ISSN 1611-3349
ISBN 978-3-642-32023-1 e-ISBN 978-3-642-32024-8
DOI 10.1007/978-3-642-32024-8
Springer Heidelberg Dordrecht London New York

Library of Congress Control Number: 2012942344

CR Subject Classification (1998): F.4.1-3, I.1, I.2.7

LNCS Sublibrary: SL 1 – Theoretical Computer Science and General Issues

Typesetting: Camera-ready by author, data conversion by Scientific Publishing Services, Chennai, India

Printed on acid-free paper

Springer is part of Springer Science+Business Media (www.springer.com)

Preface

The Formal Grammar conference series provides a forum for the presentation of new and original research on formal grammar, mathematical linguistics, and the application of formal and mathematical methods to the study of natural language.

FG 2010, the 15th conference on Formal Grammar, was held in Copenhagen, Denmark, during August 7–8, 2010. The conference consisted in 14 contributed papers (selected out of 34 submissions), and one invited talk by Mats Rooth.

FG 2011, the 16th conference on Formal Grammar, was held in Ljubljana, Slovenia, during August 6–7, 2011. The conference consisted in six contributed papers (selected out of 16 submissions), and two invited talks by Thomas Ede Zimmermann and Jeroen Groenendijk.

We would like to thank the people who made the 15th and 16th FG conferences possible: the three invited speakers, the members of the Program Committees, and the members of the ESSLLI 2010 and ESSLLI 2011 Organizing Committees.

May 2012

Philippe de Groote
Mark-Jan Nederhof

FG 2010 Organization

Program Committee

Wojciech Buszkowski	Adam Mickiewicz University, Poland
Berthold Crysmann	University of Bonn, Germany
Alexandre Dikovsky	University of Nantes, France
Denys Duchier	University of Orléans, France
Annie Foret	IRISA - University of Rennes 1, France
Nissim Francez	Technion, Israel
Makoto Kanazawa	National Institute of Informatics, Japan
Stephan Kepser	Codecentric AG, Germany
Valia Kordoni	Saarland University, Saarbrücken, Germany
Marco Kuhlmann	Uppsala University, Sweden
Glyn Morrill	UPC, Barcelona Tech, Spain
Stefan Müller	Free University of Berlin, Germany
Gerald Penn	University of Toronto, Canada
Frank Richter	University of Tübingen, Germany
Manfred Sailer	University of Göttingen, Germany
Edward Stabler	UCLA, USA
Hans-Jörg Tiede	Illinois Wesleyan University, USA
Jesse Tseng	CNRS - CLLE-ERSS, France
Shuly Wintner	University of Haifa, Israel

Standing Committee

Philippe de Groote	INRIA Nancy - Grand Est, France
Markus Egg	Humboldt-Universität Berlin, Germany
Laura Kallmeyer	University of Tübingen, Germany
Mark-Jan Nederhof	University of St. Andrews, UK

FG 2011 Organization

Program Committee

Berthold Crysmann	University of Bonn, Germany
Alexandre Dikovsky	University of Nantes, France
Denys Duchier	University of Orléans, France
Annie Foret	IRISA - University of Rennes 1, France
Nissim Francez	Technion, Israel
Laura Kallmeyer	University of Düsseldorf, Germany
Makoto Kanazawa	National Institute of Informatics, Japan
Stephan Kepser	University of Tübingen, Germany
Valia Kordoni	Saarland University, Saarbrücken, Germany
Marco Kuhlmann	Uppsala University, Sweden
Glyn Morrill	UPC, Barcelona Tech, Spain
Stefan Müller	Free University of Berlin, Germany
Gerald Penn	University of Toronto, Canada
Christian Retoré	University of Bordeaux 1, France
Manfred Sailer	University of Göttingen, Germany
Edward Stabler	UCLA, USA
Anders Søgaard	University of Copenhagen, Denmark
Jesse Tseng	CNRS - CLLE-ERSS, France

Standing Committee

Philippe de Groote	INRIA Nancy - Grand Est, France
Markus Egg	Humboldt-Universität Berlin, Germany
Mark-Jan Nederhof	University of St. Andrews, UK
Frank Richter	University of Tübingen, Germany

Table of Contents

Formal Grammar 2010: Contributed Papers

Formal Grammar 2011: Contributed Papers

Polarized Montagovian Semantics for the Lambek-Grishin Calculus

Arno Bastenhof
Utrecht University

Abstract. Grishin ([9]) proposed enriching the Lambek calculus with multiplicative disjunction (par) and coresiduals. Applications to linguistics were discussed by Moortgat ([14]), who spoke of the Lambek-Grishin calculus (**LG**). In this paper, we adapt Girard's polarity-sensitive double negation embedding for classical logic ([7]) to extract a compositional Montagovian semantics from a display calculus for focused proof search ([1]) in **LG**. We seize the opportunity to illustrate our approach alongside an analysis of extraction, providing linguistic motivation for linear distributivity of tensor over par ([3]), thus answering a question of [10]. We conclude by comparing our proposal to that of [2], where alternative semantic interpretations of **LG** are considered on the basis of call-by-name and call-by-value evaluation strategies.

Inspired by Lambek's syntactic calculus, Categorial type logics ([13]) aim at a proof-theoretic explanation of natural language syntax: syntactic categories and grammaticality are identified with formulas and provability. Typically, they show an intuitionistic bias towards asymmetric consequence, relating a structured configuration of hypotheses (a constituent) to a single conclusion (its category). The *Lambek-Grishin calculus* (**LG**, [14]) breaks with this tradition by restoring symmetry, rendering available (possibly) multiple conclusions. §1 briefly recapitulates material on **LG** from [14] and [15].

In this article, we couple **LG** with a Montagovian semantics.[1] Presented in §2, its main ingredients are focused proof search [1] and a double negation translation along the lines of [7] and [19], employing *polarities* to keep the number of negations low. In §3, we illustrate our semantics alongside an analysis of extraction inspired by linear distributivity principles ([3]). Finally, §4 compares our approach to the competing proposal of [2].

1 The Lambek-Grishin Calculus

Lambek's (non-associative) syntactic calculus ((**N**)**L**, [11], [12]) combines linguistic inquiry with the mathematical rigour of proof theory, identifying syntactic categories and derivations by formulas and proofs respectively. On the logical side, (**N**)**L** has been identified as (non-associative,)non-commutative multiplicative intuitionistic linear logic, its formulas generated as follows:

[1] Understanding *Montagovian semantics* in a broad sense, we take as its keywords *model-theoretic* and *compositional*. Our emphasis in this article lies on the latter.

P. de Groote and M.-J. Nederhof (Eds.): Formal Grammar 2010/2011, LNCS 7395, pp. 1–16, 2012.

$$\begin{array}{lr} A..E ::= p & \text{Atoms/propositional variables} \\ \quad \mid (A \otimes B) & \text{Multiplicative conjunction/tensor} \\ \quad \mid (B\backslash A) \mid (A/B) & \text{Left and right implication/division} \end{array}$$

Among **NL**'s recent offspring we find the Lambek-Grishin calculus (**LG**) of [14], inspired by Grishin's ([9]) extension of Lambek's vocabulary with a multiplicative disjunction (*par*) and *coimplications/subtractions.* Combined with the recent addition of (co)negations proposed in [15], the definition of formulas now reads

$$\begin{array}{lr} A..E ::= p & \text{Atoms} \\ \quad \mid (A \otimes B) \mid (A \oplus B) & \text{Tensor vs. par} \\ \quad \mid (A/B) \mid (B \oslash A) & \text{Right division vs. left subtraction} \\ \quad \mid (B\backslash A) \mid (A \oslash B) & \text{Left division vs. right subtraction} \\ \quad \mid {}^0B \mid B^1 & \text{Left negation vs. right conegation} \\ \quad \mid B^0 \mid {}^1B & \text{Right negation vs. left conegation} \end{array}$$

Derivability ($\leq$) satisfies the obvious preorder laws:

$$\frac{}{A \leq A}\, Refl \qquad \frac{A \leq B \quad B \leq C}{A \leq C}\, Trans$$

Logical constants group into families with independent algebraic interest. The connectives $\{\otimes, /, \backslash\}$ constitute a *residuated family* with *parent* $\otimes$, while $\{\oplus, \oslash, \oslash\}$ embodies the dual concept of a coresiduated family. Finally, $\{{}^0\cdot, \cdot^0\}$ and $\{\cdot^1, {}^1\cdot\}$ represent Galois-connected and dually Galois-connected pairs respectively.[2]

$$\begin{array}{c} B \leq A\backslash C \\ \hline\hline A \otimes B \leq C \\ \hline\hline A \leq C/B \end{array}\, r \qquad \begin{array}{c} C \oslash B \leq A \\ \hline\hline C \leq A \oplus B \\ \hline\hline A \oslash C \leq B \end{array}\, cr \qquad \begin{array}{c} A \leq {}^0B \\ \hline\hline B \leq A^0 \end{array}\, gc \qquad \begin{array}{c} A^1 \leq B \\ \hline\hline {}^1B \leq A \end{array}\, dgc$$

Among the derived rules of inference, we find the monotonicity laws:

$$\frac{A \leq B \quad C \leq D}{\begin{array}{c} A \otimes C \leq B \otimes D \\ A/D \leq B/C \\ D\backslash A \leq C\backslash B \end{array}}\, m \qquad \frac{A \leq B \quad C \leq D}{\begin{array}{c} A \oplus C \leq B \oplus D \\ A \oslash D \leq B \oslash C \\ D \oslash A \leq C \oslash B \end{array}}\, m \qquad \frac{A \leq B}{\begin{array}{c} {}^0B \leq {}^0A \\ B^0 \leq A^0 \end{array}}\, m \qquad \frac{A \leq B}{\begin{array}{c} B^1 \leq A^1 \\ {}^1B \leq {}^1A \end{array}}\, m$$

LG differs most prominently from **NL** in the existence of an *order-reversing duality*: an involution $\cdot^\infty$ on formulas s.t. $A \leq B$ iff $B^\infty \leq A^\infty$, realized by[3]

$$\frac{p \quad A \otimes B \quad A/B \quad B\backslash A \quad {}^0B \quad B^0}{p \quad B \oplus A \quad B \oslash A \quad A \oslash B \quad B^1 \quad {}^1B}\, \infty$$

A reasonable way of extending **LG** would be to allow connectives of different families to interact. Postulates licensing linear distributivity of $\otimes$ over $\oplus$ ([3], [6]) come to mind, each being self-dual under $\cdot^\infty$ (thus preserving arrow-reversal):

[2] Throughout this text, a double line indicates derivability to go in both ways. Similarly, in (m) below, the same premises are to be understood as deriving each of the inequalities listed under the horizontal line.

[3] The present formulation, adopted from [15], abbreviates a list of defining equations $(A \otimes B)^\infty = B^\infty \oplus A^\infty$, $(B \oplus A)^\infty = A^\infty \otimes B^\infty$, etc.

$$(A \oplus B) \otimes C \leq A \oplus (B \otimes C) \quad A \otimes (B \oplus C) \leq (A \otimes B) \oplus C$$
$$(A \oplus B) \otimes C \leq (A \otimes C) \oplus B \quad A \otimes (B \oplus C) \leq B \oplus (A \otimes C)$$

§3 further explores the relation **LG**/linguistics, providing as a case analysis for our Montagovian semantics of §2 a sample grammar providing linguistic support for the above linear distributivity principles. As for their proof-theoretic motivation, we note that the following generalizations of Cut become derivable:

$$\frac{A \leq E \oplus B \quad B \otimes D \leq C}{A \otimes D \leq E \oplus C} \qquad \frac{A \leq B \oplus E \quad D \otimes B \leq C}{D \otimes A \leq C \oplus E}$$

$$\frac{A \leq B \oplus E \quad B \otimes D \leq C}{A \otimes D \leq C \oplus E} \qquad \frac{A \leq E \oplus B \quad D \otimes B \leq C}{D \otimes A \leq E \oplus C}$$

For example, suppose $A \leq E \oplus B$ and $B \otimes D \leq C$. From monotonicity and linear distributivity we then deduce

$$A \otimes D \leq (E \oplus B) \otimes D \leq E \oplus (B \otimes D) \leq E \oplus C$$

2 Derivational Montagovian Semantics

We split our semantics into a *derivational* and a *lexical* component. The former is hard-wired into the grammar architecture and tells us how each inference rule builds the denotation of its conclusion (the derived constituent) from those of its premises (the direct subconstituents). The descriptive linguist gives the base of the recursion: the lexical semantics, specifying the denotations of words. Leaving lexical issues aside until §3, we define a λ-term labeled sequent calculus for simultaneously representing the proofs and the derivational semantics of **LG**. Summarized in Figure 3, its main features (and primary influences) are as follows:

1. It is, first and foremost, a *display calculus* along the lines of [8] and [14]. In particular, the notion of sequent is generalized so as to accommodate structural counterparts for each of the logical connectives. Display postulates then allow to pick out any hypothesis (conclusion) as the whole of the sequent's antecedent (consequent).
2. Our sequents are labeled by linear λ-terms for representing compositional meaning construction, adapting the polarity-sensitive double negation translations of [7] and [19].
3. We fake a one(/left)-sided sequent presentation to more closely reflect the target calculus of semantic interpretation. To this end, we adapt to our needs de Groote and Lamarche's one(/right)-sided sequents for classical **NL** ([5]).
4. In contrast with the works cited, our inference rules accommodate focused proof search ([1]), thus eliminating to a large extent the trivial rule permutations for which sequent calculi are notorious.

We proceed step by step, starting with a specification of the target language for the double negation translation.

$$\frac{}{\tau^x \vdash x : \tau} Ax$$

$$\frac{\Gamma \vdash M : \neg\tau \quad \Delta \vdash N : \tau}{\Gamma, \Delta \vdash (M\ N) : \bot} \neg E \qquad \frac{\Gamma, \tau^x \vdash M : \bot}{\Gamma \vdash \lambda x^\tau M : \neg\tau} \neg I$$

$$\frac{\Delta \vdash N : \sigma_1 \otimes \sigma_2 \quad \Gamma, \sigma_1^x, \sigma_2^y \vdash M : \tau}{\Gamma, \Delta \vdash (\mathbf{case}\ N\ \mathbf{of}\ x^{\sigma_1} \otimes y^{\sigma_2}.M) : \tau} \otimes E \qquad \frac{\Gamma \vdash M : \tau \quad \Delta \vdash N : \sigma}{\Gamma, \Delta \vdash \langle M \otimes N \rangle : \tau \otimes \sigma} \otimes I$$

$$\begin{array}{rcl}
(\lambda x M\ N) & \to_\beta & M[N/x] \\
\mathbf{case}\ \langle N_1 \otimes N_2 \rangle\ \mathbf{of}\ x \otimes y.M & \to_\beta & M[N_1/x, N_2/y] \\
((\mathbf{case}\ M_1\ \mathbf{of}\ x \otimes y.M_2)\ N) & \to_c & \mathbf{case}\ M_1\ \mathbf{of}\ x \otimes y.(M_2\ N) \\
\mathbf{case}\ (\mathbf{case}\ N_1\ \mathbf{of}\ x \otimes y.N_2)\ \mathbf{of}\ u \otimes v.M & \to_c & \mathbf{case}\ N_1\ \mathbf{of}\ x \otimes y.\mathbf{case}\ N_2\ \mathbf{of}\ u \otimes v.M
\end{array}$$

Fig. 1. Target language: typing rules and reductions. The c-conversions correspond to the obligatory commutative conversions of Prawitz ([17]).

Target Language. Instructions for meaning composition will be phrased in the linear λ-calculus of Figure 1, simply referred to as **LP**. Note that, through the Curry-Howard isomorphism, we may as well speak of a Natural Deduction presentation of multiplicative intuitionistic linear logic. Types τ, σ include multiplicative products $\tau \otimes \sigma$ and minimal negations $\neg\tau$, the latter understood as linear implications $\tau \multimap \bot$ with result a distinguished atom $\bot$:

$$\tau, \sigma ::= p \mid \bot \mid (\tau \otimes \sigma) \mid \neg\tau$$

We have understood **LP** to inherit all atoms p of **LG**. Terms M are typed relative to contexts Γ, Δ: multisets $\{\tau_1^{x_1}, \ldots, \tau_n^{x_n}\}$ of type assignments $\tau_1, \ldots, \tau_n$ to the free variables $x_1, \ldots, x_n$ in M. We often omit the braces { } and loosely write Γ, Δ for multiset union. Terms in context are represented by *sequents* $\Gamma \vdash M : \tau$, satisfying the linearity constraint that each variable in Γ is to occur free in M exactly once. We often write $\Gamma \vdash_{\mathbf{LP}} M : \tau$ to indicate $\Gamma \vdash M : \tau$ is well-typed.

From Formulas to Types: Introducing Polarities. In defining the type $[\![A]\!]$ associated with a formula A, we will parameterize over the following partitioning of non-atomic formulas, speaking of a positive/negative *polarity*:

Positive(ly polar):	$A \otimes B, A \oslash B, B \obslash A, A^1, {}^1A$	(Metavariables P, Q)
Negative(ly polar):	$A \oplus B, B \backslash A, A/B, {}^0A, A^0$	(Metavariables K, L)

Notice that the dual of a positive formula under $\cdot^\infty$ is negative and vice versa, motivating our choice of terminology. In other words, through order reversal, a positive formula on one side of the inequality sign has a negative counterpart behaving alike on the other side. In fact, we shall see that all positive formulas share proof-theoretic behavior, as do all negative formulas.

We define $[\![A]\!]$ relative to the polarities $\epsilon(\overrightarrow{B})$ of A's direct subformulae $\overrightarrow{B}$ (+ if positive, − if negative). Roughly, a connective expects its polarity to be preserved

Table 1. Interpreting **LG**'s formulas by **LP**'s types

$\epsilon(A)$	$\epsilon(B)$	$[\![A \otimes B]\!]$	$[\![B \obslash A]\!]$ $[\![A \oslash B]\!]$	$[\![A \oplus B]\!]$	$[\![A/B]\!]$ $[\![B\backslash A]\!]$	$[\![{}^0B]\!]$ $[\![B^0]\!]$	$[\![B^1]\!]$ $[\![{}^1B]\!]$
$-$	$-$	$\neg[\![A]\!] \otimes \neg[\![B]\!]$	$\neg[\![A]\!] \otimes [\![B]\!]$	$\underline{[\![A]\!] \otimes [\![B]\!]}$	$[\![A]\!] \otimes \neg[\![B]\!]$	$\neg[\![B]\!]$	$\underline{[\![B]\!]}$
$-$	$+$	$\neg[\![A]\!] \otimes [\![B]\!]$	$\neg[\![A]\!] \otimes \neg[\![B]\!]$	$[\![A]\!] \otimes \neg[\![B]\!]$	$\underline{[\![A]\!] \otimes [\![B]\!]}$	$\underline{[\![B]\!]}$	$\neg[\![B]\!]$
$+$	$-$	$[\![A]\!] \otimes \neg[\![B]\!]$	$\underline{[\![A]\!] \otimes [\![B]\!]}$	$\neg[\![A]\!] \otimes [\![B]\!]$	$\neg[\![A]\!] \otimes \neg[\![B]\!]$	$-$	$-$
$+$	$+$	$\underline{[\![A]\!] \otimes [\![B]\!]}$	$[\![A]\!] \otimes \neg[\![B]\!]$	$\neg[\![A]\!] \otimes \neg[\![B]\!]$	$\neg[\![A]\!] \otimes [\![B]\!]$	$-$	$-$

by an argument when upward monotone, while reversed when downward monotone. This is the default case, and is underlined for each connective separately in Table 1. Deviations are recorded by marking the offending argument by $\neg$. In practice, we sometimes loosely refer by $[\![A]\!]$ to some type isomorphic to it through commutativity and associativity of $\otimes$ in **LP**.

We face a choice in extending the positive/negative distinction to atoms: if assigned positive *bias* (i.e., $\epsilon(p)$ is chosen $+$), $[\![p]\!] = p$, while $[\![p]\!] = \neg p$ with negative bias ($\epsilon(p) = -$). To keep our semantics free from superfluous negations, we go with the former option.

Antecedent and Consequent Structures. Conforming to the existence of $\cdot^\infty$, we consider sequents harboring possible multitudes of both *hypotheses* (or *inputs*) and *conclusions* (*outputs*). We draw from disjoint collections of variables (written x, y, z, possiby sub- or superscripted) and covariables (ε, κ, ν) to represent in- and outputs by labeled formulas A^x and A^ε. The latter combine into (antecedent)structures Γ, Δ and co(nsequent)structures Π, Σ, as specified by[4]

$$\Gamma, \Delta ::= A^x \mid (\Gamma \bullet \Delta) \mid (\Gamma \mathbin{\bullet\!\!-} \Sigma) \mid (\Pi \mathbin{-\!\!\bullet} \Gamma) \mid \Pi^{-\bullet} \mid {}^{\bullet-}\Sigma \qquad \text{Structures}$$
$$\Pi, \Sigma ::= A^\varepsilon \mid (\Sigma \circ \Pi) \mid (\Pi \mathbin{\circ\!\!-} \Gamma) \mid (\Delta \mathbin{-\!\!\circ} \Pi) \mid \Delta^{-\circ} \mid {}^{\circ-}\Gamma \qquad \text{Costructures}$$

The various constructors involved are seen as structural counterparts for the logical connectives via the the following translation tables:

	$F(\cdot)$		$F(\cdot)$		$F(\cdot)$
A^x	A	A^ε	A	$\Delta^{-\circ}$	${}^0F(\Delta)$
$\Gamma \bullet \Delta$	$F(\Gamma) \otimes F(\Delta)$	$\Sigma \circ \Pi$	$F(\Pi) \oplus F(\Sigma)$	${}^{\circ-}\Gamma$	$F(\Gamma)^0$
$\Pi \mathbin{-\!\!\bullet} \Gamma$	$F(\Pi) \obslash F(\Gamma)$	$\Pi \mathbin{\circ\!\!-} \Gamma$	$F(\Gamma)\backslash F(\Pi)$	$\Pi^{-\bullet}$	$F(\Pi)^1$
$\Gamma \mathbin{\bullet\!\!-} \Sigma$	$F(\Gamma) \oslash F(\Sigma)$	$\Delta \mathbin{-\!\!\circ} \Pi$	$F(\Pi)/F(\Delta)$	${}^{\bullet-}\Sigma$	${}^1F(\Sigma)$

We have sided with display calculi ([8]) in rejecting the standard practice of allowing only conjunction (in the antecedent) and disjunction (consequent) to

[4] The reader eager to indulge in notational overloading may note that the symbols $\bullet, \circ$ suffice for representing each of the binary structural operations. E.g., $\Gamma \bullet \Delta, \Gamma \bullet \Sigma$ and $\Gamma \bullet \Delta$ are unambiguously recognized as $\Gamma \bullet \Delta$, $\Gamma \mathbin{\bullet\!\!-} \Sigma$ and $\Pi \mathbin{-\!\!\bullet} \Gamma$ respectively.

be understood structurally. The association of types $[\![A]\!]$ with formulae A extends to a mapping of (co)structures into **LP**-contexts. In the base case, we stipulate[5]

$$[\![A^x]\!] = \begin{cases} \{[\![A]\!]^x\} \text{ if } \epsilon(A) = + \\ \{\neg[\![A]\!]^x\} \text{ if } \epsilon(A) = - \end{cases} \qquad [\![A^\varepsilon]\!] = \begin{cases} \{\neg[\![A]\!]^\varepsilon\} \text{ if } \epsilon(A) = + \\ \{[\![A]\!]^\varepsilon\} \text{ if } \epsilon(A) = - \end{cases}$$

while structural connectives collapse into multiset union. The underlying intuition: inputs occupy the downward monotone arguments of an implication, while outputs instantiate the upward monotone ones (see also the entries for implications in Table 1).

The Display Property. We assert inequalities $F(\Gamma) \leq F(\Pi)$ through *sequents* $\Gamma, \Pi \vdash M$ or $\Pi, \Gamma \vdash M$ (the relative ordering of Γ, Π being irrelevant), where $[\![\Gamma]\!], [\![\Pi]\!] \vdash_{\mathbf{LP}} M : \bot$. Our use of structural counterparts for (co)implications scatters in- and outputs all over the sequent, as opposed to nicely partitioning them into (the yields of) the antecedent and consequent structures. Instead, the following *display postulates*, mapping to (co)residuation and (co)Galois laws under F, allow each input (output) to be displayed as the whole of the antecedent (consequent):

$$\frac{\Gamma, \Pi \vdash M}{\Pi, \Gamma \vdash M} \qquad \frac{\Gamma, \Delta^{-\circ} \vdash M}{{}^{\circ-}\Gamma, \Delta \vdash M} \qquad \frac{{}^{\bullet-}\Sigma, \Pi \vdash M}{\Sigma, \Pi^{-\bullet} \vdash M}$$

$$\frac{\Gamma \bullet \Delta, \Pi \vdash M}{\Gamma, \Delta \multimap \Pi \vdash M} \qquad \frac{\Pi, \Gamma \bullet \Delta \vdash M}{\Pi \multimapinv \Gamma, \Delta \vdash M} \qquad \frac{\Gamma, \Sigma \circ \Pi \vdash M}{\Gamma \mathrel{\bullet\!\!-} \Sigma, \Pi \vdash M} \qquad \frac{\Sigma \circ \Pi, \Gamma \vdash M}{\Sigma, \Pi \mathrel{-\!\!\bullet} \Gamma \vdash M}$$

Sequents $\Gamma, \Pi \vdash M$ and $\Delta, \Sigma \vdash M$ are declared *display-equivalent* iff they are interderivable using only the display postulates, a fact we often abbreviate

$$\frac{\Gamma, \Pi \vdash M}{\Delta, \Sigma \vdash M}\, dp$$

The *display property* now reads: for any input A^x appearing in $\Gamma, \Pi \vdash M$, there exists some Σ such that $\Sigma, A^x \vdash M$ and $\Gamma, \Pi \vdash M$ are display-equivalent, and similarly for outputs.

Focused Proof Search. We shall allow a displayed hypothesis or conclusion to inhabit the righthand zone of $\vdash$, named the *stoup*, after [7]. Thus, $\Gamma \vdash M : A$ and $\Pi \vdash M : A$ are sequents, provided

$$[\![\Gamma]\!] \vdash_{\mathbf{LP}} M : [\![A]\!] \text{ and } [\![\Pi]\!] \vdash_{\mathbf{LP}} M : \neg[\![A]\!] \text{ if } \epsilon(A) = +$$
$$[\![\Gamma]\!] \vdash_{\mathbf{LP}} M : \neg[\![A]\!] \text{ and } [\![\Pi]\!] \vdash_{\mathbf{LP}} M : [\![A]\!] \quad \text{if } \epsilon(A) = -$$

The presence of a stoup implements Andreoli's concept of *focused proof search* ([1]). That is, when working one's way backwards from the desired conclusion

[5] We assume each (co)variable of **LG** to have been uniquely associated with a variable of **LP**. The details of this correspondence are abstracted away from in our notation: the context is to differentiate between **LG**'s (co)variables and their **LP** counterparts.

to the premises, one commits to the contents of the stoup (the *focus*) as the main formula: the only (logical) inference rules deriving a sequent $\Gamma \vdash M : A$ or $\Pi \vdash M : A$ are those introducing A, and focus propagates to the subformulas of A appearing in the premises. The pay-off: a reduction of the search space.

We need structural rules for managing the contents of the stoup. As will be argued below, focusing is relevant only for the negative inputs and positive outputs. Thus, we have *decision* rules for moving the latter inside the stoup, and *reaction* rules for taking their duals out:

$$\frac{\Pi \vdash M : K}{\Pi, K^x \vdash (x\ M)} D^\bullet \quad \frac{\Gamma \vdash M : P}{\Gamma, P^\varepsilon \vdash (\varepsilon\ M)} D^\circ \quad \frac{\Pi, P^x \vdash M}{\Pi \vdash \lambda x^{[P]} M : P} R^\bullet \quad \frac{\Gamma, K^\varepsilon \vdash M}{\Gamma \vdash \lambda \varepsilon^{[K]} M : K} R^\circ$$

Logical Inferences. Each connective has its meaning defined by two types of rules: one infering it from its structural countepart (affecting the positive inputs and negative outputs) and one introducing it *alongside* its structural counterpart via monotonicity (targeting negative inputs and positive outputs). In reference to Smullyan's unified notation ([18]), we speak of rules of type α and β respectively. The former preserve provability of the conclusion in their premises and hence come for free in proof search, meaning we may apply them nondeterministically. In contrast, the order in which the available β-type rules are tried may well affect success. Not all of the non-determinism involved is meaningful, however, as witnessed by the trivial permutations of β-type rules involving disjoint active formulas. Their domain of influence we therefore restrict to the stoup.[6] It follows that we may interpret term construction under (α) and (β) by the **LP**-inferences $(\otimes E)$ and $(\otimes I)$ respectively. Using the meta-variables

$$\begin{array}{rl} \varphi, \psi, \rho & \text{for variables and covariables} \\ \text{and } \Theta, \Theta_1, \Theta_2 & \text{for antecedent and consequent structures,} \end{array}$$

we may formalize the above intuitions by the following tables and rule schemata: and for the unary connectives (overloading the α, β notation):

Θ, α^φ	α_1^ψ	α_2^ρ	$*$
$\Pi, A \otimes B^x$	A^y	B^z	$\bullet$
$\Gamma, A/B^\varepsilon$	B^y	A^ν	$\multimap$
$\Gamma, B\backslash A^\varepsilon$	A^κ	B^z	$\circ\!-$
$\Gamma, A \oplus B^\varepsilon$	B^κ	A^ν	$\circ$
$\Pi, B \oslash A^x$	B^κ	A^z	$-\!\bullet$
$\Pi, A \oslash B^x$	A^y	B^ν	$\bullet\!-$

β	Θ_1, β_1	Θ_2, β_2	$*$
$A \otimes B$	Γ, A	Δ, B	$\bullet$
A/B	Δ, B	Π, A	$\multimap$
$B\backslash A$	Π, A	Δ, B	$\circ\!-$
$A \oplus B$	Σ, B	Π, A	$\circ$
$B \oslash A$	Σ, B	Γ, A	$-\!\bullet$
$A \oslash B$	Γ, A	Σ, B	$\bullet\!-$

$$\frac{\Theta, \alpha_1^\psi * \alpha_2^\rho \vdash M}{\Theta, \alpha^\varphi \vdash \textbf{case}\ \varphi\ \textbf{of}\ \psi \otimes \rho.M} \alpha \qquad \frac{\Theta_1 \vdash M : \beta_1 \quad \Theta_2 \vdash N : \beta_2}{\Theta_1 * \Theta_2 \vdash \langle M \otimes N \rangle : \beta} \beta$$

[6] While reducing the search space, it is not immediate that completeness w.r.t. provability in **LG** is preserved. We return to this issue at the end of this section.

$$\frac{\Pi, P^y \bullet Q^z \vdash M}{\Pi, P \otimes Q^x \vdash \textbf{case}\ x\ \textbf{of}\ y^{[P]} \otimes z^{[Q]}.M}\ \alpha \qquad \frac{\Gamma \vdash M : P \quad \Delta \vdash N : Q}{\Gamma \bullet \Delta \vdash \langle M \otimes N \rangle : P \otimes Q}\ \beta$$

$$\frac{\Pi, P^y \bullet L^z \vdash M}{\Pi, P \otimes L^x \vdash \textbf{case}\ x\ \textbf{of}\ y^{[P]} \otimes z^{\neg[L]}.M}\ \alpha \qquad \frac{\Gamma \vdash M : P \quad \dfrac{\Delta, L^\nu \vdash N}{\Delta \vdash \lambda\nu^{[L]}M : L}\ R^\circ}{\Gamma \bullet \Delta \vdash \langle M \otimes \lambda\nu^{[L]}N \rangle : P \otimes L}\ \beta$$

$$\frac{\Pi, K^y \bullet Q^z \vdash M}{\Pi, K \otimes Q^x \vdash \textbf{case}\ x\ \textbf{of}\ y^{\neg[K]} \otimes z^{[Q]}.M}\ \alpha \qquad \frac{\dfrac{\Gamma, K^\kappa \vdash M}{\Gamma \vdash \lambda\kappa^{[K]}M : K}\ R^\circ \quad \Delta \vdash N : Q}{\Gamma \bullet \Delta \vdash \langle \lambda\kappa^{[K]}M \otimes N \rangle : K \otimes Q}\ \beta$$

$$\frac{\Pi, K^y \bullet L^z \vdash M}{\Pi, K \otimes L^x \vdash \textbf{case}\ x\ \textbf{of}\ y^{\neg[K]} \otimes z^{\neg[L]}.M}\ \alpha \qquad \frac{\dfrac{\Gamma, K^\kappa \vdash M}{\Gamma \vdash \lambda\kappa^{[K]}M : K}\ R^\circ \quad \dfrac{\Delta, L^\nu \vdash N}{\Delta \vdash \lambda\nu^{[L]}M : L}\ R^\circ}{\Gamma \bullet \Delta \vdash \langle \lambda\kappa^{[K]}.M \otimes \lambda\nu^{[L]}N \rangle : K \otimes L}\ \beta$$

Fig. 2. Checking all possible instantiations of $(\alpha), (\beta)$ for $\otimes$. We also mention obligatory reactions $(R^\bullet, R^\circ)$ in the premises of (β).

Θ, α^φ	α_1^ψ	$\cdot^*$
$\Gamma, {}^0A^\varepsilon$	A^y	$\cdot\multimap$
$\Gamma, A^{0\varepsilon}$	A^y	$\multimapinv\cdot$
Π, A^{1x}	A^κ	$\cdot\multimapdot$
$\Pi, {}^1A^x$	A^κ	$\multimapdotinv\cdot$

β	Θ, β_1	$\cdot^*$
${}^0A$	Δ, A	$\cdot\multimap$
A^0	Δ, A	$\multimapinv\cdot$
A^1	Σ, A	$\cdot\multimapdot$
${}^1A$	Σ, A	$\multimapdotinv\cdot$

$$\frac{\Theta, \alpha_1^{\psi *} \vdash M}{\Theta, \alpha^\varphi \vdash M[\varphi/\psi]}\ \alpha \qquad \frac{\Theta \vdash M : \beta_1}{\Theta^* \vdash M : \beta}\ \beta$$

Well-definedness is established through a case-by-case analysis. To illustrate, Figure 2 checks all possible instantiations of $(\alpha), (\beta)$ for $\otimes$. Finally, assigning positive bias to atoms implies Axioms have their conclusion placed in focus:

$$\frac{}{p^x \vdash x : p}\ Ax$$

Soundness and Completeness. In what preceded, we have already informally motivated soundness w.r.t. §1's algebraic formulation of **LG**. Completeness is demonstrated in a companion paper under preparation. Roughly, we define a syntactic phase model wherein every truth is associated with a Cut-free focused proof, similar to [16].

3 Case Analysis: Extraction

We illustrate our semantics of §2 alongside an analysis of extraction phenomena. Syntactically, their treatment in **NL** necessitates controlled associativity and commutativity ([13]). Kurtonina and Moortgat (K&M, [10]) ask whether the same results are obtainable in **LG** from having $\otimes$ and $\oplus$ interact through linear distributivity. We work out the details of such an approach, after first having

$$\frac{}{p^x \vdash x : p}\ Ax \qquad \frac{\Gamma, \Pi \vdash M}{\Pi, \Gamma \vdash M} \qquad \frac{\Gamma \bullet \Delta, \Pi \vdash M}{\Gamma, \Delta \multimap \Pi \vdash M}$$

$$\frac{\Pi, \Gamma \bullet \Delta \vdash M}{\Pi \circ\!- \Gamma, \Delta \vdash M} \qquad \frac{\Gamma, \Sigma \circ \Pi \vdash M}{\Gamma \bullet\!- \Sigma, \Pi \vdash M} \qquad \frac{\Sigma \circ \Pi, \Gamma \vdash M}{\Sigma, \Pi -\!\bullet \Gamma \vdash M}$$

$$\frac{\Pi \vdash M : K}{\Pi, K^x \vdash \lambda x^{[K]} M}\ D^{\bullet} \qquad \frac{\Gamma \vdash M : P}{\Gamma, P^{\varepsilon} \vdash \lambda \varepsilon^{[P]} M}\ D^{\circ}$$

$$\frac{\Pi, P^x \vdash M}{\Pi \vdash (x\ M) : P}\ R^{\bullet} \qquad \frac{\Gamma, K^{\varepsilon} \vdash M}{\Gamma \vdash (\varepsilon\ M) : K}\ R^{\circ}$$

$$\frac{\Theta, \alpha_1^{\psi} * \alpha_2^{\rho} \vdash M}{\Theta, \alpha^{\varphi} \vdash \textbf{case}\ \varphi\ \textbf{of}\ \psi \otimes \rho . M}\ \alpha \qquad \frac{\Theta_1 \vdash M : \beta_1 \quad \Theta_2 \vdash N : \beta_2}{\Theta_1 * \Theta_2 \vdash \langle M \otimes N \rangle : \beta}\ \beta$$

$$\frac{\Theta, \alpha_1^{\psi *} \vdash M}{\Theta, \alpha^{\varphi} \vdash M[\varphi / \psi]}\ \alpha \qquad \frac{\Theta \vdash M : \beta_1}{\Theta^{*} \vdash M : \beta}\ \beta$$

Fig. 3. An overview of the term-labeled sequent calculus for **LG**. An easy induction shows that terms labeling derivable sequents are in (β-)normal form.

pointed out a flaw in an alternative proposal by K&M. As illustration, we work out the derivational and lexical semantics of several sample sentences.

The Good. We first consider a case that already works fine in **NL**. The following complex noun demonstrates extraction out of (subordinate) subject position:

(1) (the) mathematician who founded intuitionism

We analyze (1) into a binary branching tree, categorizing the words as follows:

mathematician	[who	[invented	intuitionism]]
n	$(n\backslash n)/(np\backslash s)$	$(np\backslash s)/np$	np

employing atoms s (categorizing sentences), *np* (noun phrases) and n (nouns). Note the directionality in the category $np\backslash s$ assigned to the gapped clause (as selected for by *who*), seeing as the *np* gap occurs in a left branch. Figure 6 demonstrates (1), with bracketing as indicated above, to be categorizable by n, referring to the 'macro' from Figure 4 for deriving transitive clauses.

We proceed with a specification of the lexical semantics. Linearity no longer applies at this stage, as our means of referring to the world around us is not so restricted. Thus, we allow access to the full repertoire of the simply-typed λ-calculus, augmented with logical constants for the propositional connectives. Concretely, lexical denotations are built over types

$$\tau, \sigma ::= e \mid t \mid (\tau \times \sigma) \mid (\tau \rightarrow \sigma)$$

$$\dfrac{\dfrac{\dfrac{\dfrac{\dfrac{\dfrac{\dfrac{}{s^{u} \vdash u : s}\,Ax}{s^{\nu}, s^{u} \vdash (\nu\ u)}\,D^{\circ}}{s^{\nu} \vdash \lambda u(\nu\ u) : s}\,R^{\bullet} \quad \dfrac{}{np^{x} \vdash x : np}\,Ax}{s^{\nu} \mathbin{\circ\!\!-} np^{x} \vdash \langle \lambda u(\nu\ u) \otimes x \rangle : np \backslash s}\,\beta \quad \dfrac{}{np^{z} \vdash z : np}\,Ax}{np^{z} \multimap (s^{\nu} \mathbin{\circ\!\!-} np^{x}) \vdash \langle z \otimes \langle \lambda u(\nu\ u) \otimes x \rangle\rangle : (np\backslash s)/np}\,\beta}{np^{z} \multimap (s^{\nu} \mathbin{\circ\!\!-} np^{x}), (np\backslash s)/np^{z} \vdash (y\ \langle z \otimes \langle \lambda u(\nu\ u) \otimes x \rangle\rangle)}\,D^{\bullet}}{np^{x} \bullet ((np\backslash s)/np^{y} \bullet np^{z}), s^{\nu} \vdash (y\ \langle z \otimes \langle \lambda u(\nu\ u) \otimes x \rangle\rangle)}\,dp$$

Fig. 4. Derivation of a transitive clause in an SVO language

where e, t interpret (a fixed set of) entities and (Boolean) truth values respectively. The linear types and terms of §2 carry over straightforwardly: interpret $\bot$, $\tau \otimes \sigma$ and $\neg\tau$ by t, $\tau \times \sigma$ and $\tau \to t$, with terms $\langle M \otimes N \rangle$ and **case** N **of** $x \otimes y.M$ being replaced by pairs $\langle M, N \rangle$ and projections $M[\pi_1(N)/x, \pi_2(N)/y]$. The remaining atoms s, np and n we interpret by t (sentences denote truth values), e (noun phrases denote entities) and $e \to t$ (nouns denote first-order properties) respectively. Abbreviating $\lambda x^{\tau \times \sigma} M[\pi_1(x)/y, \pi_2(x)/z]$ by $\lambda\langle y, z\rangle M$ and types $\tau \to t$ by $\neg\tau$, the linear terms of §2 remain practically unchanged. For instance, delinearization of the term found in Figure 6 for (1) gives

$$(w\ \langle \lambda\langle \kappa, b\rangle (f\ \langle i, \langle \lambda z(\kappa\ z), b\rangle\rangle), \langle \lambda y(\nu\ y), m\rangle\rangle)$$

the free variables w, f, i and m ranging over the denotations of *who*, *founded*, *intuitionism* and *mathematician*. Since words act as inputs, those categorized P (K) are interpreted by a closed term M of type $[\![P]\!]$ ($\neg[\![K]\!]$). These remarks motivate the following lexical entries, conveniently written as nonlogical axioms:

$$\begin{aligned}
\text{mathematician} &\vdash \textsc{mathematician} : n \\
\text{who} &\vdash \lambda\langle Q, \langle \nu, P\rangle\rangle (\nu\ \lambda x((P\ x) \wedge (Q\ \langle \lambda pp, x\rangle))) : (n\backslash n)/(np\backslash s) \\
\text{founded} &\vdash \lambda\langle y, \langle q, x\rangle\rangle (q\ ((\textsc{founded}\ y)\ x)) : (np\backslash s)/np \\
\text{intuitionism} &\vdash \textsc{intuitionism} : np
\end{aligned}$$

We applied the familiar trick of switching fonts to abstract away from certain interpretations, yielding constants MATHEMATICIAN (type $\neg e$), FOUNDED ($e \to \neg e$) and INTUITIONISM (e). If we take the nonlogical axiom perspective seriously, lexical substitution proceeds via Cut. Simplifying, we directly substitute terms for the free variables m, w, f and i, yielding, after β-reduction, the term (with free variable γ corresponding to the assigned category n)

$$(\gamma\ \lambda x((\textsc{mathematician}\ x) \wedge ((\textsc{founded}\ \textsc{intuitionism})\ x))),$$

The Bad. Cases of non-subject extraction are illustrated in (2) and (3) below:

(2) (the) law that Brouwer rejected
(3) (the) mathematician whom TNT pictured on a post stamp

While tempting to categorize *that* and *whom* by $(n\backslash n)/(s/np)$ (noticing the gap now occurs in a right branch), we find that derivability of (2) then necessitates rebracketing (mentioning also the dual concept for reasons of symmetry):

$$\begin{array}{ll} (A \otimes B) \otimes C \leq A \otimes (B \otimes C) & A \oplus (B \oplus C) \leq (A \oplus B) \oplus C \\ A \otimes (B \otimes C) \leq (A \otimes B) \otimes C & (A \oplus B) \oplus C \leq A \oplus (B \oplus C) \end{array}$$

Worse yet, (3) requires (weak) commutativity for its derivability:

$$\begin{array}{ll} (A \otimes B) \otimes C \leq (A \otimes C) \otimes B & (A \oplus C) \oplus B \leq (A \oplus B) \oplus C \\ A \otimes (B \otimes C) \leq B \otimes (A \otimes C) & B \oplus (A \oplus C) \leq A \oplus (B \oplus C) \end{array}$$

Said principles, however, contradict the resource sensitive nature of linguistic reality. Kurtonina and Moortgat (K&M, [10]), working in **LG**, questioned the viability of a different solution: revise the categorization of *whom* such that recourse need be made only to linear distributivity of $\otimes$ over $\oplus$ (or *mixed* associativity and commutativity, if you will):

$$\begin{array}{ll} (A \oplus B) \otimes C \leq A \oplus (B \otimes C) & (A \oplus B) \otimes C \leq (A \otimes C) \oplus B \\ A \otimes (B \oplus C) \leq (A \otimes B) \oplus C & A \otimes (B \oplus C) \leq B \oplus (A \otimes C) \end{array}$$

As observed by Moortgat (using a slightly different syntax), the presence of (co)implications allows a presentation in the following rule format:

$$\frac{\Pi \rightharpoondown \Gamma, \Delta \multimap \Sigma \vdash M}{\Gamma \bullet \Delta, \Sigma \circ \Pi \vdash M}\,(\oslash, /) \qquad \frac{\Delta \leftharpoondown \Sigma, \Pi \multimapinv \Gamma \vdash M}{\Gamma \bullet \Delta, \Sigma \circ \Pi \vdash M}\,(\oslash, \backslash)$$

$$\frac{\Pi \rightharpoondown \Delta, \Sigma \multimapinv \Gamma \vdash M}{\Gamma \bullet \Delta, \Sigma \circ \Pi \vdash M}\,(\oslash, \backslash) \qquad \frac{\Gamma \rightharpoondown \Sigma, \Delta \multimap \Pi \vdash M}{\Gamma \bullet \Delta, \Sigma \circ \Pi \vdash M}\,(\oslash, /)$$

K&M suggested in particular to categorize *whom* by $(n\backslash n)/((s \oslash s) \oplus (s/np))$. However, their analysis assumes $(\oslash, /)$, $(\oslash, \backslash)$, $(\oslash, \backslash)$ and $(\oslash, /)$ to be *invertible*, thereby seriously compromising the resource sensitivity of **LG**, as illustrated by the derivable inferences of Figure 5 (and similar ones under $\cdot^{\infty}$).

The Analysis. We propose a solution to K&M's challenge by categorizing *whom* using not only the (co)residuated families of connectives, but also the Galois connected pair ${}^{0}\cdot, \cdot^{0}$. In particular, we have in mind the following lexicon for (2):

$$\begin{array}{rl} \text{law} \vdash & \text{LAW} : n \\ \text{that} \vdash & \lambda\langle Q, \langle \nu, P\rangle\rangle(\nu\ \lambda x((P\ x) \wedge (Q\ \langle \lambda pp, x\rangle))) : (n\backslash n)/(s \oplus {}^{0}np) \\ \text{Brouwer} \vdash & \text{BROUWER} : np \\ \text{rejected} \vdash & \lambda\langle y, \langle q, x\rangle\rangle(q\ ((\text{REJECTED}\ y)\ x)) : (np\backslash s)/np \end{array}$$

employing constants LAW, BROUWER and REJECTED of types $\neg e$, e and $e \to \neg e$. Note the formula $s \oplus {}^{0}np$ (selected for by *that*) categorizing the gapped clause; had the gap occurred in a left branch, we would have used $np^{0} \oplus s$ instead. Figure 6 gives the derivation. Lexical substitution and β-reduction yield

$$\begin{array}{cl} \varGamma_2 \bullet (\varGamma_1 \bullet \varGamma_3), \varSigma \circ \varPi \vdash M & \\ \hline\hline \varPi \multimapdot (\varGamma_1 \bullet \varGamma_3), \varSigma \multimapinv \varGamma_2 \vdash M & (\obslash, \backslash) \\ \hline\hline \varGamma_1 \bullet \varGamma_3, (\varSigma \multimapinv \varGamma_2) \circ \varPi \vdash M & dp \\ \hline\hline \varGamma_3 \multimapdotinv (\varSigma \multimapinv \varGamma_2), \varPi \multimapdotinv \varGamma_1 \vdash M & (\oslash, \backslash) \\ \hline\hline (\varPi \multimapinv \varGamma_1) \multimapdot \varGamma_3, \varSigma \multimapinv \varGamma_2 \vdash M & dp \\ \hline\hline \varGamma_2 \bullet \varGamma_3, \varSigma \circ (\varPi \multimapinv \varGamma_1) \vdash M & (\obslash, \backslash) \\ \hline\hline (\varGamma_2 \bullet \varGamma_3) \multimapdotinv \varSigma, \varPi \multimapinv \varGamma_1 \vdash M & dp \\ \hline\hline \varGamma_1 \bullet (\varGamma_2 \bullet \varGamma_3), \varSigma \circ \varPi \vdash M & (\oslash, \backslash) \end{array} \qquad \begin{array}{cl} (\varGamma_1 \bullet \varGamma_2) \bullet \varGamma_3, \varSigma \circ \varPi \vdash M & \\ \hline\hline \varPi \multimapdot (\varGamma_1 \bullet \varGamma_2), \varGamma_3 \multimap \varSigma \vdash M & (\obslash, /) \\ \hline\hline \varGamma_1 \bullet \varGamma_2, (\varGamma_3 \multimap \varSigma) \circ \varPi \vdash M & dp \\ \hline\hline \varGamma_2 \multimapdotinv (\varGamma_3 \multimap \varSigma), \varPi \multimapinv \varGamma_1 \vdash M & (\obslash, \backslash) \\ \hline\hline (\varPi \multimapinv \varGamma_1) \multimapdot \varGamma_2, \varGamma_3 \multimap \varSigma \vdash M & dp \\ \hline\hline \varGamma_2 \bullet \varGamma_3, \varSigma \circ (\varPi \multimapinv \varGamma_1) \vdash M & (\obslash, /) \\ \hline\hline (\varGamma_2 \bullet \varGamma_3) \multimapdotinv \varSigma, \varPi \multimapinv \varGamma_1 \vdash M & dp \\ \hline\hline \varGamma_1 \bullet (\varGamma_2 \bullet \varGamma_3), \varSigma \circ \varPi \vdash M & (\oslash, \backslash) \end{array}$$

Fig. 5. Illustrating the structural collapse induced by making $(\oslash, /)$, $(\oslash, \backslash)$, $(\obslash, /)$ and $(\obslash, \backslash)$ invertible

$$(\gamma\ \lambda x((\textsc{law}\ x) \wedge ((\textsc{rejected}\ x)\ \textsc{brouwer})))$$

Like K&M, we have, in Figure 6, not relied exclusively on linear distributivity: the (dual) Galois connected pairs now go 'halfway De Morgan', as explicated by the following three equivalent groups of axioms

$$\begin{array}{ccc} (A \otimes B)^1 \leq {}^0B \oplus {}^0A & A \oslash {}^0B \leq A \otimes B & A/B \leq A \oplus {}^0B \\ {}^1(A \otimes B) \leq B^0 \oplus A^0 & B^0 \obslash A \leq B \otimes A & B \backslash A \leq B^0 \oplus A \\ A^1 \otimes B^1 \leq {}^0(B \oplus A) & B \oplus A \leq B^1 \backslash A & B^1 \otimes A \leq B \obslash A \\ {}^1A \otimes {}^1B \leq (B \oplus A)^0 & B \oplus A \leq B/{}^1A & A \otimes {}^1B \leq A \oslash B \end{array}$$

Note their independence of their converses (i.e., with $\leq$ turned around). The following equivalent presentation in rule format is adapted from [15]:

$$\frac{\varGamma \bullet \varDelta, \varPi \vdash M}{\varGamma \multimapdotinv \varPi, \varDelta^{\multimap} \vdash M}\,(\oslash, {}^0\cdot) \qquad \frac{\varGamma \bullet \varDelta, \varPi \vdash M}{\varPi \multimapdot \varDelta, {}^{\multimapinv}\varGamma \vdash M}\,(\obslash, \cdot^0)$$

$$\frac{\varGamma \bullet \varDelta, \varPi \vdash M}{\varDelta \multimapdotinv \varPi, {}^{\multimapinv}\varGamma \vdash M}\,(\oslash, \cdot^0) \qquad \frac{\varGamma \bullet \varDelta, \varPi \vdash M}{\varPi \multimapdot \varGamma, \varDelta^{\multimap} \vdash M}\,(\obslash, {}^0\cdot)$$

The intuition behind our analysis is as follows. If we were to also adopt the converses of the above De Morgan axioms (turning $\leq$ around), same-sort associativity and weak commutativity would find equivalent presentations as

$$\begin{array}{cc} (A \otimes B) \oslash {}^0C \leq A \otimes (B \oslash {}^0C) & (A \otimes B) \oslash {}^0C \leq (A \oslash {}^0C) \otimes B \\ A^0 \obslash (B \otimes C) \leq (A^0 \obslash B) \otimes C & A^0 \obslash (B \otimes C) \leq B \otimes (A^0 \obslash C) \end{array}$$

Going only halfway with De Morgan, however, the above inferences remain derivable (by virtue of linear distributivity) and useful (by composing with $A \oslash {}^0B \leq A \otimes B$ and $B^0 \obslash A \leq B \otimes A$), but without inducing a collapse. Indeed, none of the derivabilities of Figure 5 carry over, and neither do the variations

$$\begin{array}{c} (\varGamma_1 \bullet \varGamma_2) \bullet \varGamma_3, \varSigma \circ {}^{\multimapinv}\varPi \vdash M \\ \hline\hline \varGamma_1 \bullet (\varGamma_2 \bullet \varGamma_3), \varSigma \circ {}^{\multimapinv}\varPi \vdash M \end{array} \qquad \begin{array}{c} \varGamma_2 \bullet (\varGamma_1 \bullet \varGamma_3), \varSigma \circ {}^{\multimapinv}\varPi \vdash M \\ \hline\hline \varGamma_1 \bullet (\varGamma_2 \bullet \varGamma_3), \varSigma \circ {}^{\multimapinv}\varPi \vdash M \end{array}$$

$$\dfrac{}{np^b \bullet ((np\backslash s)/np^f \bullet np^i), s^\kappa \vdash (f\ \langle i \otimes \langle \lambda z(\kappa\ z) \otimes b\rangle\rangle)}\ tv$$

$$\dfrac{np^b \bullet ((np\backslash s)/np^f \bullet np^i), s^\kappa \vdash (f\ \langle i \otimes \langle \lambda z(\kappa\ z) \otimes b\rangle\rangle)}{(np\backslash s)/np^f \bullet np^i, s^\kappa \mathbin{\circ\!\!-} np^b \vdash (f\ \langle i \otimes \langle \lambda z(\kappa\ z) \otimes b\rangle\rangle)}\ dp$$

$$\dfrac{(np\backslash s)/np^f \bullet np^i, s^\kappa \mathbin{\circ\!\!-} np^b \vdash (f\ \langle i \otimes \langle \lambda z(\kappa\ z) \otimes b\rangle\rangle)}{(np\backslash s)/np^f \bullet np^i, np\backslash s^\delta \vdash \mathbf{case}\ \delta\ \mathbf{of}\ \kappa \otimes b.(f\ \langle i \otimes \langle \lambda z(\kappa\ z) \otimes b\rangle\rangle)}\ \alpha$$

$$\dfrac{(np\backslash s)/np^f \bullet np^i, np\backslash s^\delta \vdash \mathbf{case}\ \delta\ \mathbf{of}\ \kappa \otimes b.(f\ \langle i \otimes \langle \lambda z(\kappa\ z) \otimes b\rangle\rangle)}{(np\backslash s)/np^f \bullet np^i \vdash \lambda\langle \kappa \otimes b\rangle(f\ \langle i \otimes \langle \lambda z(\kappa\ z) \otimes b\rangle\rangle) : np\backslash s}\ R^\circ$$

$$\dfrac{\dfrac{\dfrac{\dfrac{}{n^y \vdash y : n}\ Ax}{n^\nu, n^y \vdash (\nu\ y)}\ D^\circ}{n^\nu \vdash \lambda y(\nu\ y) : n}\ R^\bullet \quad \dfrac{}{n^m \vdash m : n}\ Ax}{n^\nu \mathbin{\circ\!\!-} n^m \vdash \langle \lambda y(\nu\ y) \otimes m\rangle : n\backslash n :}\ \beta$$

$$\dfrac{(np\backslash s)/np^f \bullet np^i \vdash \lambda\langle \kappa \otimes b\rangle(\ldots) : np\backslash s \quad n^\nu \mathbin{\circ\!\!-} n^m \vdash \langle \lambda y(\nu\ y) \otimes m\rangle : n\backslash n :}{((np\backslash s)/np^f \bullet np^i) \multimap (n^\nu \mathbin{\circ\!\!-} n^m) \vdash \langle \lambda\langle \kappa \otimes b\rangle(f\ \langle i \otimes \langle \lambda z(\kappa\ z) \otimes b\rangle\rangle) \otimes \langle \lambda y(\nu\ y) \otimes m\rangle\rangle : (n\backslash n)/(np\backslash s)}\ \beta$$

$$\dfrac{((np\backslash s)/np^f \bullet np^i) \multimap (n^\nu \mathbin{\circ\!\!-} n^m) \vdash \langle \ldots \rangle : (n\backslash n)/(np\backslash s)}{((np\backslash s)/np^f \bullet np^i) \multimap (n^\nu \mathbin{\circ\!\!-} n^m), (n\backslash n)/(np\backslash s)^w \vdash (w\ \langle \lambda\langle \kappa \otimes b\rangle(f\ \langle i \otimes \langle \lambda z(\kappa\ z) \otimes b\rangle\rangle) \otimes \langle \lambda y(\nu\ y) \otimes m\rangle\rangle)}\ D^\bullet$$

$$\dfrac{((np\backslash s)/np^f \bullet np^i) \multimap (n^\nu \mathbin{\circ\!\!-} n^m), (n\backslash n)/(np\backslash s)^w \vdash (w\ \langle \ldots \rangle)}{n^m \bullet ((n\backslash n)/(np\backslash s)^w \bullet ((np\backslash s)/np^f \bullet np^i)), n^\nu \vdash (w\ \langle \lambda\langle \kappa \otimes b\rangle(f\ \langle i \otimes \langle \lambda z(\kappa\ z) \otimes b\rangle\rangle) \otimes \langle \lambda y(\nu\ y) \otimes m\rangle\rangle)}\ dp$$

$$\dfrac{}{np^b \bullet ((np\backslash s)/np^r \bullet np^e), s^\kappa \vdash (r\ \langle e \otimes \langle \lambda z(\kappa\ z) \otimes b\rangle\rangle)}\ tv$$

$$\dfrac{np^b \bullet ((np\backslash s)/np^r \bullet np^e), s^\kappa \vdash (r\ \langle e \otimes \langle \lambda z(\kappa\ z) \otimes b\rangle\rangle)}{(s^\kappa \mathbin{\circ\!\!-} np^b) \mathbin{-\!\!\bullet} (np\backslash s)/np^r, (np^e)^{\multimap} \vdash (r\ \langle e \otimes \langle \lambda z(\kappa\ z) \otimes b\rangle\rangle)}\ (\oslash, {}^0\cdot), dp$$

$$\dfrac{(s^\kappa \mathbin{\circ\!\!-} np^b) \mathbin{-\!\!\bullet} (np\backslash s)/np^r, (np^e)^{\multimap} \vdash (r\ \langle e \otimes \langle \lambda z(\kappa\ z) \otimes b\rangle\rangle)}{(s^\kappa \mathbin{\circ\!\!-} np^b) \mathbin{-\!\!\bullet} (np\backslash s)/np^r, {}^0np^\varepsilon \vdash (r\ \langle \varepsilon \otimes \langle \lambda z(\kappa\ z) \otimes b\rangle\rangle)}\ \alpha$$

$$\dfrac{(s^\kappa \mathbin{\circ\!\!-} np^b) \mathbin{-\!\!\bullet} (np\backslash s)/np^r, {}^0np^\varepsilon \vdash (r\ \langle \varepsilon \otimes \langle \lambda z(\kappa\ z) \otimes b\rangle\rangle)}{np^b \bullet (np\backslash s)/np^r, {}^0np^\varepsilon \circ s^\kappa \vdash \mathbf{case}\ \delta\ \mathbf{of}\ \varepsilon \otimes \kappa.(r\ \langle \varepsilon \otimes \langle \lambda z(\kappa\ z) \otimes b\rangle\rangle)}\ (\oslash, \backslash), dp$$

$$\dfrac{np^b \bullet (np\backslash s)/np^r, {}^0np^\varepsilon \circ s^\kappa \vdash \mathbf{case}\ \delta\ \mathbf{of}\ \varepsilon \otimes \kappa.(r\ \langle \varepsilon \otimes \langle \lambda z(\kappa\ z) \otimes b\rangle\rangle)}{np^b \bullet (np\backslash s)/np^r, s \oplus {}^0np^\delta \vdash \mathbf{case}\ \delta\ \mathbf{of}\ \varepsilon \otimes \kappa.(r\ \langle \varepsilon \otimes \langle \lambda z(\kappa\ z) \otimes b\rangle\rangle)}\ \alpha$$

$$\dfrac{np^b \bullet (np\backslash s)/np^r, s \oplus {}^0np^\delta \vdash \mathbf{case}\ \delta\ \mathbf{of}\ \varepsilon \otimes \kappa.(r\ \langle \varepsilon \otimes \langle \lambda z(\kappa\ z) \otimes b\rangle\rangle)}{np^b \bullet (np\backslash s)/np^r \vdash \lambda\langle \varepsilon \otimes \kappa\rangle(r\ \langle \varepsilon \otimes \langle \lambda z(\kappa\ z) \otimes b\rangle\rangle) : s \oplus {}^0np}\ R^\circ$$

$$\dfrac{\dfrac{\dfrac{\dfrac{}{n^y \vdash y : n}\ Ax}{n^\nu, n^y \vdash (\nu\ y)}\ D^\circ}{n^\nu \vdash \lambda y(\nu\ y) : n}\ R^\bullet \quad \dfrac{}{n^l \vdash l : n}\ Ax}{n^\nu \mathbin{\circ\!\!-} n^l \vdash \langle \lambda y(\nu\ y) \otimes l\rangle : n\backslash n :}\ \beta$$

$$\dfrac{np^b \bullet (np\backslash s)/np^r \vdash \lambda\langle \varepsilon \otimes \kappa\rangle(\ldots) : s \oplus {}^0np \quad n^\nu \mathbin{\circ\!\!-} n^l \vdash \langle \lambda y(\nu\ y) \otimes l\rangle : n\backslash n :}{(np^b \bullet (np\backslash s)/np^r) \multimap (n^\nu \mathbin{\circ\!\!-} n^l) \vdash \langle \lambda\langle \varepsilon \otimes \kappa\rangle(r\ \langle \varepsilon \otimes \langle \lambda z(\kappa\ z) \otimes b\rangle\rangle) \otimes \langle \lambda y(\nu\ y) \otimes l\rangle\rangle : (n\backslash n)/(s \oplus {}^0np)}\ \beta$$

$$\dfrac{(np^b \bullet (np\backslash s)/np^r) \multimap (n^\nu \mathbin{\circ\!\!-} n^l) \vdash \langle \ldots \rangle : (n\backslash n)/(s \oplus {}^0np)}{(np^b \bullet (np\backslash s)/np^r) \multimap (n^\nu \mathbin{\circ\!\!-} n^l), (n\backslash n)/(s \oplus {}^0np)^t \vdash (t\ \langle \lambda\langle \varepsilon \otimes \kappa\rangle(r\ \langle \varepsilon \otimes \langle \lambda z(\kappa\ z) \otimes b\rangle\rangle) \otimes \langle \lambda y(\nu\ y) \otimes l\rangle\rangle)}\ D^\bullet$$

$$\dfrac{(np^b \bullet (np\backslash s)/np^r) \multimap (n^\nu \mathbin{\circ\!\!-} n^l), (n\backslash n)/(s \oplus {}^0np)^t \vdash (t\ \langle \ldots \rangle)}{n^l \bullet ((n\backslash n)/(s \oplus {}^0np)^t \bullet (np^b \bullet (np\backslash s)/np^r)), n^\nu \vdash (t\ \langle \lambda\langle \varepsilon \otimes \kappa\rangle(r\ \langle \varepsilon \otimes \langle \lambda z(\kappa\ z) \otimes b\rangle\rangle) \otimes \langle \lambda y(\nu\ y) \otimes l\rangle\rangle)}\ dp$$

Fig. 6. Derivations of complex nouns demonstrating (peripheral) subject and non-subject extraction respectively. Words (or rather, the formulas representing their categories) appear as hypotheses, grouping together into binary branching tree structures via the structural counterpart $\bullet$ of $\otimes$. The chosen variable names are meant to be suggestive of the words they represent. Applications of (tv) refer to Figure 4.

as an exhaustive exploration of the search space will tell, noting we need only consider structural rules.

By virtue of the mixed commutativity involved in some of the linear distributivity postulates, it should be clear our formula $(n\backslash n)/(s \oplus {}^0 np)$ for *that* in (2) also applies to *whom* in (3), the latter example involving non-peripheral extraction. For reasons of space, we leave its analysis as an exercise.

LGT. (Call-by-name)

$$\frac{}{p^{\varepsilon} \vdash \varepsilon : p}\ Ax \qquad \frac{\Pi \vdash M : A}{\Pi, A^{x} \vdash (x\ M)}\ D$$

$$\frac{\Pi, A^{x} \bullet B^{y} \vdash M}{\Pi \vdash \lambda\langle x \otimes y\rangle M : A \otimes B}\ \otimes^{\bullet} \qquad \frac{\Gamma, A^{\varepsilon} \vdash M \quad \Delta, B^{\kappa} \vdash N}{\Gamma \bullet \Delta, A \otimes B^{\nu} \vdash (\nu\ \langle \lambda\varepsilon M \otimes \lambda\kappa N\rangle)}\ \otimes^{\circ}$$

$$\frac{\Gamma, A^{\varepsilon} \multimap B^{y} \vdash M}{\Gamma, B\backslash A^{\nu} \vdash \textbf{case}\ \nu\ \textbf{of}\ \varepsilon \otimes y.M}\ \backslash^{\circ} \qquad \frac{\Delta, B^{\varepsilon} \vdash N \quad \Pi \vdash M : A}{\Pi \multimap \Delta \vdash \langle \lambda\varepsilon N \otimes M\rangle : B\backslash A}\ \backslash^{\bullet}$$

$$\frac{\Gamma, B^{\kappa} \circ A^{\varepsilon} \vdash M}{\Gamma, A \oplus B^{\nu} \vdash \textbf{case}\ \gamma\ \textbf{of}\ \beta \otimes \alpha.M}\ \oplus^{\circ} \qquad \frac{\Sigma \vdash N : B \quad \Pi \vdash M : A}{\Sigma \circ \Pi \vdash \langle N \otimes M\rangle : A \oplus B}\ \oplus^{\bullet}$$

$$\frac{\Pi, A^{y} \mathrel{\bullet\!\!-} B^{\nu} \vdash M}{\Pi \vdash \lambda\langle y \otimes \nu\rangle M : A \oslash B}\ \oslash^{\bullet} \qquad \frac{\Sigma \vdash N : B \quad \Gamma, A^{\varepsilon} \vdash M}{\Gamma \mathrel{\bullet\!\!-} \Sigma, A \oslash B^{\nu} \vdash (\nu\ \langle N \otimes \lambda\varepsilon M\rangle)}\ \oslash^{\circ}$$

LGQ. (Call-by-value)

$$\frac{}{p^{x} \vdash x : p}\ Ax \qquad \frac{\Gamma \vdash M : A}{\Gamma, A^{\varepsilon} \vdash (\varepsilon\ M)}\ D$$

$$\frac{\Pi, A^{y} \bullet B^{z} \vdash M}{\Pi, A \otimes B^{x} \vdash \textbf{case}\ x\ \textbf{of}\ y \otimes z.M}\ \otimes^{\bullet} \qquad \frac{\Gamma \vdash M : A \quad \Delta \vdash N : B}{\Gamma \bullet \Delta \vdash \langle M \otimes N\rangle : A \otimes B}\ \otimes^{\circ}$$

$$\frac{\Gamma, A^{\varepsilon} \multimapinv B^{y} \vdash M}{\Gamma \vdash \lambda\langle \varepsilon \otimes y\rangle M : B\backslash A}\ \backslash^{\circ} \qquad \frac{\Delta \vdash N : B \quad \Pi, A^{x} \vdash M}{\Pi \multimapinv \Delta, B\backslash A^{z} \vdash (z\ \langle N \otimes \lambda x M\rangle)}\ \backslash^{\bullet}$$

$$\frac{\Gamma, B^{\kappa} \circ A^{\varepsilon} \vdash M}{\Gamma \vdash \lambda\langle \kappa \otimes \varepsilon\rangle M : A \oplus B}\ \oplus^{\circ} \qquad \frac{\Sigma, B^{y} \vdash N \quad \Pi, A^{x} \vdash M}{\Sigma \circ \Pi, A \oplus B^{z} \vdash (z\ \langle \lambda y N \otimes \lambda x M\rangle)}\ \oplus^{\bullet}$$

$$\frac{\Pi, A^{x} \mathrel{\bullet\!\!-} B^{\kappa} \vdash M}{\Pi, A \oslash B^{z} \vdash \textbf{case}\ z\ \textbf{of}\ x \otimes \kappa.M}\ \oslash^{\bullet} \qquad \frac{\Sigma, B^{z} \vdash N \quad \Gamma \vdash M : A}{\Gamma \mathrel{-\!\!\bullet} \Sigma \vdash \langle \lambda z N \otimes M\rangle : A \oslash B}\ \oslash^{\circ}$$

Fig. 7. Explicating the CBN and CBV interpretations of ([2]) through the display calculi **LGT** and **LGQ**. For reasons of space, we discuss only the binary connectives and have refrained from mentioning the display postulates (see Figure 3). In addition, only rules for $\backslash, \oslash$ are explicated, those for $/, \otimes$ being similar.

4 Comparison

Bernardi and Moortgat (B&M, [2]) alternatively propose designing a Montagovian semantics for **LG** on the assumption that all formulae are of equal polarity: either all negative, inducing a call-by-name translation (CBN), or all positive, corresponding to call-by-value (CBV). Thus, the corresponding maps $\lfloor\cdot\rfloor$ and $\lceil\cdot\rceil$ restrict to the top- and bottom levels respectively of the polarity table in §2, inserting additional negations for positives in CBN and negatives in CBV:

	$\lfloor\cdot\rfloor$ (CBN)	$\lceil\cdot\rceil$ (CBV)
p	$\neg p$	p
$A/B, B\backslash A$	$\neg\lfloor B\rfloor \otimes \lfloor A\rfloor$	$\neg(\lceil B\rceil \otimes \neg\lceil A\rceil)$
$B \oslash A, A \oslash B$	$\neg(\lfloor B\rfloor \otimes \neg\lfloor A\rfloor)$	$\neg\lceil B\rceil \otimes \lceil A\rceil$

B&M code their derivations inside a variation of Curien and Herbelin's $\bar{\lambda}\mu\tilde{\mu}$-calculus. However, to facilitate comparison with our own approach, we express in Figure 7 B&M's CBN and CBV translations by 'stouped' display calculi **LGT** and **LGQ**, named after their obvious sources of inspiration [4]. Sequents, as well as their display equivalences, carry over straightforwardly from §2, their interpretations being as before. In particular, atomic (co)structures are interpreted

$$\begin{array}{rclcrcl}\lfloor A^x\rfloor &=& \{\neg\lfloor A\rfloor^x\} & \quad & \lfloor A^\varepsilon\rfloor &=& \{\lfloor A\rfloor^\varepsilon\}\\ \lceil A^x\rceil &=& \{\lceil A\rceil^x\} & & \lceil A^\varepsilon\rceil &=& \{\neg\lceil A\rceil^\varepsilon\}\end{array}$$

The differences between the various display calculi of Figures 3 and 7 are now reduced to the maintenance of the stoup. In particular, **LGT**, considering all formulas negative, allows only hypotheses inside, whereas **LGQ** restricts the contents of the stoup to conclusions.

In comparing the various proposals at the level of the lexical semantics, the polarized approach often amounts to the more economic one. For instance, a ditransitive verb like *offered*, categorized $((np\backslash s)/np)/np$ (abbreviated dtv), receives denotations of types $\neg\lfloor dtv\rfloor$ (CBN), $\lceil dtv\rceil$ (CBV) and $\neg[\![dtv]\!]$ (polarized):

$$\begin{array}{rl}\text{CBN:} & \lambda\langle Z,\langle Y,\langle X,q\rangle\rangle\rangle(Z\ \lambda z(Y\ \lambda y(X\ \lambda x(q\ (((\textsc{offered}\ z)\ y)\ x)))))\\ \text{CBV:} & \lambda\langle z,Y\rangle(Y\ \lambda\langle y,X\rangle(X\ \lambda\langle x,q\rangle(q\ (((\textsc{offered}\ z)\ y)\ x))))\\ \text{polarized:} & \lambda\langle z,\langle y,\langle q,x\rangle\rangle\rangle(q\ (((\textsc{offered}\ z)\ y)\ x))\end{array}$$

Acknowledgements. This work has benefited from discussions with Michael Moortgat, Jeroen Bransen and Vincent van Oostrom, as well as from comments from two anonymous referees. All remaining errors are my own.

References

1. Andreoli, J.-M.: Logic programming with focusing proofs in linear logic. Journal of Logic and Computation 2(3), 297–347 (1992)
2. Bernardi, R., Moortgat, M.: Continuation Semantics for Symmetric Categorial Grammar. In: Leivant, D., de Queiroz, R. (eds.) WoLLIC 2007. LNCS, vol. 4576, pp. 53–71. Springer, Heidelberg (2007)

3. Cockett, J.R.B., Seely, R.A.G.: Weakly distributive categories. Journal of Pure and Applied Algebra, 45–65 (1991)
4. Danos, V., Joinet, J.-B., Schellinx, H.: LKQ and LKT: Sequent calculi for second order logic based upon dual linear decompositions of classical implication. In: Proceedings of the workshop on Advances in Linear Logic, New York, NY, USA, pp. 211–224. Cambridge University Press (1995)
5. De Groote, P., Lamarche, F.: Classical Non Associative Lambek Calculus. Studia Logica 71, 355–388 (2002)
6. Došen, K., Petrić, Z.: Proof-theoretical Coherence. King's College Publications (2004)
7. Girard, J.-Y.: A new constructive logic: Classical logic. Mathematical Structures in Computer Science 1(3), 255–296 (1991)
8. Goré, R.: Substructural logics on display. Logic Journal of the IGPL 6(3), 451–504 (1998)
9. Grishin, V.N.: On a generalization of the Ajdukiewicz-Lambek system. In: Mikhailov, A.I. (ed.) Studies in Nonclassical Logics and Formal Systems, Nauka, Moscow, pp. 315–334 (1983)
10. Kurtonina, N., Moortgat, M.: Relational semantics for the Lambek-Grishin calculus. Mathematics of Language. Citeseerx (2007), doi:10.1.1.92.3297
11. Lambek, J.: The mathematics of sentence structure. American Mathematical Monthly 65, 154–169 (1958)
12. Lambek, J.: On the calculus of syntactic types. In: Jakobson, R. (eds.) tructure of Language and its Mathematical Aspects, Proceedings of the Twelfth Symposium in Applied Mathematics (1961)
13. Moortgat, M.: Categorial type logics. In: Handbook of Logic and Language, pp. 93–177. Elsevier (1997)
14. Moortgat, M.: Symmetric categorial grammar. Journal of Philosophical Logic 38(6), 681–710 (2009)
15. Moortgat, M.: Symmetric categorial grammar: residuation and Galois connections. Linguistic Analysis. Special Issue Dedicated to Jim Lambek 36(1-4) (2010)
16. Okada, M.: A uniform semantic proof for cut-elimination and completeness of various first and higher order logics. Theoretical Computer Science 281(1-2), 471–498 (2002)
17. Prawitz, D.: Natural Deduction. Dover Publications (2006)
18. Smullyan, R.M.: First–Order Logic. Springer (1968); Revised edn. Dover Press, NY (1994)
19. Reus, B., Lafont, Y., Streichter, T.: Continuation semantics or expressing implication by negation. Technical Report 93-21. University of Munich (1993)

Two Models of Learning Iterated Dependencies

Denis Béchet[1], Alexander Dikovsky[1], and Annie Foret[2]

[1] LINA UMR CNRS 6241, Université de Nantes, France
{Denis.Bechet,Alexandre.Dikovsky}@univ-nantes.fr
[2] IRISA, Université de Rennes1, France
Annie.Foret@irisa.fr

Abstract. We study the learnability problem in the family of Categorial Dependency Grammars (CDG), a class of categorial grammars defining unlimited dependency structures. CDG satisfying a reasonable condition on iterated (i.e., repeatable and optional) dependencies are shown to be incrementally learnable in the limit.

1 Introduction

The idea of grammatical inference is as follows. A class of languages defined using a class of grammars $\mathcal{G}$ is learnable if there exists a learning algorithm ϕ from finite sets of words generated by the target grammar $G_0 \in \mathcal{G}$ to hypothetical grammars in $\mathcal{G}$, such that (i) the sequence of languages generated by the output grammars converges to the target language $L(G_0)$ and (ii) this is true for any increasing enumeration of finite sublanguages of $L(G_0)$.

This concept due to E.M. Gold [8] is also called `learning from strings`. More generally, the hypothetical grammars may be generated from finite sets of structures defined by the target grammar. This kind of learning is called `learning from structures`. Both concepts were intensively studied (see excellent surveys in [1] and [10]). Most results are pessimistic. In particular, any family of grammars generating all finite languages and at least one infinite language (as it is the case of all classical grammars) is not learnable from strings. Nevertheless, due to several sufficient conditions of learnability, such as `finite elasticity` [15,12] and `finite thickness` [14], some interesting positive results were obtained. In particular, k-rule string and term generating grammars are learnable from strings for every k [14] and k-`rigid` (i.e. assigning no more than k types per word) classical categorial grammars (CG) are learnable from so called "function-argument" structures and also from strings [4,10].

In this paper we study the learnability problem in the family of Categorial Dependency Grammars (CDG) introduced in [7]. CDG is a class of categorial grammars defining unlimited dependency structures. In [3] it is shown that, in contrast with the classical categorial grammars, the `rigid` (i.e. 1-rigid) CDG are not learnable. This negative effect is due to the use of iterated subtypes which express the *iterated* dependencies i.e. unlimited repeatable optional dependencies (those of noun modifiers and of verb circumstantials). On the other hand, it is also shown that the k-rigid CDG with iteration-free types are learnable

P. de Groote and M.-J. Nederhof (Eds.): Formal Grammar 2010/2011, LNCS 7395, pp. 17–32, 2012.

from the so called "dependency nets" (an analogue of the function-argument structures adapted to CDG) and also from strings. However, the iteration-free CDG cannot be considered as an acceptable compromise because the linguistically relevant dependency grammars must express the iterated dependencies. Below we propose a pragmatic solution of the learnability problem for CDG with iterated dependency subtypes. It consists in limiting the family of CDG to the grammars satisfying a strong condition on the iterated dependencies. Intuitively, in the grammars satisfying this condition, the iterated dependencies and the dependencies repeatable at least K times for some fixed K are indiscernible. This constraint, called below *K-star-revealing*, is more or less generally accepted in the traditional dependency syntax (cf. [11], where $K = 2$). For the class of K-star-revealing CDG, we show an algorithm which incrementally learns the target CDG from the dependency structures in which the iteration is not marked. We compare this new model of learning grammars from structures with the traditional model as applied to iterated dependencies. As one might expect, the CDG with unlimited iterated dependencies are not learnable from input functor/argument-like structures. Moreover, this is true even for the rigid CDG.

2 Background

2.1 Categorial Dependency Grammars

Categorial dependency grammars [6] may be seen as an assignment to words of first order dependency types of the form: $t = [l_m\backslash \ldots \backslash l_1\backslash g/r_1/\ldots/r_n]^P$. Intuitively, $w \mapsto [\alpha\backslash d\backslash\beta]^P$ means that the word w has a left subordinate through dependency d (similar for the right subtypes $[\alpha/d/\beta]^P$). The *head subtype* g in $w \mapsto [\alpha\backslash g/\beta]^P$ intuitively means that w is governed through dependency g. In this way t defines all local (projective) dependencies of a word.

Example 1. For instance, the assignment:
$in \mapsto [c{-}copul/prepos{-}in]$, $the \mapsto [det]$, $Word \mapsto [det\backslash pred]$
$beginning \mapsto [det\backslash prepos{-}in]$, $was \mapsto [c{-}copul\backslash S/pred]$
determines the projective dependency structure in 1.

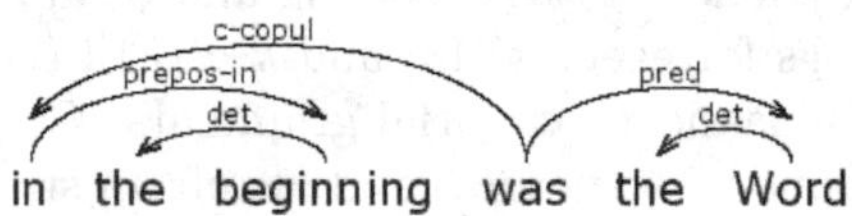

Fig. 1. Projective dependency structure

The intuitive meaning of subtype P, called *potential*, is that it defines the distant (non-projective, discontinuous) dependencies of the word w. P is a string of *polarized valencies*, i.e. of symbols of four kinds: $\swarrow d$ (*left negative valency* d), $\searrow d$ (*right negative valency* d), $\nwarrow d$ (*left positive valency* d), $\nearrow d$ (*right positive* *valency d*). Intuitively, $v = \nwarrow d$ requires a subordinate through dependency d situated *somewhere* on the left, whereas the *dual* valency $\breve{v} = \swarrow d$ requires a governor through the same dependency d situated somewhere on the right. So together they describe the discontinuous dependency d. Similar for the other pairs of dual valencies. For negative valencies $\swarrow d, \searrow d$ are provided a special kind of subtypes $\#(\swarrow d)$, $\#(\searrow d)$. Intuitively, they serve to check the adjacency of a distant word subordinate through discontinuous dependency d to a *host word*. The dependencies of these types are called *anchor*. A *primitive dependency type* is either a *local dependency name* d or its *iteration* d^* or an anchor type $\#(v)$.

Example 2. For instance, the assignment:
$elle \mapsto [pred]$, $la \mapsto [\#(\swarrow clit{-}a{-}obj)]^{\swarrow clit\text{-}a\text{-}obj}$,
$lui \mapsto [\#(\swarrow clit{-}3d{-}obj)]^{\swarrow clit\text{-}3d\text{-}obj}$, $donnée \mapsto [aux]^{\nwarrow clit\text{-}3d\text{-}obj \nwarrow clit\text{-}a\text{-}obj}$,
$a \mapsto [\#(\swarrow clit{-}3d{-}obj)\backslash\#(\swarrow clit{-}a{-}obj)\backslash pred\backslash S/aux{-}a{-}d]$
determines the non projective DS in Fig. 2.[1]

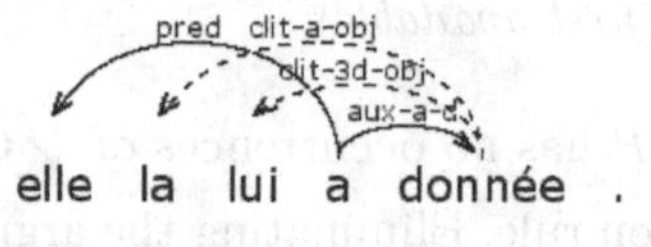

(*fr.* **she it*$_{\mathbf{g}=fem}$ *to him has given*)

Fig. 2. Non-projective dependency structure

Definition 1. *Let $w = a_1 \ldots a_n$ be a string, W be the set of all occurrences of symbols in w and $C = \{d_1, \ldots, d_m\}$ be a set of* dependency names. *A graph $D = (W, E)$ with labeled arcs is a* dependency structure (DS) *of w if it has a* root, *i.e. a node $a_0 \in W$ such that (i) for any node $a \in W$, $a \neq a_0$, there is a path from a_0 to a and (ii) there is no arc (a', d, a_0).*[2] *An arc $(a_1, d, a_2) \in E$ is called* dependency d *from a_1 to a_2. The linear order on W induced by w is the* precedence order *on D.*

Definition 2. *Let* $\mathbf{C}$ *be a set of* local dependency names *and* $\mathbf{V}$ *be a set of* valency names.

The expressions of the form $\swarrow v$, $\nwarrow v$, $\searrow v$, $\nearrow v$, where $v \in \mathbf{V}$, are called polarized valencies. $\nwarrow v$ *and* $\nearrow v$ *are* positive, $\swarrow v$ *and* $\searrow v$ *are* negative; $\nwarrow v$ *and* $\swarrow v$ *are* left, $\nearrow v$ *and* $\searrow v$ *are* right. *Two polarized valencies with the same valency name and orientation, but with the opposite signs are* dual.

An expression of one of the forms $\#(\swarrow v)$, $\#(\searrow v)$, $v \in \mathbf{V}$, is called anchor type *or just* anchor. *An expression of the form d^* where $d \in \mathbf{C}$, is called* iterated dependency type.

[1] Anchors are not displayed for a better readability.

[2] Evidently, every DS is connected and has a unique root.

Local dependency names, iterated dependency types and anchor types are primitive types.

An expression of the form $t = [l_m\backslash \ldots \backslash l_1\backslash H/ \ldots /r_1 \ldots /r_n]$ *in which* $m, n \geq 0$, $l_1, \ldots, l_m, r_1, \ldots, r_n$ *are primitive types and* H *is either a local dependency name or an anchor type, is called* basic dependency type. $l_1, \ldots, l_m$ *and* $r_1, \ldots, r_n$ *are respectively left and right* argument subtypes *of* t. H *is called* head subtype *of* t *(or* head type *for short).*

A (possibly empty) string P *of polarized valencies is called* potential.[3]

A dependency type *is an expression* B^P *in which* B *is a basic dependency type and* P *is a potential.* $\mathbf{CAT(C, V)}$ *and* $\mathbf{B(C)}$ *will denote respectively the set of all dependency types over* $\mathbf{C}$ *and* $\mathbf{V}$ *and the set of all basic dependency types over* $\mathbf{C}$.

CDG are defined using the following calculus of dependency types [4]

$\mathbf{L^l}$. $C^{P_1}[C\backslash\beta]^{P_2} \vdash [\beta]^{P_1P_2}$

$\mathbf{I^l}$. $C^{P_1}[C^*\backslash\beta]^{P_2} \vdash [C^*\backslash\beta]^{P_1P_2}$

$\mathbf{\Omega^l}$. $[C^*\backslash\beta]^P \vdash [\beta]^P$

$\mathbf{D^l}$. $\alpha^{P_1(\swarrow C)P(\nwarrow C)P_2} \vdash \alpha^{P_1PP_2}$, if the potential $(\swarrow C)P(\nwarrow C)$ satisfies the following pairing rule **FA** (*first available*):

$$\textbf{FA}: \quad P \text{ has no occurrences of } \swarrow C, \nwarrow C.$$

$\mathbf{L^l}$ is the classical elimination rule. Eliminating the argument subtype $C \neq \#(\alpha)$ it constructs the (*projective*) dependency C and concatenates the potentials. $C = \#(\alpha)$ creates the *anchor dependency.* $\mathbf{I^l}$ derives $k > 0$ instances of C. $\mathbf{\Omega^l}$ serves for the case $k = 0$. $\mathbf{D^l}$ creates *discontinuous dependencies.* It pairs and eliminates dual valencies with name C satisfying the rule **FA** to create the discontinuous dependency C.

Definition 3. *A* categorial dependency grammar (CDG) *is a system* $G = (W, \mathbf{C}, S, \lambda)$, *where* W *is a finite set of words,* $\mathbf{C}$ *is a finite set of local dependency names containing the selected name* S *(an axiom), and* λ, *called* lexicon, *is a finite substitution on* W *such that* $\lambda(a) \subset \mathbf{CAT(C, V)}$ *for each word* $a \in W$.

For a DS D *and a string* x, *let* $G(D, x)$ *denote the relation:* D *is constructed in a proof* $\Gamma \vdash S$ *for some* $\Gamma \in \lambda(x)$. *Then the* language *generated by* G *is the set* $L(G) =_{df} \{w \mid \exists D\ G(D, w)\}$ *and the* DS-language *generated by* G *is the set* $\Delta(G) =_{df} \{D \mid \exists w\ G(D, w)\}$. $\mathcal{D}(CDG)$ *and* $\mathcal{L}(CDG)$ *will denote the families of DS-languages and languages generated by these grammars.*

Example 3. For instance, the proof in Fig. 3 shows that the DS in Fig. 2 belongs to the DS-language generated by a grammar containing the type assignments shown above for the french sentence *Elle la lui a donnée*.

---

[3] In fact, the potentials should be defined as multi-sets. We define them as strings in order to simplify definitions and notation. Nevertheless, to make the things clear, below we will present potentials in the normal form, where all left valencies precede all right valencies.

[4] We show left-oriented rules. The right-oriented are symmetrical.

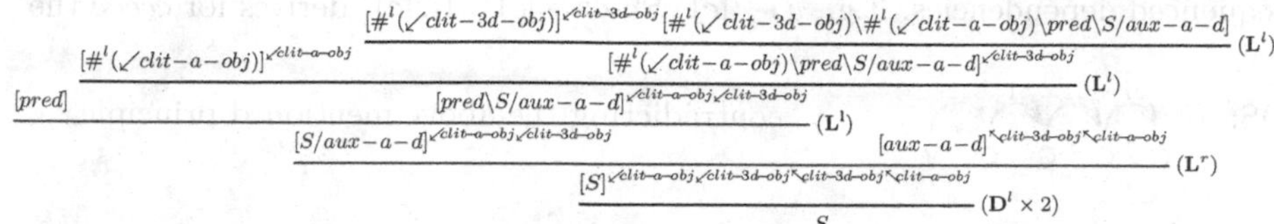

Fig. 3. Dependency structure correctness proof

CDG are very expressive. Evidently, they generate all CF-languages. They can also generate non-CF languages.

Example 4. [7]. The CDG:
$a \mapsto A^{\swarrow A}, [A\backslash A]^{\swarrow A}$, $b \mapsto [B/C]^{\nwarrow A}, [A\backslash S/C]^{\nwarrow A}$, $c \mapsto C, [B\backslash C]$
generates the language $\{a^n b^n c^n \mid n > 0\}$.[5]

Seemingly, the family $\mathcal{L}(CDG)$ of CDG-languages is different from that of the mildly context sensitive languages [9,13] generated by multi-component TAG, linear CF rewrite systems and some other grammars. $\mathcal{L}(CDG)$ contains non-TAG languages, e.g. $L^{(m)} = \{a_1^n a_2^n ... a_m^n \mid n \geq 1\}$ for all $m > 0$. In particular, it contains the language $MIX = \{w \in \{a, b, c\}^+ \mid |w|_a = |w|_b = |w|_c\}$ [2], for which E. Bach has conjectured that it is not mildly CS. On the other hand, [5] conjectures that this family does not contain the TAG language $L_{copy} = \{xx \mid x \in \{a, b\}^*\}$. This comparison shows a specific nature of the valencies' pairing rule **FA**. It can be expressed in terms of valencies' bracketing. For this, one should interpret $\swarrow d$ and $\nearrow d$ as *left brackets* and $\nwarrow d$ and $\searrow d$ as *right brackets*. A potential is *balanced* if it is well bracketed in the usual sense.

CDG have an important property formulated in terms of two images of sequences of types γ: the *local projection* $\|\gamma\|_l$ and the *valency projection* $\|\gamma\|_v$:
1. $\|\varepsilon\|_l = \|\varepsilon\|_v = \varepsilon$; $\|\alpha\gamma\|_l = \|\alpha\|_l\|\gamma\|_l$ and $\|\alpha\gamma\|_v = \|\alpha\|_v\|\gamma\|_v$ for a type α.
2. $\|C^P\|_l = C$ et $\|C^P\|_v = P$ for every type C^P.

Theorem 1. *[5,6] For a CDG G with lexicon λ and a string x, $x \in L(G)$ iff there is $\Gamma \in \lambda(x)$ such that $\|\Gamma\|_l$ is reduced to S without the rule* **D** *and $\|\Gamma\|_v$ is balanced.*

On this property resides a polynomial time parsing algorithm for CDG [5,6].

It is important to understand why the iterated subtypes are unavoidable in dependency grammars. This is due to one of the basic principles of dependency syntax, which concerns the optional repeatable dependencies (cf. [11]): all modifiers of a noun n share n as their *governor* and, similar, all circonstants of a verb v share v as their *governor*. For instance, in the dependency structure in Figure 4 there are three circonstants dependent on the same verb *fallait* (fr. *had to*). In particular, this means that the iterated dependencies cannot be simulated through recursive types. Indeed, $a \mapsto [\alpha\backslash d]$ and $b \mapsto [d\backslash\beta]$ derives the dependency $a \xleftarrow{d} b$ for ab. Therefore, the recursive types derive

[5] One can see that the DS may be not trees.

sequenced dependencies. E.g., $v \mapsto [c1\backslash S]$, $c \mapsto [c1\backslash c1], [c1]$ derives for *ccccv* the

DS: c c c c v . (dependencies c1 c1 c1 c1) contradicting the above mentioned principle.

2.2 Learnability, Finite Elasticity and Limit Points

With every grammar $G \in \mathcal{C}$ is related an **observation set** $\Phi(G)$ of G. This may be the generated language $L(G)$ or an image of the constituent or dependency structures generated by G. Below we call **training sequence** for G an enumeration of $\Phi(G)$. An algorithm $\mathcal{A}$ is an **inference algorithm** for $\mathcal{C}$ if, for every grammar $G \in \mathcal{C}$, $\mathcal{A}$ applies to its training sequences σ of $\Phi(G)$ and, for every initial subsequence $\sigma[i] = \{s_1, \ldots, s_i\}$ of σ, it returns a **hypothetical grammar** $\mathcal{A}(\sigma[i]) \in \mathcal{C}$. $\mathcal{A}$ **learns** a **target grammar** $G \in \mathcal{C}$ if on any training sequence σ for G $\mathcal{A}$ stabilizes on a grammar $\mathcal{A}(\sigma[T]) \equiv G$.[6] The grammar $\lim\limits_{i\to\infty} \mathcal{A}(\sigma[i]) = \mathcal{A}(\sigma[T])$ returned at the stabilization step is the **limit grammar**. $\mathcal{A}$ **learns** $\mathcal{C}$ if it learns every grammar in $\mathcal{C}$. $\mathcal{C}$ is **learnable** if there is an inference algorithm learning $\mathcal{C}$.

Learnability and unlearnability properties have been widely studied from a theoretical point of view. In particular, in [15,12] was introduced finite elasticity, a property of classes of languages implying their learnability. The following elegant presentation of this property is cited from [10].

Definition 4 (Finite Elasticity). *A class $\mathcal{L}$ of languages has* infinite elasticity *iff $\exists(e_i)_{i\in\mathbb{N}}$ an infinite sequence of sentences, $\exists(L_i)_{i\in\mathbb{N}}$ an infinite sequence of languages of $\mathcal{L}$ such that $\forall i \in \mathbb{N}$: $e_i \notin L_i$ and $\{e_0, \ldots, e_{i-1}\} \subseteq L_i$. A class has* finite elasticity *iff it has not infinite elasticity.*

Theorem 2. [**Wright 1989**] *A class that is not learnable has infinite elasticity.*

Corollary 1. *A class that has finite elasticity is learnable.*

The finite elasticity can be extended from a class to every class obtained by a *finite-valued relation*[7]. We use here a version of the theorem that has been proved in [10] and is useful for various kinds of languages (strings, structures, nets) that can be described by lists of elements over some alphabets.

Theorem 3. [**Kanazawa 1998**] *Let $\mathcal{L}$ be a class of languages over Γ that has finite elasticity, and let $R \subseteq \Sigma^* \times \Gamma^*$ be a finite-valued relation. Then the class of languages $\{R^{-1}[L] = \{s \in \Sigma^* \mid \exists u \in L \wedge (s,u) \in R\} \mid L \in \mathcal{L}\}$ has finite elasticity.*

---

6 $\mathcal{A}$ **stabilizes** on σ on step T means that T is the minimal number t for which there is no $t_1 > t$ such that $\mathcal{A}(\sigma[t_1]) \neq \mathcal{A}(\sigma[t])$.

7 A relation $R \subseteq \Sigma^* \times \Gamma^*$ is finite-valued iff for every $s \in \Sigma^*$, there are at most finitely many $u \in \Gamma^*$ such that $(s,u) \in R$.

Definition 5 (Limit Points). *A class $\mathcal{L}$ of languages has* a limit point *iff there exists an infinite sequence $(L_n)_{n\in N}$ of languages in $\mathcal{L}$ and a language $L \in \mathcal{L}$ such that: $L_0 \subsetneq L_1 \ldots \subsetneq \ldots \subsetneq L_n \subsetneq \ldots$ and $L = \bigcup_{n\in N} L_n$ (L is* a limit point *of $\mathcal{L}$).*

Limit Points Imply Unlearnability. If the languages of the grammars in a class $\mathcal{G}$ have a limit point then the class $\mathcal{G}$ is *unlearnable.* [8]

2.3 Limit Points for CDGs with Iterated Subtypes

In [3] it is shown that, in contrast with the classical categorial grammars, the `rigid` (i.e. 1-rigid) CDG are not learnable. This negative effect is due to the use of iterated subtypes. We recall the limit point construction of [3] concerning iterative subtypes and discuss it later.

Definition 6. *Let S, A, B be local dependency names. We define G'_n, G'_* by:*

$$C'_0 = S \qquad G'_0 = \{a \mapsto A, b \mapsto B, c \mapsto C'_0\}$$
$$C'_{n+1} = C'_n \;/\; A^* \;/\; B^* \qquad G'_n = \{a \mapsto A, b \mapsto B, c \mapsto [C'_n]\}$$
$$G'_* = \{a \mapsto A, b \mapsto A, c \mapsto [S \;/\; A^*]\}$$

Theorem 4. *These constructions yield a limit point as follows [3]:*
$L(G'_n) = \{c(b^*a^*)^k \mid k \le n\}$ and $L(G'_*) = c\{b,a\}^*$

Corollary 2. *The constructions show the non-learnability from strings for the classes of (rigid) grammars allowing iterative subtypes (A^*).*

We observe that in these constructions, the number of iterative subtypes (A^*) is not bound.

3 Incremental Learning

Below we show an incremental algorithm `strongly` learning CDG from DS. This means that $\Delta(G)$ serves as the observation set $\Phi(G)$ and that the limit grammar is `strongly` equivalent to the target grammar. From the very beginning, it should be clear that, in contrast with the constituent structure grammars and also with the classical CG, the existence of such learning algorithm is not guaranteed because, due to the iterated subtypes, the straightforward arguments of subtypes' set cardinality do not work. In particular, even the rigid CDG (monotonic with respect to the subgrammar partial order (PO)) do not satisfy the finite thickness condition. On the other hand, the learning algorithm $\mathcal{A}$ below is `incremental` in the sense that every next hypothetical CDG $\mathcal{A}(\sigma[i+1])$ is an "extension" of the preceding grammar $\mathcal{A}(\sigma[i])$ and it is so `without any rigidity constraint`. Incremental learning algorithms are rare. Those we know, are unification based and apply only to *rigid* grammars (cf. [4] and [3]). They cannot be considered as practical (at least for the NLP) because the real application grammars are never rigid. In the cases when the k-rigid learnability is a consequence of the rigid

[8] This implies that the class has infinite elasticity.

learnability, it is only of a theoretical interest because the existence of a learning algorithm is based on the Kanazawa's finite-valued-relation reduction [10].

Our notion of incrementality is based on a partial "flexibility" order $\preceq$ on CDGs. Basically, the order corresponds to grammar expansion in the sense that $G_1 \preceq G_2$ means that G_2 defines no less dependency structures than G_1 and at least as precise dependency structures as G_1. This PO is the reflexive-transitive closure of the following preorder $<$.

Definition 7. *For a type* $t = [l_m\backslash \cdots l_1\backslash g/r_1 \cdots /r_n]^P$ *a dependency name* c, $i \geq 0$, $0 \leq j \leq m$, *let* $t_c^{(i\backslash,j)} = [l_m\backslash \cdots \backslash l_j\backslash c \cdots \backslash c\backslash l_{j-1}\backslash \cdots l_1\backslash g/r_1 \cdots /r_n]^P$ *(i times) and* $t_c^{(*\backslash,j)} = [l_m\backslash \cdots \backslash l_j\backslash c* \backslash l_{j-1}\backslash \cdots l_1\backslash g/r_1 \cdots /r_n]^P$. *Respectively, for* $0 \leq k \leq n$ $t_c^{(i/,k)} = [l_m\backslash \cdots l_1\backslash g/r_1 \cdots /r_{k-1}/c \cdots /c/r_k/ \cdots /r_n]^P$ *and* $t_c^{(*/,k)} = [l_m\backslash \cdots l_1\backslash g/r_1 \cdots /r_{k-1}/c* /r_k/ \cdots /r_n]^P$. *Then:*
1. $t_c^{(i\backslash,j)} < t_c^{(*\backslash,j)}$ *and* $t_c^{(i/,k)} < t_c^{(*/,k)}$ *for all* $i \geq 0$, $0 \leq j \leq m$ *and* $0 \leq k \leq n$
2. $\tau < \tau'$ *for sets of types* τ, τ', *if either:*
  (i) $\tau' = \tau \cup \{t\}$ *for a type* $t \notin \tau$ *or*
  (ii) $\tau = \tau_0 \cup \{t'\}$ *and* $\tau' = \tau_0 \cup \{t''\}$
for a set of types τ_0 *and some types* t', t'' *such that* $t' < t''$.
3. $\lambda < \lambda'$ *for two type assignments* λ *and* λ', *if* $\lambda(w') < \lambda'(w')$ *for a word* w' *and* $\lambda(w) = \lambda'(w)$ *for all words* $w \neq w'$.
4. $\preceq$ *is the PO which is the reflexive-transitive closure of the preorder* $<$.

It is not difficult to prove that the expressive power of CDG monotonically grows with respect to this PO.

Proposition 1. *Let* G_1 *and* G_2 *be two CDG such that* $G_1 \preceq G_2$. *Then* $\Delta(G_1) \subseteq \Delta(G_2)$ *and* $\mathcal{L}(G_1) \subseteq \mathcal{L}(G_2)$.

The flexibility PO $\preceq$ serves to define the following main notion of `incremental learning`.

Definition 8. *Let* $\mathcal{A}$ *be an inference algorithm for* $\mathcal{CDG}$ *from DS and* σ *be a training sequence for a CDG* G.
1. $\mathcal{A}$ *is* `monotonic` *on* σ *if* $\mathcal{A}(\sigma[i]) \preceq \mathcal{A}(\sigma[j])$ *for all* $i \leq j$.
2. $\mathcal{A}$ *is* `faithful` *on* σ *if* $\Delta(\mathcal{A}(\sigma[i])) \subseteq \Delta(G)$ *for all* i.
3. $\mathcal{A}$ *is* `expansive` *on* σ *if* $\sigma[i] \subseteq \Delta(\mathcal{A}(\sigma[i]))$ *for all* i.

Definition 9. *Let* G_1 *and* G_2 *be two CDG,* $G_1 \equiv_s G_2$ *iff* $\Delta(G_1) = \Delta(G_2)$.

Theorem 5. *Let* σ *be a training sequence for a CDG* G. *If an inference algorithm* $\mathcal{A}$ *is monotonic, faithful, and expansive on* σ, *and if* $\mathcal{A}$ *stabilizes on* σ *then* $\lim_{i\to\infty} \mathcal{A}(\sigma[i]) \equiv_s G$.

Proof. Indeed, stabilization implies that $\lim_{i\to\infty} \mathcal{A}(\sigma[i]) = \mathcal{A}(\sigma[T])$ for some T. Then $\Delta(\mathcal{A}(\sigma[T])) \subseteq \Delta(G)$ because of faithfulness. At the same time, by expansiveness and monotonicity, $\Delta(G) = \sigma = \bigcup_{i=1}^{\infty} \sigma[i] \subseteq \bigcup_{i=1}^{\infty} \Delta(\mathcal{A}(\sigma[i])) \subseteq \bigcup_{i=1}^{T} \Delta(\mathcal{A}(\sigma[i])) \subseteq \Delta(\mathcal{A}(\sigma[T]))$.

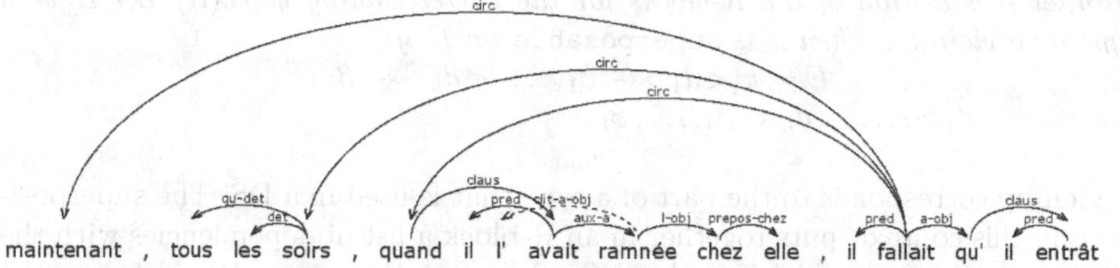

(fr. *now all the evenings when he took her home he had to enter* [*M.Proust*])

Fig. 4. Iterated circumstantial dependency

As we explain it in Section 4, the unlearnability of rigid or k-rigid CDG is due to the use of iterated types. Such types are unavoidable in real grammars (cf. the iterated dependency *circ* in Fig. 4). But in particular in the real application grammars, the iterated types have very special properties. Firstly, the discontinuous dependencies are never iterated. Secondly, in natural languages, the optional constructions repeated successively several times (two or more) are exactly those iterated. This is the resource we use to resolve the learnability problem. To formalize these properties we need some notations and definitions. The main definition concerns a restriction on the class of grammars that is learned. This class corresponds to grammars where an argument that is used at least K times in a DS must be an iterated argument. Such grammars are called `K-star-revealing` grammars.

Definition 10
1. `Repetition blocks` *(R-blocks) : For* $d \in \mathbf{C}$,

$$LB_d = \{t_1\backslash \cdots \backslash t_i \mid i > 0, t_1, \ldots, t_i \in \{d\} \cup \{x^* \mid x \in \mathbf{C}\}\}$$

and symmetrically for RB_d.
2. `Patterns`*: Patterns are defined exactly as types, but in the place of* $\mathbf{C}$*, we use* $\mathbf{G}$*, where* $\mathbf{G}$ *is the set of* gaps $\mathbf{G} = \{<d> \mid d \in \mathbf{C}\}$*. Moreover, for any* α, β, P *and* d*,* $[\alpha\backslash <d> \backslash <d> \backslash\beta]^P$ *and* $[\alpha/ <d> / <d> /\beta]^P$ *are not patterns.*
3. `Vicinity`*: Let* D *be a DS in which an occurrence of a word* w *has : the incoming local dependency* h *(or the axiom* S*), the left projective dependencies or anchors* $l_k, \ldots, l_1$ *(in this order), the right projective dependencies or anchors* $r_1, \ldots, r_m$ *(in this order), and discontinuous dependencies* $p_1(d_1), \ldots, p_n(d_n)$*, where* $p_1, \ldots, p_n$ *are polarities and* $d_1, \ldots, d_n \in \mathbf{V}$ *are valency names. Then the* `vicinity` *of* w *in* D *is the type*

$$V(w, D) = [l_1\backslash \ldots \backslash l_k\backslash h/r_m/ \ldots /r_1]^P,$$

in which P *is a permutation of* $p_1(d_1), \ldots, p_n(d_n)$ *in a standard lexicographical order, for instance, compatible with the polarity order* $\nwarrow < \searrow < \swarrow < \nearrow$.
4. `Superposition` *and* `indexed occurrences` *of R-blocks :*
(i) Let π *be a pattern,* $\beta_1, \ldots, \beta_k$ *be R-blocks and* $< d_1 >, \ldots, < d_k >$ *be gaps. Then* $\pi(< d_1 > \leftarrow \beta_1, \ldots, < d_k > \leftarrow \beta_k)$ *is the expression resulting from* π *by the*

parallel substitution of the R-blocks for the corresponding gaps.(ii) Let E be a type or a vicinity. Then π is **superposable** *on E if:*

$$E = \pi(<d_1>\leftarrow \beta_1,\ldots,<d_k>\leftarrow \beta_k)$$

for some $<d_1>,\ldots,<d_k>,\beta_1,\ldots,\beta_k$.

A vicinity corresponds to the part of a type that is used in a DS. The superposition, in this context, puts together in an R-block a list of dependencies with the same name some of which may be defined by iterative types. For instance, the verb *fallait* in the DS in Fig. 4 has the vicinity $[pred\backslash circ\backslash circ\backslash circ\backslash S/a{-}obj]$. The pattern superposable on this vicinity is $\pi = [<pred>\backslash <circ>\backslash S/<a{-}obj>]$ and the corresponding type is obtained through the following substitution:

$$\pi(<pred>\leftarrow pred, <circ>\leftarrow circ\backslash circ\backslash circ, <a{-}obj>\leftarrow a{-}obj).$$

The vicinity of the participle *ramenée* is $[aux{-}a/l{-}obj]^{\nwarrow clit\text{-}a\text{-}obj}$. It is the same as the type:

$$[aux{-}a/<l{-}obj>]^{\nwarrow clit\text{-}a\text{-}obj}(<l{-}obj>\leftarrow l{-}obj).$$

Proposition 2. *For every vicinity V there is a single pattern π superposable on V and a single decomposition (*R-decomposition*)*

$$V = \pi(<d_1>\leftarrow \beta_1,\ldots,<d_k>\leftarrow \beta_k)^P$$

Proposition 3. *For $D \in \Delta(G)$ and an occurrence w of a word in D, let $V(w,D) = \pi(<d_1>\leftarrow \beta_1,\ldots,<d_k>\leftarrow \beta_k)^P$ be the R-decomposition of the vicinity of w in D. Then, for every type $t\in\lambda(w)$ which can be used in a proof of D for w, there exists a permutation P' of P such that $\pi^{P'}$ is superposable on t.*

Notation. Let G be a CDG with lexicon λ, w be a word and t be a type. Then G^t_w denotes the CDG with lexicon $\lambda \cup \{w \mapsto t\}$.

Definition 11. *Let $K > 1$ be an integer. We define a CDG $\mathcal{C}^K(G)$, the* **K-star-generalization** *of G, by recursively adding for every word w and every local dependency name d the types*

$$[l_1\backslash\cdots\backslash l_a\backslash d^*\backslash m_1\backslash\cdots\backslash m_b\backslash h/r_1/\cdots/r_c]^P$$

and

$$[l_1\backslash\cdots\backslash l_a\backslash m_1\backslash\cdots\backslash m_b\backslash h/r_1/\cdots/r_c]^P$$

when w has a type assignment $w \mapsto t$, where

$$t = [l_1\backslash\cdots\backslash l_a\backslash t_1\backslash\cdots\backslash t_k\backslash m_1\backslash\cdots\backslash m_b\backslash h/r_1/\cdots/r_c]^P,$$

every $t_1,\ldots,t_k$ is either d or some iterated dependency type x^* and among $t_1,\ldots,t_k$ there are **at least** K **occurrences of** d **or at least one occurrence of** d^*. *Symmetrically, we also add the corresponding types if $t_1,\ldots,t_k$ appear in the right part of t.*

For instance, with $K = 2$, for the type $[a\backslash b^*\backslash a\backslash S/a^*]$, we add $[a\backslash a\backslash S/a^*]$ and $[a\backslash b^*\backslash a\backslash S]$ but also $[a^*\backslash S/a^*]$ and $[S/a^*]$. Recursively, we also add $[a\backslash a\backslash S]$, $[a^*\backslash S]$ and $[S]$. The size of $\mathcal{C}^K(G)$ can be exponential with respect to the size of G.

Definition 12. *Let $K > 1$ be an integer. CDG G is K-***star-revealing** *if $\mathcal{C}^K(G) \equiv_s G$*

For instance, if we define the grammar $G(t)$ by $A \mapsto [a], B \mapsto [b], C \mapsto t$, where t is a type, then we can prove that:

- $G([a^*\backslash S/a^*])$, $G([a^*\backslash b^*\backslash a^*\backslash S])$ and $G([a^*\backslash b\backslash a^*\backslash S])$ are all 2-star-revealing,
- $G([a^*\backslash a\backslash S])$, $G([a^*\backslash b^*\backslash a\backslash S])$ and $G([a\backslash b^*\backslash a\backslash S])$ are not 2-star-revealing.

We see that in a K-star-revealing grammar, one and the same iterated subtype d^* may be used in a type several times. Usually, each occurrence is not in the same block as the local dependency name d. Besides this, there should be less than K occurrences of d in a block if there is no occurrence of d^* and this block is separated from other blocks by types that are not iterated.

Theorem 6. *The class $\mathcal{CDG}^{K\to *}$ of K-star-revealing CDG is (incrementally) learnable from DS.*

To prove the theorem, we present an inference algorithm $\mathbf{TGE}^{(K)}$ (see Fig. 5) which, for every next DS in a training sequence, transforms the observed local, anchor and discontinuous dependencies of every word into a type with repeated local dependencies by introducing iteration for each group of at least K local dependencies with the same name. $\mathbf{TGE}^{(K)}$ is learning $\mathcal{CDG}^{K\to *}$ due to the following two statements.

Lemma 1. *The inference algorithm $\mathbf{TGE}^{(K)}$ is monotonic, faithful and expansive on every training sequence σ of a K-star-revealing CDG.*

Proof. By definition, the algorithm $\mathbf{TGE}^{(K)}$ is `monotonic` (the lexicon is always extended). It is `expansive` because for $\sigma[i]$, we add types to the grammar that are based on the vicinities of the words of $\sigma[i]$. Thus, $\sigma[i] \subseteq \Delta(\mathbf{TGE}^{(K)}(\sigma[i]))$. To prove that $\mathbf{TGE}^{(K)}$ is `faithful` for $\sigma[i]$ of $\Delta(G) = \Delta(\mathcal{C}^K(G))$, we have to remark that $\mathbf{TGE}^{(K)}(\sigma[i]) \preceq \mathcal{C}^K(G)$.

Lemma 2. *The inference algorithm $\mathbf{TGE}^{(K)}$ stabilizes on every training sequence σ of a K-star-revealing CDG.*

Proof. Because $\mathcal{C}^K(G)$ has a finite number of types, the number of corresponding patterns is also finite. Thus the number of patterns that correspond to the DS in $\Delta(\mathcal{C}^K(G))$ (and of course in σ) is also finite. Because the R-blocks are generalized using $*$ by $\mathbf{TGE}^{(K)}$ when their length is greater or equal to K, the number of R-blocks used by $\mathbf{TGE}^{(K)}$ is finite. Thus the number of generated types is finite and the algorithm certainly stabilizes.

```
Algorithm TGE^(K) (type-generalize-expand):
Input: σ[i] (σ being a training sequence).
Output: CDG TGE^(K)(σ[i]).
let G_H = (W_H, C_H, S, λ_H) where
W_H := ∅; C_H := {S}; λ_H := ∅; k := 0

(loop) for i ≥ 0                    //Infinite loop on σ
  let σ[i+1] = σ[i] · D;
  let (x, E) = D;
  (loop) for every w ∈ x
      W_H := W_H ∪ {w};
      let V(w, D) = π(<d_1>← β_1, ..., <d_k>← β_k)^P
      (loop) for j := 1, ..., k
          if β_j ∈ LD_d ∪ RD_d and length(β_j) ≥ K
              then γ_j := d*          // generalization
              else γ_j := β_j end end
      let t_w := π(<d_1>← γ_1, ..., <d_k>← γ_k)^P    // typing
      λ_H(w) := λ_H(w) ∪ {t_w};          // expansion
      end end
```

Fig. 5. *Inference algorithm* $\mathbf{TGE}^{(K)}$

4 Learnability from Positive Examples

Below we study the problem of learning CDG from positive examples of structures analogous to the FA-structures used for learning of categorial grammars.

4.1 Original Algorithm on Functor-Argument Data

An *FA structure* over an alphabet Σ is a binary tree where each leaf is an element of Σ and each internal node is labelled by the name of the binary rule.

Background - RG Algorithm. We recall Buszkowski's Algorithm called RG as in [10] it is defined for *AB* grammars, based on $/_e$ and $\backslash_e$ (binary elimination rules, like the local rules of CDG $\mathbf{L^r}$ and $\mathbf{L^l}$, without potentials) :

$$/_e : A \,/\, B, B \Rightarrow A \qquad \text{and} \qquad \backslash_e : B, B \,\backslash\, A \Rightarrow A$$

The RG algorithm takes a set D of functor-argument structures as positive examples and returns a rigid grammar $RG(D)$ compatible with the input if there is one (compatible means that D is in the set of functor-argument structures generated by the grammar).

Sketch of RG-Algorithm, Computing $RG(D)$:

1. assign S to the root of each structure
2. assign distinct variables to argument nodes
3. compute the other types on functor nodes according to $/_e$ and $\backslash_e$
4. collect the types assigned to each symbol, this provides $GF(D)$

5. unify (classical unification) the types assigned to the same symbol in $GF(D)$, and compute the most general unifier σ_{mgu} of this family of types.
6. The algorithm fails if unification fails, otherwise the result is the application of σ_{mgu} to the types of $GF(D)$: $RG(D) = \sigma_{mgu}(GF(D))$.

4.2 Functor-Argument Structures for CDG with Iterated Subtypes

Definition 13. *Let $\mathcal{D}$ be a dependency structure proof, ending in a type t. The* **labelled functor-argument structure** *associated to $\mathcal{D}$, $lfa_{iter}(\mathcal{D})$, is defined by induction on the length of the dependency proof $\mathcal{D}$ considering its last rule :*

- if $\mathcal{D}$ has no rule, it is a type t assigned to a word w, let $lfa_{iter}(\mathcal{D}) = w$;
- if the last rule is: $c^{P_1}\ [c \setminus \beta]^{P_2} \vdash [\beta]^{P_1P_2}$, by induction let $\mathcal{D}_1$ be a dependency structure proof for c^{P_1} and $\mathcal{T}_1 = lfa_{iter}(\mathcal{D}_1)$; and let $\mathcal{D}_2$ be a dependency structure proof for $[c \setminus \beta]^{P_2}$ and $\mathcal{T}_2 = lfa_{iter}(\mathcal{D}_2)$: then $lfa_{iter}(\mathcal{D})$ is the tree with root labelled by $\mathbf{L^l}_{[c]}$ and subtrees $\mathcal{T}_1$, $\mathcal{T}_2$;
- if the last rule is: $[c^* \setminus \beta]^{P_2} \vdash [\beta]^{P_2}$, by induction let $\mathcal{D}_2$ be a dependency structure proof for $[c^* \setminus \beta]^{P_2}$ and $\mathcal{T}_2 = lfa_{iter}(\mathcal{D}_2)$: then $lfa_{iter}(\mathcal{D})$ is $\mathcal{T}_2$;
- if the last rule is: $c^{P_1}\ [c^* \setminus \beta]^{P_2} \vdash [c^* \setminus \beta]^{P_1P_2}$, by induction let $\mathcal{D}_1$ be a dependency structure proof for c^{P_1} and $\mathcal{T}_1 = lfa_{iter}(\mathcal{D}_1)$ and let $\mathcal{D}_2$ be a dependency structure proof for $[c^* \setminus \beta]^{P_2}$ and $\mathcal{T}_2 = lfa_{iter}(\mathcal{D}_2)$: $lfa_{iter}(\mathcal{D})$ is the tree with root labelled by $\mathbf{L^l}_{[c]}$ and subtrees $\mathcal{T}_1$, $\mathcal{T}_2$;
- we define similarly the function lfa_{iter} when the last rule is on the right, using $/$ and $\mathbf{L^r}$ instead of $\setminus$ and $\mathbf{L^l}$;
- if the last rule is the one with potentials, $lfa_{iter}(\mathcal{D})$ is taken as the image of the proof above.

The functor-argument structure $fa_{iter}(\mathcal{D})$ is the one obtained from $lfa_{iter}(\mathcal{D})$ (the labelled one) by erasing the labels $[c]$.

Example 5. Let $\lambda(John) = N$, $\lambda(ran) = [N \setminus S / A^*]$, $\lambda(fast) = \lambda(yesterday) = A$, then $s'_3 = \mathbf{L^l}_{[N]}(John, \mathbf{L^r}_{[A]}(\mathbf{L^r}_{[A]}(ran, fast), yesterday)$ (labelled structure) and $s_3 = \mathbf{L^l}(John, \mathbf{L^r}(\mathbf{L^r}(ran, fast), yesterday)$ are associated to $\mathcal{D}_1$ below :

$$\mathcal{D}_1 : \quad \dfrac{N \qquad \dfrac{\dfrac{\dfrac{[N \setminus S / A^*]\ A}{[N \setminus S / A^*]}\mathbf{I^r} \qquad A}{[N \setminus S / A^*]}\mathbf{I^r}}{[N \setminus S]}\mathbf{\Omega^r}}{S}\mathbf{L^l}$$

John ran fast yesterday
(dependency structure)

4.3 On RG-Like Algorithms and Iteration

Example 6. We consider the following functor-argument structures :

$s_1 = \mathbf{L^l}(John, ran)$
$s_2 = \mathbf{L^l}(John, \mathbf{L^r}(ran, fast))$

$s_3 = \mathbf{L^l}(John, \mathbf{L^r}(\mathbf{L^r}(ran, fast), yesterday)$
$s_4 = \mathbf{L^l}(John, \mathbf{L^r}(\mathbf{L^r}(\mathbf{L^r}(ran, fast), yesterday), nearby)$

An RG-like algorithm could compute the following assignments and grammar from $\{s_1, s_2, s_3\}$:

$\mathbf{L^l}(John : X_1, ran : X_1 \setminus S) : S$
$\mathbf{L^l}(John : X_1', \mathbf{L^r}(ran : X_1' \setminus S \,/\, X_2, fast : X_2) : X_1' \setminus S) : S$
$\mathbf{L^l}(John : X"_1, \mathbf{L^r}(\, \mathbf{L^r}(ran : X"_1 \setminus S \,/\, X"_2 \,/\, X_2', fast : X_2') : X"_1 \setminus S \,/\, X"_2,$
$yesterday : X"_2) : X"_1 \setminus S) : S$

	general form	unification	flat rigid grammar for 2-iteration
John	$X_1, X_1', X"_1$	$X_1 = X_1' = X"_1$	X_1
ran	$X_1 \setminus S$ $X_1' \setminus S \,/\, X_2$ $X"_1 \setminus S \,/\, X"_2 \,/\, X_2'$	*fails*	$X_1 \setminus S \,/\, X_2^*$ with $X_2 = X_2' = X"_2$
fast	X_2, X_2'	X_2	X_2
yesterday	$X"_2$	$X"_2$	X_2

Notice that the next example s_4 would not change the type of *ran*.

In fact, such an RG-like algorithm, when the class of grammars is restricted to rigid grammars, when positive examples are functor-argument structures (without dependency names), cannot converge (in the sense of Gold).

This can be seen, as explained below, using the same grammars as in the limit point construction for string languages in [3], involving iterated dependency types. In fact, the functor-argument structures are all flat structures, with only / operators.

$$C_0' = S \qquad G_0' = \{a \mapsto A, b \mapsto B, c \mapsto C_0'\}$$
$$C_{n+1}' = C_n' \,/\, A^* \,/\, B^* \qquad G_n' = \{a \mapsto A, b \mapsto B, c \mapsto [C_n']\}$$
$$G_*' = \{a \mapsto A, b \mapsto A, c \mapsto [S \,/\, A^*]\}$$

Positive structured examples are then of the form :

$$c,\ \mathbf{L^r}(c, b),\ \mathbf{L^r}(\mathbf{L^r}(c, b), b),\ \mathbf{L^r}(c, a), \mathbf{L^r}(\mathbf{L^r}(c, a), a),\ \mathbf{L^r}(\mathbf{L^r}(c, b), a), \ldots$$

Definition 14. *We define $flat_{\mathbf{L^r}}$ and $flat_{\mathbf{L^r}_{[A]}}$ on words by : $flat_{\mathbf{L^r}}(x1) = x1$ $= flat_{\mathbf{L^r}_{[A]}}(x1)$ for words of length 1, and $flat_{\mathbf{L^r}}(x1.w1) = \mathbf{L^r}(x, flat_{\mathbf{L^r}}(w1))$; $flat_{\mathbf{L^r}_{[A]}}(x1.w1) = \mathbf{L^r}_{[A]}(x, flat_{\mathbf{L^r}_{[A]}}(w1))$; we extend the notation $flat_{\mathbf{L^r}}$ and $flat_{\mathbf{L^r}_{[A]}}$ to sets of words (as the set of word images).*

Let $FL(G)$ denote the language of functor-arguments structures of G.

Theorem 7. $FL(G_n') = flat_{\mathbf{L^r}}(\{c(b^*a^*)^k \mid k \leq n\})$ and $FL(G_*') = flat_{\mathbf{L^r}}(c\{b, a\}^*)$

Corollary 3. *The limit point establishes the non-learnability from functor-argument structures for the underlying classes of (rigid) grammars: those allowing iterated dependency types (A^*).*

A Limit Point, for Labelled Functor-Arguments Structures. If we drop restrictions such as k-rigid, and consider learnability from labelled functor-arguments structures, we have a limit point as follows :

$$\begin{array}{ll} C_0 = S & G_0 = \{a \mapsto A, c \mapsto C_0\} \\ C_{n+1} = (C_n \,/\, A) & G_n = \{a \mapsto A, c \mapsto [C_n], c \mapsto [C_{n-1}], \ldots c \mapsto C_0\} \\ & G_* = \{a \mapsto [A], c \mapsto [S \,/\, A^*]\} \end{array}$$

In fact, the functor-argument structures are all flat structures, with only $/$ operators and always the same label A.

Let $LFL(G)$ denote the language of labelled functor-argument structures of G.

Theorem 8. $LFL(G_n) = flat_{\mathbf{L^r}[A]}(\{c \;\; a^k \mid k \leq n\})$ *and* $LFL(G_*) = flat_{\mathbf{L^r}[A]}(c\, a^*)$

Corollary 4. *The limit point establishes the non-learnability from labelled functor-argument structures for the underlying classes of grammars: those allowing iterated dependency types (A^*).*

The similar question for rigid or k-rigid CDG with iteration is left open.

4.4 Bounds and String Learnability

A List-Like Simulation. In order to simulate an iterated type such that :

$$[\beta \,/\, a^*]^{P_0} a^{P_1} \ldots a^{P_n} \vdash [\beta]^{P_0 P_1 \ldots P_n}$$

we can distinguish two types, one type a for a first use in a sequence and one type $a \setminus a$ for next uses in a sequence of elements of type a, as in :

$$\begin{array}{cccccc} John & ran & fast & yesterday & nearby \\ n & n \setminus s \,/\, a & a & a \setminus a & a \setminus a \end{array}$$

Bounds. As a corollary, for a class of CDG *without potentials* for which the number of iterated types is bound by a fixed N, the simulation leads to a class of grammars without iterated types, which is also k-rigid: the number of assignments per word is bound by a large but fixed number ($k = 2^N$). This means that the class of rigid CDG allowing at most N iterated types is learnable from strings. This fact also extends to k-rigid CDG, not only to rigid (1-rigid) CDG.

5 Conclusion

In this paper, we propose a new model of incremental learning of categorial dependency grammars with unlimited iterated types from input dependency structures without marked iteration. The model reflects the real situation of deterministic inference of a dependency grammar from a dependency treebank. The learnability sufficient condition of K-star-revealing we use, is widely accepted in traditional linguistics for small K, which makes this model interesting for practical purposes. As shows our study, the more traditional unification based learning from function-argument structures fails even for rigid categorial dependency grammars with unlimited iterated types.

On the other hand, in this paper, the K-star-revealing condition is defined in "semantic" terms. It is an interesting question, whether one can find a simple syntactic formulation.

References

1. Angluin, D.: Inductive inference of formal languages from positive data. Information and Control 45, 117–135 (1980)
2. Béchet, D., Dikovsky, A., Foret, A.: Dependency Structure Grammars. In: Blache, P., Stabler, E.P., Busquets, J.V., Moot, R. (eds.) LACL 2005. LNCS (LNAI), vol. 3492, pp. 18–34. Springer, Heidelberg (2005)
3. Béchet, D., Dikovsky, A., Foret, A., Moreau, E.: On learning discontinuous dependencies from positive data. In: Proc. of the 9th Intern. Conf. "Formal Grammar 2004" (FG 2004), Nancy, France, pp. 1–16 (August 2004)
4. Buszkowski, W., Penn, G.: Categorial grammars determined from linguistic data by unification. Studia Logica 49, 431–454 (1990)
5. Dekhtyar, M., Dikovsky, A.: Categorial dependency grammars. In: Proc. of Intern. Conf. on Categorial Grammars, Montpellier, pp. 76–91 (2004)
6. Dekhtyar, M., Dikovsky, A.: Generalized Categorial Dependency Grammars. In: Avron, A., Dershowitz, N., Rabinovich, A. (eds.) Pillars of Computer Science. LNCS, vol. 4800, pp. 230–255. Springer, Heidelberg (2008)
7. Dikovsky, A.: Dependencies as categories. In: Recent Advances in Dependency Grammars (COLING 2004) Workshop, pp. 90–97 (2004)
8. Gold, E.M.: Language identification in the limit. Information and Control 10, 447–474 (1967)
9. Joshi, A.K., Shanker, V.K., Weir, D.J.: The convergence of mildly context-sensitive grammar formalisms. In: Foundational Issues in Natural Language Processing, Cambridge, MA, pp. 31–81 (1991)
10. Kanazawa, M.: Learnable classes of categorial grammars. Studies in Logic, Language and Information. FoLLI & CSLI (1998)
11. Mel'čuk, I.: Dependency Syntax. SUNY Press, Albany (1988)
12. Motoki, T., Shinohara, T., Wright, K.: The correct definition of finite elasticity: Corrigendum to identification of unions. In: The fourth Annual Workshop on Computational Learning Theory, San Mateo, Calif, p. 375 (1991)
13. Shanker, V.K., Weir, D.J.: The equivalence of four extensions of context-free grammars. Mathematical Systems Theory 27, 511–545 (1994)
14. Shinohara, T.: Inductive inference of monotonic formal systems from positive data. New Generation Computing 8(4), 371–384 (1991)
15. Wright, K.: Identifications of unions of languages drawn from an identifiable class. In: The 1989 Workshop on Computational Learning Theory, San Mateo, Calif, pp. 328–333 (1989)

The Lambek-Grishin Calculus Is NP-Complete

Jeroen Bransen

Utrecht University, The Netherlands

Abstract. The Lambek-Grishin calculus **LG** is the symmetric extension of the non-associative Lambek calculus **NL**. In this paper we prove that the derivability problem for **LG** is NP-complete.

1 Introduction

In his 1958 and 1961 papers, Lambek formulated two versions of the *Syntactic Calculus*: in (Lambek, 1958), types are assigned to *strings*, which are then combined by an *associative* operation; in (Lambek, 1961), types are assigned to *phrases* (bracketed strings), and the composition operation is non-associative. We refer to these two versions as **L** and **NL** respectively.

As for generative power, Kandulski (1988) proved that **NL** defines exactly the context-free languages. Pentus (1993) showed that this also holds for associative **L**. As for the complexity of the derivability problem, de Groote (1999) showed that for **NL** this belongs to `PTIME`; for **L**, Pentus (2003) proves that the problem is NP-complete and Savateev (2009) shows that NP-completeness also holds for the product-free fragment of **L**.

It is well known that some natural language phenomena require generative capacity beyond context-free. Several extensions of the Syntactic Calculus have been proposed to deal with such phenomena. In this paper we look at the Lambek-Grishin calculus **LG** (Moortgat, 2007, 2009). **LG** is a *symmetric* extension of the nonassociative Lambek calculus **NL**. In addition to $\otimes, \backslash, /$ (product, left and right division), **LG** has dual operations $\oplus, \oslash, \oslash$ (coproduct, left and right difference). These two families are related by linear distributivity principles. Melissen (2009) shows that all languages which are the intersection of a context-free language and the permutation closure of a context-free language are recognizable in **LG**. This places the lower bound for **LG** recognition beyond LTAG. The upper bound is still open.

The key result of the present paper is a proof that the derivability problem for **LG** is NP-complete. This will be shown by means of a reduction from SAT.[1]

2 Lambek-Grishin Calculus

We define the formula language of **LG** as follows.

[1] This paper has been written as a result of my Master thesis supervised by Michael Moortgat. I would like to thank him, Rosalie Iemhoff and Arno Bastenhof for comments and I acknowledge that any errors are my own.

P. de Groote and M.-J. Nederhof (Eds.): Formal Grammar 2010/2011, LNCS 7395, pp. 33–49, 2012.

Let Var be a set of *primitive types*, we use lowercase letters to refer to an element of Var. Let *formulas* be constructed using primitive types and the binary connectives $\otimes$, $/$, $\backslash$, $\oplus$, $\oslash$ and $\obslash$ as follows:

$$A, B ::= p \mid A \otimes B \mid A/B \mid B\backslash A \mid A \oplus B \mid A \oslash B \mid B \obslash A$$

The sets of *input* and *output structures* are constructed using formulas and the binary structural connectives $\cdot \otimes \cdot$, $\cdot / \cdot$, $\cdot \backslash \cdot$, $\cdot \oplus \cdot$, $\cdot \oslash \cdot$ and $\cdot \obslash \cdot$ as follows:

$$\text{(input)} \qquad X, Y ::= A \mid X \cdot \otimes \cdot Y \mid X \cdot \oslash \cdot P \mid P \cdot \obslash \cdot X$$

$$\text{(output)} \qquad P, Q ::= A \mid P \cdot \oplus \cdot Q \mid P \cdot / \cdot X \mid X \cdot \backslash \cdot P$$

The *sequents* of the calculus are of the form $X \to P$, and as usual we write $\vdash_{LG} X \to P$ to indicate that the sequent $X \to P$ is derivable in **LG**. The axioms and inference rules are presented in Figure 1, where we use the *display logic* from (Goré, 1998), but with different symbols for the *structural connectives.*

It has been proven by Moortgat (2007) that we have *Cut admissibility* for **LG**. This means that for every derivation using the *Cut*-rule, there exists a corresponding derivation that is *Cut-free.* Therefore we will assume that the Cut-rule is not needed anywhere in a derivation.

3 Preliminaries

3.1 Derivation Length

We will first show that for every derivable sequent there exists a Cut-free derivation that is polynomial in the length of the sequent. The length of a sequent φ, denoted as $|\varphi|$, is defined as the number of (formula and structural) connectives used to construct this sequent. A subscript will be used to indicate that we count only certain connectives, for example $|\varphi|_{\otimes}$.

Lemma 1. *If* $\vdash_{LG} \varphi$ *there exists a derivation with exactly* $|\varphi|$ logical rules.

Proof. If $\vdash_{LG} \varphi$ then there exists a Cut-free derivation for φ. Because every logical rule removes one logical connective and there are no rules that introduce logical connectives, this derivation contains $|\varphi|$ logical rules. □

Lemma 2. *If* $\vdash_{LG} \varphi$ *there exists a derivation with at most* $\frac{1}{4}|\varphi|^2$ Grishin interactions.

Proof. Let us take a closer look at the Grishin interaction principles. First of all, it is not hard to see that the interactions are irreversible. Also note that the interactions happen between the families of input connectives $\{\otimes, /, \backslash\}$ and output connectives $\{\oplus, \oslash, \obslash\}$ and that the Grishin interaction principles are the only rules of inference that apply on both families. So, on any pair of one input and one output connective, at most one Grishin interaction principle can be applied.

$$\frac{}{p \to p}\ Ax$$

$$\frac{X \to A \quad A \to P}{X \to P}\ Cut$$

$$\frac{\dfrac{Y \to X \cdot \backslash \cdot P}{X \cdot \otimes \cdot Y \to P}\ r}{X \to P \cdot / \cdot Y}\ r \qquad \frac{\dfrac{X \cdot \oslash \cdot Q \to P}{X \to P \cdot \oplus \cdot Q}\ dr}{P \cdot \obslash \cdot X \to Q}\ dr$$

(a) Display rules

$$\frac{X \cdot \otimes \cdot Y \to P \cdot \oplus \cdot Q}{X \cdot \oslash \cdot Q \to P \cdot / \cdot Y}\ d\oslash/ \qquad \frac{X \cdot \otimes \cdot Y \to P \cdot \oplus \cdot Q}{Y \cdot \oslash \cdot Q \to X \cdot \backslash \cdot P}\ d\oslash\backslash$$

$$\frac{X \cdot \otimes \cdot Y \to P \cdot \oplus \cdot Q}{P \cdot \obslash \cdot X \to Q \cdot / \cdot Y}\ d\obslash/ \qquad \frac{X \cdot \otimes \cdot Y \to P \cdot \oplus \cdot Q}{P \cdot \obslash \cdot Y \to X \cdot \backslash \cdot Q}\ d\obslash\backslash$$

(b) Distributivity rules (Grishin interaction principles)

$$\frac{A \cdot \otimes \cdot B \to P}{A \otimes B \to P}\ \otimes L \qquad \frac{X \to B \cdot \oplus \cdot A}{X \to B \oplus A}\ \oplus R$$

$$\frac{X \to A \cdot / \cdot B}{X \to A/B}\ /R \qquad \frac{B \cdot \obslash \cdot A \to P}{B \obslash A \to P}\ \obslash L$$

$$\frac{X \to B \cdot \backslash \cdot A}{X \to B\backslash A}\ \backslash R \qquad \frac{A \cdot \oslash \cdot B \to P}{A \oslash B \to P}\ \oslash L$$

$$\frac{X \to A \quad Y \to B}{X \cdot \otimes \cdot Y \to A \otimes B}\ \otimes R \qquad \frac{B \to P \quad A \to Q}{B \oplus A \to P \cdot \oplus \cdot Q}\ \oplus L$$

$$\frac{X \to A \quad B \to P}{B/A \to P \cdot / \cdot X}\ /L \qquad \frac{X \to B \quad A \to P}{P \cdot \obslash \cdot X \to A \obslash B}\ \obslash R$$

$$\frac{X \to A \quad B \to P}{A\backslash B \to X \cdot \backslash \cdot P}\ \backslash L \qquad \frac{X \to B \quad A \to P}{X \cdot \oslash \cdot P \to B \oslash A}\ \oslash R$$

(c) Logical rules

Fig. 1. The Lambek-Grishin calculus inference rules

If $\vdash_{LG} \varphi$ there exists a Cut-free derivation of φ. The maximum number of possible Grishin interactions in 1 Cut-free derivation is reached when a Grishin interaction is applied on every pair of one input and one output connective. Thus, the maximum number of Grishin interactions in one Cut-free derivation is $|\varphi|_{\{\otimes,/,\backslash\}} \cdot |\varphi|_{\{\oplus,\oslash,\obslash\}}$.

By definition, $|\varphi|_{\{\otimes,/,\backslash\}} + |\varphi|_{\{\oplus,\oslash,\obslash\}} = |\varphi|$, so the maximum value of $|\varphi|_{\{\otimes,/,\backslash\}} \cdot |\varphi|_{\{\oplus,\oslash,\obslash\}}$ is reached when $|\varphi|_{\{\otimes,/,\backslash\}} = |\varphi|_{\{\oplus,\oslash,\obslash\}} = \frac{|\varphi|}{2}$. Then the total number of Grishin interactions in 1 derivation is $\frac{|\varphi|}{2} \cdot \frac{|\varphi|}{2} = \frac{1}{4}|\varphi|^2$, so any Cut-free derivation of φ will contain at most $\frac{1}{4}|\varphi|^2$ Grishin interactions. □

Lemma 3. *In a derivation of sequent φ at most $2|\varphi|$ display rules are needed to display any of the structural parts.*

Proof. A structural part in sequent φ is nested under at most $|\varphi|$ structural connectives. For each of these connectives, one or two *r* or *dr* rules can display the desired part, after which the next connective is visible. Thus, at most $2|\varphi|$ display rules are needed to display any of the structural parts.

Lemma 4. *If $\vdash_{LG} \varphi$ there exists a Cut-free derivation of length $O(|\varphi|^3)$.*

Proof. ¿From Lemma 1 and Lemma 2 we know that there exists a derivation with at most $|\varphi|$ logical rules and $\frac{1}{4}|\varphi|^2$ Grishin interactions. Thus, the derivation consists of $|\varphi| + \frac{1}{4}|\varphi|^2$ rules, with between each pair of consecutive rules the display rules. From Lemma 3 we know that at most $2|\varphi|$ display rules are needed to display any of the structural parts. So, at most $2|\varphi| \cdot (|\varphi| + \frac{1}{4}|\varphi|^2) = 2|\varphi|^2 + \frac{1}{2}|\varphi|^3$ derivation steps are needed in the shortest possible Cut-free derivation for this sequent, and this is in $O(|\varphi|^3)$. □

3.2 Additional Notations

Let us first introduce some additional notations to make the proofs shorter and easier readable.

Let us call an input structure X which does not contain any structural operators except for $\cdot \otimes \cdot$ a *$\otimes$-structure*. A $\otimes$-structure can be seen as a binary tree with $\cdot \otimes \cdot$ in the internal nodes and formulas in the leafs. Formally we define $\otimes$-structures U and V as:

$$U, V ::= A \mid U \cdot \otimes \cdot V$$

We define $X[]$ and $P[]$ as the input and output structures X and P with a hole in one of their leafs. Formally:

$$X[] ::= [] \mid X[] \cdot \otimes \cdot Y \mid Y \cdot \otimes \cdot X[] \mid X[] \cdot \oslash \cdot Q \mid Y \cdot \oslash \cdot P[] \mid Q \cdot \obslash \cdot X[] \mid P[] \cdot \obslash \cdot Y$$

$$P[] ::= [] \mid P[] \cdot \oplus \cdot Q \mid Q \cdot \oplus \cdot P[] \mid P[] \cdot / \cdot Y \mid Q \cdot / \cdot X[] \mid Y \cdot \backslash \cdot P[] \mid X[] \cdot \backslash \cdot Q$$

This notation is similar to the one of de Groote (1999) but with structures. If $X[]$ is a structure with a hole, we write $X[Y]$ for $X[]$ with its hole filled with structure Y. We will write $X^{\otimes}[]$ for a $\otimes$-structure with a hole.

Furthermore, we extend the definition of hole to formulas, and define $A[]$ as a *formula* A with a hole in it, in a similar manner as for structures. Hence, by $A[B]$ we mean the formula $A[]$ with its hole filled by formula B.

In order to distinguish between input and output polarity formulas, we write $A^{\bullet}$ for a formula with *input* polarity and A° for a formula with *output* polarity. Note that for structures this is already defined by using X and Y for input polarity and P and Q for output polarity. This can be extended to formulas in a similar way, and we will use this notation only in cases where the polarity is not clear from the context.

3.3 Derived Rules of Inference

Now we will show and prove some derived rules of inference of **LG**.

Lemma 5. *If* $\vdash_{LG} A \to B$ *and we want to derive* $X^{\otimes}[A] \to P$*, we can* replace *A by B in* $X^{\otimes}[]$*. We have the inference rule below:*

$$\frac{A \to B \quad X^{\otimes}[B] \to P}{X^{\otimes}[A] \to P}\ Repl$$

Proof. We consider three cases:

1. If $X^{\otimes}[A] = A$, it is simply the cut-rule:

$$\frac{A \to B \quad B \to P}{A \to P}\ Cut$$

2. If $X^{\otimes}[A] = Y^{\otimes}[A] \cdot \otimes \cdot V$, we can move V to the righthand-side and use induction to prove the sequent:

$$\cfrac{A \to B \quad \cfrac{Y^{\otimes}[B] \cdot \otimes \cdot V \to P}{Y^{\otimes}[B] \to P \cdot / \cdot V}\ r}{\cfrac{Y^{\otimes}[A] \to P \cdot / \cdot V}{Y^{\otimes}[A] \cdot \otimes \cdot V \to P}\ r}\ Repl$$

3. If $X^{\otimes}[A] = U \cdot \otimes \cdot Y^{\otimes}[A]$, we can move U to the righthand-side and use induction to prove the sequent:

$$\cfrac{A \to B \quad \cfrac{U \cdot \otimes \cdot Y^{\otimes}[B] \to P}{Y^{\otimes}[B] \to U \cdot \backslash \cdot P}\ r}{\cfrac{Y^{\otimes}[A] \to U \cdot \backslash \cdot P}{U \cdot \otimes \cdot Y^{\otimes}[A] \to P}\ r}\ Repl$$

□

Lemma 6. *If we want to derive $X^{\otimes}[A \oslash B] \rightarrow P$, then we can move the expression $\oslash B$ out of the $\otimes$-structure. We have the inference rule below:*

$$\frac{X^{\otimes}[A] \cdot \oslash \cdot B \rightarrow P}{X^{\otimes}[A \oslash B] \rightarrow P}\ \mathit{Move}$$

Proof. We consider three cases:

1. If $X^{\otimes}[A \oslash B] = A \oslash B$, then this is simply the $\oslash L$-rule:

$$\frac{A \cdot \oslash \cdot B \rightarrow Y}{A \oslash B \rightarrow Y}\ \oslash L$$

2. If $X^{\otimes}[A \oslash B] = Y^{\otimes}[A \oslash B] \cdot \otimes \cdot V$, we can move V to the righthand-side and use induction together with the Grishin interaction principles to prove the sequent:

$$\cfrac{\cfrac{\cfrac{\cfrac{(Y^{\otimes}[A] \cdot \otimes \cdot V) \cdot \oslash \cdot B \rightarrow P}{Y^{\otimes}[A] \cdot \otimes \cdot V \rightarrow P \cdot \oplus \cdot B}\ dr}{Y^{\otimes}[A] \cdot \oslash \cdot B \rightarrow P \cdot / \cdot V}\ d\oslash /}{Y^{\otimes}[A \oslash B] \rightarrow P \cdot / \cdot V}\ \mathit{Move}}{Y^{\otimes}[A \oslash B] \cdot \otimes \cdot V \rightarrow P}\ r$$

3. If $X^{\otimes}[A \oslash B] = U \cdot \otimes \cdot Y^{\otimes}[A \oslash B]$, we can move U to the righthand-side and use induction together with the Grishin interaction principles to prove the sequent:

$$\cfrac{\cfrac{\cfrac{\cfrac{(U \cdot \otimes \cdot Y^{\otimes}[A]) \cdot \oslash \cdot B \rightarrow P}{U \cdot \otimes \cdot Y^{\otimes}[A] \rightarrow P \cdot \oplus \cdot B}\ dr}{Y^{\otimes}[A] \cdot \oslash \cdot B \rightarrow U \cdot \backslash \cdot P}\ d\oslash \backslash}{Y^{\otimes}[A \oslash B] \rightarrow U \cdot \backslash \cdot P}\ \mathit{Move}}{U \cdot \otimes \cdot Y^{\otimes}[A \oslash B] \rightarrow P}\ r$$

□

Lemma 7. $\vdash_{LG} A_1 \otimes (A_2 \otimes \ldots (A_{n-1} \otimes A_n)) \rightarrow P$ *iff* $\vdash_{LG} A_1 \cdot \otimes \cdot (A_2 \cdot \otimes \cdot \ldots (A_{n-1} \cdot \otimes \cdot A_n)) \rightarrow P$

Proof. The *if*-part can be derived by the application of $n-1$ times the $\otimes L$ rule together with the r rule:

$$\cfrac{\cfrac{\cfrac{\cfrac{\cfrac{\cfrac{\cfrac{A_1 \cdot \otimes \cdot (A_2 \cdot \otimes \cdot \ldots (A_{n-1} \cdot \otimes \cdot A_n)) \rightarrow P}{A_{n-1} \cdot \otimes \cdot A_n \rightarrow \ldots \cdot \backslash \cdot (A_2 \cdot \backslash \cdot (A_1 \cdot \backslash \cdot P))}\ r^*}{A_{n-1} \otimes A_n \rightarrow \ldots \cdot \backslash \cdot (A_2 \cdot \backslash \cdot (A_1 \cdot \backslash \cdot P))}\ \otimes L}{\ldots (A_{n-1} \otimes A_n) \rightarrow A_2 \cdot \backslash \cdot (A_1 \cdot \backslash \cdot P)}\ \ldots}{A_2 \cdot \otimes \cdot \ldots (A_{n-1} \otimes A_n) \rightarrow A_1 \cdot \backslash \cdot P}\ r}{A_2 \otimes \ldots (A_{n-1} \otimes A_n) \rightarrow A_1 \cdot \backslash \cdot P}\ \otimes L}{A_1 \cdot \otimes \cdot (A_2 \otimes \ldots (A_{n-1} \otimes A_n)) \rightarrow P}\ r}{A_1 \otimes (A_2 \otimes \ldots (A_{n-1} \otimes A_n)) \rightarrow P}\ \otimes L$$

The *only-if*-part can be derived by application of $n-1$ times the $\otimes R$ rule followed by a *Cut*:

$$
\dfrac{\dfrac{A_1 \to A_1 \quad \dfrac{A_2 \to A_2 \quad \dfrac{\dfrac{A_{n-1} \to A_{n-1} \quad A_n \to A_n}{A_{n-1} \cdot \otimes \cdot A_n \to A_{n-1} \otimes A_n} \otimes R}{\ldots (A_{n-1} \cdot \otimes \cdot A_n) \to \ldots (A_{n-1} \otimes A_n)} \ldots}{A_2 \cdot \otimes \cdot \ldots (A_{n-1} \cdot \otimes \cdot A_n) \to A_2 \otimes \ldots (A_{n-1} \otimes A_n)} \otimes R}{A_1 \cdot \otimes \cdot (A_2 \cdot \otimes \cdot \ldots (A_{n-1} \cdot \otimes \cdot A_n)) \to A_1 \otimes (A_2 \otimes \ldots (A_{n-1} \otimes A_n))} \otimes R \quad A_1 \otimes (A_2 \otimes \ldots (A_{n-1} \otimes A_n)) \to P}{A_1 \cdot \otimes \cdot (A_2 \cdot \otimes \cdot \ldots (A_{n-1} \cdot \otimes \cdot A_n)) \to P} \; Cut
$$

Note that because of the Cut elimination theorem, there exists a cut-free derivation for this sequent. □

3.4 Type Similarity

The type simililarity relation $\sim$, introduced by Lambek (1958), is the reflexive transitive symmetric closure of the derivability relation. Formally we define this as:

Definition 1. $A \sim B$ iff *there exists a sequence* $C_1 \ldots C_n (1 \leq i \leq n)$ *such that* $C_1 = A$, $C_n = B$ *and* $C_i \to C_{i+1}$ *or* $C_{i+1} \to C_i$ *for all* $1 \leq i < n$.

It was proved by Lambek that $A \sim B$ *iff* one of the following equivalent statements holds (the so-called *diamond property*):

$$\exists C \text{ such that } A \to C \text{ and } B \to C \quad \text{(join)}$$

$$\exists D \text{ such that } D \to A \text{ and } D \to B \quad \text{(meet)}$$

This diamond property will be used in the reduction from SAT to create a choice for a truthvalue of a variable.

Definition 2. *If* $A \sim B$ *and* C *is the* join *type of* A *and* B *so that* $A \to C$ *and* $B \to C$, *we define* $A \overset{C}{\sqcap} B = (A/((C/C)\backslash C)) \otimes ((C/C)\backslash B)$ *as the meet type of* A *and* B.

This is also the solution given by Lambek (1958) for the associative system **L**, but in fact this is the shortest solution for the non-associative system **NL** (Foret, 2003).

Lemma 8. *If* $A \sim B$ *with join-type* C *and* $\vdash_{LG} A \to P$ *or* $\vdash_{LG} B \to P$, *then we also have* $\vdash_{LG} A \overset{C}{\sqcap} B \to P$. *We can write this as a derived rule of inference:*

$$\frac{A \to P \quad \text{or} \quad B \to P}{A \overset{C}{\sqcap} B \to P} \; Meet$$

Proof

1. If $A \to P$:

$$\dfrac{\dfrac{\dfrac{\dfrac{\dfrac{\dfrac{C \to C \quad C \to C}{C/C \to C \cdot / \cdot C}\,/L}{C/C \to C/C}\,/R \qquad B \to C}{(C/C)\backslash B \to (C/C) \cdot \backslash \cdot C}\,\backslash L}{(C/C)\backslash B \to (C/C)\backslash C}\,\backslash R \qquad A \to P}{A/((C/C)\backslash C) \to P \cdot / \cdot ((C/C)\backslash B)}\,/L}{\dfrac{(A/((C/C)\backslash C)) \cdot \otimes \cdot ((C/C)\backslash B) \to P}{(A/((C/C)\backslash C)) \otimes ((C/C)\backslash B) \to P}\,\otimes L}\,r$$

2. If $B \to P$:

$$\dfrac{\dfrac{\dfrac{\dfrac{A \to C \qquad \dfrac{\dfrac{\dfrac{\dfrac{C \to C \quad C \to C}{C/C \to C \cdot / \cdot C}\,/L}{(C/C) \cdot \otimes \cdot C \to C}\,r}{C \to (C/C) \cdot \backslash \cdot C}\,r}{C \to (C/C)\backslash C}\,\backslash R}{A/((C/C)\backslash C) \to C \cdot / \cdot C}\,/L}{A/((C/C)\backslash C) \to C/C}\,/R \qquad B \to P}{\dfrac{\dfrac{(C/C)\backslash B \to (A/((C/C)\backslash C)) \cdot \backslash \cdot P}{(A/((C/C)\backslash C)) \cdot \otimes \cdot ((C/C)\backslash B) \to P}\,r}{(A/((C/C)\backslash C)) \otimes ((C/C)\backslash B) \to P}\,\otimes L}\,\backslash L$$

□

The following lemma is the key lemma of this paper, and its use will become clear to the reader in the construction of Section 4.

Lemma 9. *If* $\vdash_{LG} A \overset{C}{\sqcap} B \to P$ *then* $\vdash_{LG} A \to P$ *or* $\vdash_{LG} B \to P$, *if it is not the case that:*

- $P = P'[A'[(A_1 \otimes A_2)^\circ]]$
- $\vdash_{LG} A/((C/C)\backslash C) \to A_1$
- $\vdash_{LG} (C/C)\backslash B \to A_2$

Proof. We have that $\vdash_{LG} (A/((C/C)\backslash C)) \otimes ((C/C)\backslash B) \to P$, so from Lemma 7 we know that $\vdash_{LG} (A/((C/C)\backslash C)) \cdot \otimes \cdot ((C/C)\backslash B) \to P$. Remark that this also means that there exists a cut-free derivation for this sequent. By case analysis on the derivation we will show that *if* $\vdash_{LG} (A/((C/C)\backslash C)) \cdot \otimes \cdot ((C/C)\backslash B) \to P$, *then* $\vdash_{LG} A \to P$ or $\vdash_{LG} B \to P$, under the assumption that P is not of the form that is explicitly excluded in this lemma.

We will look at the derivations in a top-down way, and we will do case analysis on the rules that are applied to the sequent. We will show that in all cases we could change the derivation in such way that the meet-type $A \overset{C}{\sqcap} B$ in the conclusion could have immediately been replaced by either A or B.

The first case is where a logical rule is applied on the lefthand-side of the sequent. At a certain point in the derivation, possibly when P is an atom, one of the following three rules must be applied:

1. The $\otimes R$ rule, but then $P = A_1 \otimes A_2$ and in order to come to a derivation it must be the case that $\vdash_{LG} A/((C/C)\backslash C) \to A_1$ and $\vdash_{LG} (C/C)\backslash B \to A_2$. However, this is explicitly excluded in this lemma so this can never be the case.
2. The $/L$ rule, in this case first the r rule is applied so that we have $\vdash_{LG} A/((C/C)\backslash C) \to P \cdot / \cdot ((C/C)\backslash B)$. Now if the $/L$ rule is applied, we must have that $\vdash_{LG} A \to P$.
3. The $\backslash L$ rule, in this case first the r rule is applied so that we have $\vdash_{LG} (C/C)\backslash B \to (A/((C/C)\backslash C)) \cdot \backslash \cdot P$. Now if the $\backslash L$ rule is applied, we must have that $\vdash_{LG} B \to P$.

The second case is where a logical rule is applied on the righthand-side of the sequent. Let $\delta = \{r, dr, d\oslash/, d\oslash\backslash, d\otimes/, d\otimes\backslash\}$ and let δ^* indicate a (possibly empty) sequence of structural residuation steps and Grishin interactions. For example for the $\oslash R$ rule there are two possibilities:

– The lefthand-side ends up in the first premisse of the $\oslash R$ rule:

$$\dfrac{\dfrac{\dfrac{(A/((C/C)\backslash C)) \cdot \otimes \cdot ((C/C)\backslash B) \to P''[A']}{P'[(A/((C/C)\backslash C)) \cdot \otimes \cdot ((C/C)\backslash B)] \to A'}\,\delta^* \quad B' \to Q}{P'[(A/((C/C)\backslash C)) \cdot \otimes \cdot ((C/C)\backslash B)] \cdot \oslash \cdot Q \to A' \oslash B'}\,\oslash R}{(A/((C/C)\backslash C)) \cdot \otimes \cdot ((C/C)\backslash B) \to P[A' \oslash B']}\,\delta^*$$

In order to be able to apply the $\oslash R$ rule, we need to have a formula of the form $A' \oslash B'$ on the righthand-side. In the first step all structural rules are applied to display this formula in the righthand-side, and we assume that in the lefthand-side the meet-type ends up in the first structural part (inside a structure with the remaining parts from P that we call P'). After the $\oslash R$ rule has been applied, we can again display our meet-type in the lefthand-side of the formula by moving all other structural parts from P' back to the righthand-side (P'').

In this case it must be that $\vdash_{LG} (A/((C/C)\backslash C)) \cdot \otimes \cdot ((C/C)\backslash B) \to P''[A']$, so from this lemma we know that in this case also $\vdash_{LG} A \to P''[A']$ or $\vdash_{LG} B \to P''[A']$. In the case that $\vdash_{LG} A \to P''[A']$, we can show that $\vdash_{LG} A \to P[A' \oslash B']$ as follows:

$$\dfrac{\dfrac{\dfrac{A \to P''[A']}{P'[A] \to A'}\,\delta^* \quad B' \to Q}{P'[A] \cdot \oslash \cdot Q \to A' \oslash B'}\,\oslash R}{A \to P[A' \oslash B']}\,\delta^*$$

The case for B is similar.

– The lefthand-side ends up in the second premisse of the $\oslash R$ rule:

$$\dfrac{Q \to A' \quad \dfrac{(A/((C/C)\backslash C)) \cdot \otimes \cdot ((C/C)\backslash B) \to P''[B']}{B' \to P'[(A/((C/C)\backslash C)) \cdot \otimes \cdot ((C/C)\backslash B)]}\delta^*}{\dfrac{Q \cdot \oslash \cdot P'[(A/((C/C)\backslash C)) \cdot \otimes \cdot ((C/C)\backslash B)] \to A' \oslash B'}{(A/((C/C)\backslash C)) \cdot \otimes \cdot ((C/C)\backslash B) \to P[A' \oslash B']}\delta^*}\oslash R$$

This case is similar to the other case, except that the meet-type ends up in the other premisse. Note that, although in this case it is temporarily moved to the righthand-side, the meet-type will still be in an input polarity position and can therefore be displayed in the lefthand-side again.

In this case it must be that $\vdash_{LG} (A/((C/C)\backslash C)) \cdot \otimes \cdot ((C/C)\backslash B) \to P''[B']$, and from this lemma we know that in this case also $\vdash_{LG} A \to P''[B']$ or $\vdash_{LG} B \to P''[B']$. In the case that $\vdash_{LG} A \to P''[B']$, we can show that $\vdash_{LG} A \to P[A' \oslash B']$ as follows:

$$\dfrac{Q \to A' \quad \dfrac{A \to P''[B']}{B' \to P'[A]}\delta^*}{\dfrac{Q \cdot \oslash \cdot P'[A] \to A' \oslash B'}{A \to P[A' \oslash B']}\delta^*}\oslash R$$

The case for B is similar.

The cases for the other logical rules are similar. □

4 Reduction from SAT to LG

In this section we will show that we can reduce a Boolean formula in conjunctive normal form to a sequent of the *Lambek-Grishin calculus*, so that the corresponding **LG** sequent is provable *if and only if* the CNF formula is satisfiable. This has already been done for the associative system **L** by Pentus (2003) with a similar construction.

Let $\varphi = c_1 \wedge \ldots \wedge c_n$ be a Boolean formula in conjunctive normal form with clauses $c_1 \ldots c_n$ and variables $x_1 \ldots x_m$. For all $1 \leq j \leq m$ let $\neg_0 x_j$ stand for the literal $\neg x_j$ and $\neg_1 x_j$ stand for the literal x_j. Now $\langle t_1, \ldots, t_m\rangle \in \{0,1\}^m$ is a satisfying assignment for φ if and only if for every $1 \leq i \leq n$ there exists a $1 \leq j \leq m$ such that the literal $\neg_{t_j} x_j$ appears in clause c_i.

Let p_i (for $1 \leq i \leq n$) be distinct primitive types from Var. We now define the following families of types:

$$
\begin{array}{lll}
E_j^i(t) \leftrightharpoons \begin{cases} p_i \oslash (p_i \otimes p_i) & \text{if } \neg_t x_j \text{ appears in clause } c_i \\ p_i & \text{otherwise} \end{cases} & & \text{if } 1 \leq i \leq n,\ 1 \leq j \leq m \text{ and } t \in \{0,1\} \\
E_j(t) \leftrightharpoons E_j^1(t) \otimes (E_j^2(t) \otimes (\ldots (E_j^{n-1}(t) \otimes E_j^n(t)))) & & \text{if } 1 \leq j \leq m \text{ and } t \in \{0,1\} \\
H_j \leftrightharpoons p_1 \otimes (p_2 \otimes (\ldots (p_{n-1} \otimes p_n))) & & \text{if } 1 \leq j \leq m \\
F_j \leftrightharpoons E_j(1) \overset{H_j}{\sqcap} E_j(0) & & \text{if } 1 \leq j \leq m \\
G_0 \leftrightharpoons H_1 \otimes (H_2 \otimes (\ldots (H_{m-1} \otimes H_m))) & & \\
G_i \leftrightharpoons G_{i-1} \oslash (p_i \otimes p_i) & & \text{if } 1 \leq i \leq n
\end{array}
$$

Let $\bar{\varphi} = F_1 \otimes (F_2 \otimes (\ldots (F_{m-1} \otimes F_m))) \rightarrow G_n$ be the **LG** sequent corresponding to the Boolean formula φ. We now claim that the $\models \varphi$ *if and only if* $\vdash_{LG} \bar{\varphi}$.

4.1 Example

Let us take the Boolean formula $(x_1 \vee \neg x_2) \wedge (\neg x_1 \vee \neg x_2)$ as an example. We have the primitive types $\{p_1, p_2\}$ and the types as shown in Figure 2. The formula is satisfiable (for example with the assignment $\langle 1, 0 \rangle$), thus $\vdash_{LG} F_1 \otimes F_2 \rightarrow G_2$. A sketch of the derivation is given in Figure 2, some parts are proved in lemma's later on.

4.2 Intuition

Let us give some intuitions for the different parts of the construction, and a brief idea of why this would work. The basic idea is that on the lefthand-side we create a type for each literal (F_j is the formula for literal j), which will in the end result in the base type H_j, so $F_1 \otimes (F_2 \otimes (\ldots (F_{m-1} \otimes F_m)))$ will result in G_0. However, on the righthand-side we have an occurence of the expression $\oslash(p_i \otimes p_i)$ for each clause i, so in order to come to a derivation, we need to apply the $\oslash R$ rule for every clause i.

Each literal on the lefthand-side will result in either $E_j(1)$ (x_j is *true*) or $E_j(0)$ (x_j is *false*). This choice is created using a *join type* H_j such that $\vdash_{LG} E_j(1) \rightarrow H_j$ and $\vdash_{LG} E_j(0) \rightarrow H_j$, which we use to construct the *meet type* F_j. It can be shown that in this case $\vdash_{LG} F_j \rightarrow E_j(1)$ and $\vdash_{LG} F_j \rightarrow E_j(0)$, i.e. in the original formula we can replace F_j by either $E_j(1)$ or $E_j(0)$, giving us a choice for the truthvalue of x_j.

Let us assume that we need $x_1 = true$ to satisfy the formula, so on the lefthand-side we need to replace F_j by $E_1(1)$. $E_1(1)$ will be the product of exactly n parts, one for each clause ($E_1^1(1) \ldots E_1^n(1)$). Here $E_1^i(1)$ is $p_i \oslash (p_i \otimes p_i)$ *iff* x_1 does appear in clause i, and p_i otherwise. The first thing that should be noticed is that $\vdash_{LG} p_i \oslash (p_i \otimes p_i) \rightarrow p_i$, so we can rewrite all $p_i \oslash (p_i \otimes p_i)$ into p_i so that $\vdash_{LG} E_1(1) \rightarrow H_1$.

However, we can also use the type $p_i \oslash (p_i \otimes p_i)$ to facilitate the application of the $\oslash R$ rule on the occurrence of the expression $\oslash(p_i \otimes p_i)$ in the righthand-side. From Lemma 6 we know that $\vdash_{LG} X^{\otimes}[p_i \oslash (p_i \otimes p_i)] \rightarrow G_i$ if $\vdash_{LG} X^{\otimes}[p_i] \cdot \oslash \cdot$

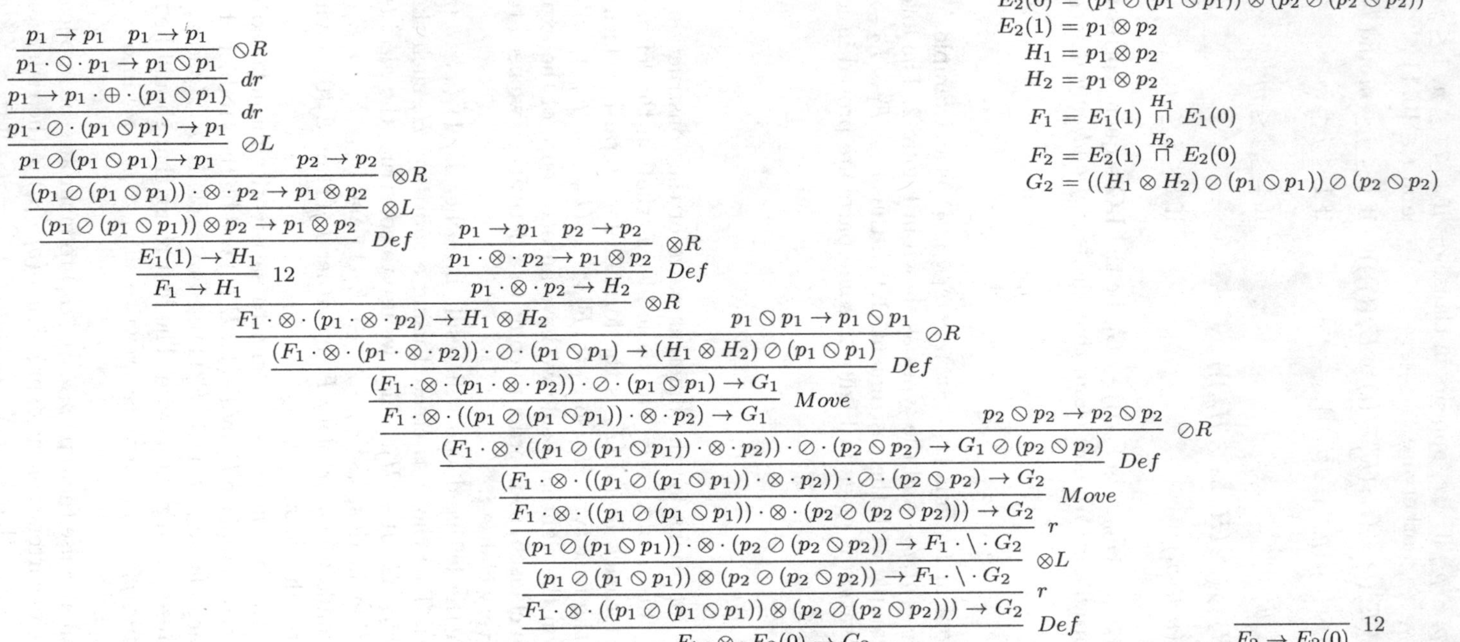

Fig. 2. Sketch proof for LG sequent corresponding to $(x_1 \vee \neg x_2) \wedge (\neg x_1 \vee \neg x_2)$

$(p_i \oslash p_i) \to G_i$, so if the expression $\oslash Y$ occurs somewhere in a $\otimes$-structure we can move it to the outside. Hence, from the occurrence of $p_i \oslash (p_i \oslash p_i)$ on the lefthand-side we can move $\oslash(p_i \oslash p_i)$ to the outside of the $\otimes$-structure and p_i will be left behind within the original structure (just as if we rewrote it to p_i). However, the sequent is now of the form $X^{\otimes}[p_i] \cdot \oslash \cdot (p_i \oslash p_i) \to G_{i-1} \oslash (p_i \oslash p_i)$, so after applying the $\oslash R$ rule we have $X^{\otimes}[p_i] \to G_{i-1}$.

Now if the original CNF formula is satisfiable, we can use the meet types on the lefthand-side to derive the correct value of $E_j(1)$ or $E_j(0)$ for all j. If this assignment indeed satisfies the formula, then for each i the formula $p_i \oslash (p_i \oslash p_i)$ will appear at least once. Hence, for all occurrences of the expression $\oslash(p_i \oslash p_i)$ on the righthand-side we can apply the $\oslash R$ rule, after which the rest of the $p_i \oslash (p_i \oslash p_i)$ can be rewritten to p_i in order to derive the base type.

If the formula is not satisfiable, then there will be no way to have the $p_i \oslash (p_i \oslash p_i)$ types on the lefthand-side for *all* i, so there will be at least one occurence of $\oslash(p_i \oslash p_i)$ on the righthand-side where we cannot apply the $\oslash R$ rule. Because the $\oslash$ will be the main connective we cannot apply any other rule, and we will never come to a valid derivation.

Note that the meet type F_j provides an *explicit* switch, so we first have to replace it by *either* $E_j(1)$ *or* $E_j(0)$ before we can do anything else with it. This guarantees that if $\vdash_{LG} \bar{\varphi}$, there also must be some assignment $\langle t_1, \ldots, t_m \rangle \in \{0,1\}^m$ such that $\vdash_{LG} E_1(t_1) \otimes (E_2(t_2) \otimes (\ldots (E_{m-1}(t_{m-1}) \otimes E_m(t_m)))) \to G_n$, which means that $\langle t_1, \ldots, t_m \rangle$ is a satisfying assigment for φ.

5 Proof

We will now prove the main claim that $\models \varphi$ *if and only if* $\vdash_{LG} \bar{\varphi}$. First we will prove that *if* $\models \varphi$, *then* $\vdash_{LG} \bar{\varphi}$.

5.1 Only If-Part

Let us assume that $\models \varphi$, so there is an assignment $\langle t_1, \ldots, t_m \rangle \in \{0,1\}^m$ that satisfies φ.

Lemma 10. *If* $1 \leq i \leq n$, $1 \leq j \leq m$ *and* $t \in \{0,1\}$ *then* $\vdash_{LG} E_j^i(t) \to p_i$.

Proof. We consider two cases:

1. If $E_j^i(t) = p_i$ this is simply the axiom rule.
2. If $E_j^i(t) = p_i \oslash (p_i \oslash p_i)$ we can prove it as follows:

$$\cfrac{\cfrac{\cfrac{\cfrac{p_i \to p_i \quad p_i \to p_i}{p_i \cdot \oslash \cdot p_i \to p_i \oslash p_i}\ \oslash R}{p_i \to p_i \cdot \oplus \cdot (p_i \oslash p_i)}\ dr}{p_i \cdot \oslash \cdot (p_i \oslash p_i) \to p_i}\ dr}{p_i \oslash (p_i \oslash p_i) \to p_i}\ \oslash L$$

□

Lemma 11. *If $1 \leq j \leq m$ and $t \in \{0,1\}$, then $\vdash_{LG} E_j(t) \rightarrow H_j$.*

Proof. ¿From Lemma 7 we know that we can turn $E_j(t)$ into a $\otimes$-structure. From Lemma 10 we know that $\vdash_{LG} E_j^i(t) \rightarrow p_i$, so using Lemma 5 we can replace all $E_j^i(t)$ by p_i in $E_j(t)$ after which we can apply the $\otimes R$ rule $n-1$ times to prove the lemma. □

Lemma 12. *If $1 \leq j \leq m$, then $\vdash_{LG} F_j \rightarrow E_j(t_j)$*

Proof. ¿From Lemma 11 we know that $\vdash_{LG} E_j(1) \rightarrow H_j$ and $\vdash_{LG} E_j(0) \rightarrow H_j$, so $E_j(1) \sim E_j(0)$ with join-type H_j. Now from Lemma 8 we know that $\vdash_{LG} E_j(1) \overset{H_j}{\sqcap} E_j(0) \rightarrow E_j(1)$ and $\vdash_{LG} E_j(1) \overset{H_j}{\sqcap} E_j(0) \rightarrow E_j(0)$. □

Lemma 13. *We can replace each F_j in $\bar{\varphi}$ by $E_j(t_j)$, so:*

$$\frac{E_1(t_1) \cdot\otimes\cdot (E_2(t_2) \cdot\otimes\cdot (\ldots (E_{m-1}(t_{m-1}) \cdot\otimes\cdot E_m(t_m)))) \rightarrow G_n}{F_1 \otimes (F_2 \otimes (\ldots (F_{m-1} \otimes F_m))) \rightarrow G_n}$$

Proof. This can be proven by using Lemma 7 to turn it into a $\otimes$-structure, and then apply Lemma 12 in combination with Lemma 5 m times. □

Lemma 14. *In $E_1(t_1) \cdot\otimes\cdot (E_2(t_2) \cdot\otimes\cdot (\ldots (E_{m-1}(t_{m-1}) \cdot\otimes\cdot E_m(t_m)))) \rightarrow G_n$, there is at least one occurrence of $p_i \oslash (p_i \obslash p_i)$ in the lefthand-side for every $1 \leq i \leq n$.*

Proof. This sequence of $E_1(t_1), \ldots, E_m(t_m)$ represents the truthvalue of all variables, and because this is a satisfying assignment, for all i there is at least one index k such that $\neg_{t_k} x_k$ appears in clause i. By definition we have that $E_k^i(t_k) = p_i \oslash (p_i \obslash p_i)$. □

Definition 3. $Y_j^i \leftrightharpoons E_j(t_j)$ *with every occurrence of $p_k \oslash (p_k \obslash p_k)$ replaced by p_k for all $i < k \leq n$*

Lemma 15. $\vdash_{LG} Y_1^0 \cdot\otimes\cdot (Y_2^0 \cdot\otimes\cdot (\ldots (Y_{m-1}^0 \cdot\otimes\cdot Y_m^0))) \rightarrow G_0$

Proof. Because $Y_j^0 = H_j$ by definition for all $1 \leq j \leq m$ and $G_0 = H_1 \otimes (H_2 \otimes (\ldots (H_{m-1} \otimes H_m)))$, this can be proven by applying the $\otimes R$ rule $m-1$ times. □

Lemma 16. *If $\vdash_{LG} Y_1^{i-1} \cdot\otimes\cdot (Y_2^{i-1} \cdot\otimes\cdot (\ldots (Y_{m-1}^{i-1} \cdot\otimes\cdot Y_m^{i-1}))) \rightarrow G_{i-1}$, then $\vdash_{LG} Y_1^i \cdot\otimes\cdot (Y_2^i \cdot\otimes\cdot (\ldots (Y_{m-1}^i \cdot\otimes\cdot Y_m^i))) \rightarrow G_i$*

Proof. ¿From Lemma 14 we know that $p_i \oslash (p_i \obslash p_i)$ occurs in $Y_1^i \cdot\otimes\cdot (Y_2^i \cdot\otimes\cdot (\ldots (Y_{m-1}^i \cdot\otimes\cdot Y_m^i)))$ (because the Y_j^i parts are $E_j(t_j)$ but with $p_k \oslash (p_k \obslash p_k)$ replaced by p_k only for $k > i$). Using Lemma 6 we can move the expression $\oslash(p_i \obslash p_i)$ to the outside of the lefthand-side of the sequent, after which we can apply the $\oslash R$-rule. After this we can replace all other occurrences of $p_i \oslash (p_i \obslash p_i)$ by p_i using Lemma 10 and Lemma 5. This process can be summarized as:

$$\cfrac{\cfrac{\cfrac{Y_1^{i-1} \cdot \otimes \cdot (Y_2^{i-1} \cdot \otimes \cdot (\ldots (Y_{m-1}^{i-1} \cdot \otimes \cdot Y_m^{i-1}))) \to G_{i-1} \quad p_i \oslash p_i \to p_i \oslash p_i}{(Y_1^{i-1} \cdot \otimes \cdot (Y_2^{i-1} \cdot \otimes \cdot (\ldots (Y_{m-1}^{i-1} \cdot \otimes \cdot Y_m^{i-1})))) \cdot \oslash \cdot (p_i \oslash p_i) \to G_{i-1} \oslash (p_i \oslash p_i)} \oslash R}{Y_1^{i-1} \cdot \otimes \cdot (Y_2^{i-1} \cdot \otimes \cdot (\ldots (Y_{m-1}^{i-1} \cdot \otimes \cdot Y_m^{i-1}))) \cdot \oslash \cdot (p_i \oslash p_i) \to G_i} Def}{Y_1^{i} \cdot \otimes \cdot (Y_2^{i} \cdot \otimes \cdot (\ldots (Y_{m-1}^{i} \cdot \otimes \cdot Y_m^{i}))) \to G_i} 14, 6, 10, 5$$

□

Lemma 17. $\vdash_{LG} Y_1^n \cdot \otimes \cdot (Y_2^n \cdot \otimes \cdot (\ldots (Y_{m-1}^n \cdot \otimes \cdot Y_m^n))) \to G_n$

Proof. We can prove this using induction with Lemma 15 as base and Lemma 16 as induction step. □

Lemma 18. *If* $\models \varphi$, *then* $\vdash_{LG} \bar{\varphi}$,

Proof. ¿From Lemma 17 we know that $\vdash_{LG} Y_1^n \cdot \otimes \cdot (Y_2^n \cdot \otimes \cdot (\ldots (Y_{m-1}^n \cdot \otimes \cdot Y_m^n))) \to G_n$, and because by definition $Y_j^n = E_j(t_j)$, we also have that $\vdash_{LG} E_1(t_1) \cdot \otimes \cdot (E_2(t_2) \cdot \otimes \cdot (\ldots (E_{m-1}(t_{m-1}) \cdot \otimes \cdot E_m(t_m)))) \to G_n$. Finally combining this with Lemma 13 we have that $\vdash_{LG} \bar{\varphi} = F_1 \otimes (F_2 \otimes (\ldots (F_{m-1} \otimes F_m))) \to G_n$, using the assumption that $\models \varphi$. □

5.2 If-Part

For the if part we will need to prove that *if* $\vdash_{LG} \bar{\varphi}$, *then* $\models \varphi$. Let us now assume that $\vdash_{LG} \bar{\varphi}$.

Lemma 19. *If* $\vdash_{LG} X \to P'[(P \oslash Y)^\circ]$, *then there exist a* Q *such that* Q *is part of* X *or* P' *(possibly inside a formula in* X *or* P'*) and* $\vdash_{LG} Y \to Q$.

Proof. The only rule that matches a $\oslash$ in the righthand-side is the $\oslash R$ rule, so somewhere in the derivation this rule must be applied on the occurrence of $P \oslash Y$. Because this rule needs a $\cdot \oslash \cdot$ connective in the lefthand-side, we know that if $\vdash_{LG} X \to P'[(P \oslash Y)^\circ]$ it must be the case that we can turn this into $X' \cdot \oslash \cdot Q \to P \oslash Y$ such that $\vdash_{LG} Y \to Q$. □

Lemma 20. *If* $\vdash_{LG} E_1(t_1) \cdot \otimes \cdot (E_2(t_2) \cdot \otimes \cdot (\ldots (E_{m-1}(t_{m-1}) \cdot \otimes \cdot E_m(t_m))) \to G_n$, *then there is an occurrence* $p_i \oslash (p_i \oslash p_i)$ *on the lefthand-side at least once for all* $1 \leq i \leq n$.

Proof. G_n by definition contains an occurrence of the expression $\oslash(p_i \oslash p_i)$ for all $1 \leq i \leq n$. From Lemma 19 we know that somewhere in the sequent we need an occurrence of a structure Q such that $\vdash_{LG} p_i \oslash p_i \to Q$. From the construction it is obvious that the only possible type for Q is in this case $p_i \oslash p_i$, and it came from the occurrence of $p_i \oslash (p_i \oslash p_i)$ on the lefthand-side. □

Lemma 21. *If* $\vdash_{LG} E_1(t_1) \cdot \otimes \cdot (E_2(t_2) \cdot \otimes \cdot (\ldots (E_{m-1}(t_{m-1}) \cdot \otimes \cdot E_m(t_m))) \to G_n$, *then* $\langle t_1, t_2, \ldots, t_{m-1}, t_m \rangle$ *is a satisfying assignment for the CNF formula.*

Proof. ¿From Lemma 20 we know that there is a $p_i \oslash (p_i \otimes p_i)$ in the lefthand-side of the formula for all $1 \leq i \leq n$. From the definition we know that for each i there is an index j such that $E^i_j(t_j) = p_i \oslash (p_i \otimes p_i)$, and this means that $\neg_{t_j} x_j$ appears in clause i, so all clauses are satisfied. Hence, this choice of $t_1 \dots t_m$ is a satisfying assignment. □

Lemma 22. *If* $1 \leq j \leq m$ *and* $\vdash_{LG} X^{\otimes}[F_j] \to G_n$*, then* $\vdash_{LG} X^{\otimes}[E_j(0)] \to G_n$ *or* $\vdash_{LG} X^{\otimes}[E_j(1)] \to G_n$.

Proof. We know that $X^{\otimes}[F_j]$ is a $\otimes$-structure, so we can apply the r rule several times to move all but the F_j-part to the righthand-side. We then have that $\vdash_{LG} F_j \to \dots \cdot \backslash \cdot G_n \cdot / \cdot \dots$. From Lemma 9 we know that we now have that $\vdash_{LG} E_j(0) \to \dots \cdot \backslash \cdot G_n \cdot / \cdot \dots$ or $\vdash_{LG} E_j(1) \to \dots \cdot \backslash \cdot G_n \cdot / \cdot \dots$. Finally we can apply the r rule again to move all parts back to the lefthand-side, to show that $\vdash_{LG} X^{\otimes}[E_j(0)] \to G_n$ or $\vdash_{LG} X^{\otimes}[E_j(1)] \to G_n$.

Note that, in order for Lemma 9 to apply, we have to show that this sequent satisfies the constraints. G_n does contain $A_1 \otimes A_2$ with output polarity, however the only connectives in A_1 and A_2 are $\otimes$. Because no rules apply on $A/((C/C)\backslash C) \to A'_1 \otimes A''_1$, we have that $\not\vdash_{LG} A/((C/C)\backslash C) \to A_1$. In $X^{\otimes}[]$, the only $\otimes$ connectives are within other F_k, however these have an input polarity and do not break the constraints either.

So, in all cases F_j provides an *explicit switch*, which means that the truthvalue of a variable can only be changed in all clauses simultanously. □

Lemma 23. *If* $\vdash_{LG} \bar{\varphi}$*, then* $\models \varphi$.

Proof. ¿From Lemma 22 we know that all derivations will first need to replace each F_j by *either* $E_j(1)$ *or* $E_j(0)$. This means that if $\vdash_{LG} F_1 \otimes (F_2 \otimes (\dots (F_{m-1} \otimes F_m))) \to G_n$, then also $\vdash_{LG} E_1(t_1) \cdot \otimes \cdot (E_2(t_2) \cdot \otimes \cdot (\dots (E_{m-1}(t_{m-1}) \cdot \otimes \cdot E_m(t_m))) \to G_n$ for some $\langle t_1, t_2, \dots, t_{m-1}, t_m \rangle \in \{0,1\}^m$. From Lemma 21 we know that this is a satisfying assignment for φ, so if we assume that $\vdash_{LG} \bar{\varphi}$, then $\models \varphi$. □

6 Conclusion

Theorem 1. ***LG*** *is NP-complete.*

Proof. ¿From Lemma 4 we know that for every derivable sequent there exists a proof that is of polynomial length, so the derivability problem for **LG** is in NP. From Lemma 18 and Lemma 23 we can conclude that we can reduce SAT to **LG**. Because SAT is a known NP-hard problem (Garey and Johnson, 1979), and our reduction is polynomial, we can conclude that derivability for **LG** is also NP-hard.

Combining these two facts we conclude that the derivability problem for **LG** is NP-complete. □

References

de Groote, P.: The Non-Associative Lambek Calculus with Product in Polynomial Time. In: Murray, N.V. (ed.) TABLEAUX 1999. LNCS (LNAI), vol. 1617, pp. 128–139. Springer, Heidelberg (1999)

Foret, A.: On the computation of joins for non associative Lambek categorial grammars. In: Proceedings of the 17th International Workshop on Unification (UNIF 2003), Valencia, Spain, June 8-9 (2003)

Garey, M.R., Johnson, D.S.: Computers and Intractability: A Guide to the Theory of NP-Completeness. W. H. Freeman & Co., New York (1979)

Goré, R.: Substructural logics on display. Logic Jnl IGPL 6(3), 451–504 (1998)

Kandulski, M.: The non-associative Lambek calculus. Categorial Grammar, Linguistic and Literary Studies in Eastern Europe (LLSEE) 25, 141–151 (1988)

Lambek, J.: The Mathematics of Sentence Structure. American Mathematical Monthly 65, 154–170 (1958)

Lambek, J.: On the calculus of syntactic types. In: Structure of Language and Its Mathematical Aspects, pp. 166–178 (1961)

Melissen, M.: The Generative Capacity of the Lambek–Grishin Calculus: A New Lower Bound. In: de Groote, P., Egg, M., Kallmeyer, L. (eds.) Formal Grammar. LNCS, vol. 5591, pp. 118–132. Springer, Heidelberg (2011)

Moortgat, M.: Symmetries in Natural Language Syntax and Semantics: The Lambek-Grishin Calculus. In: Leivant, D., de Queiroz, R. (eds.) WoLLIC 2007. LNCS, vol. 4576, pp. 264–284. Springer, Heidelberg (2007)

Moortgat, M.: Symmetric categorial grammar. Journal of Philosophical Logic 38(6), 681–710 (2009)

Pentus, M.: Lambek grammars are context free. In: Proceedings of the 8th Annual IEEE Symposium on Logic in Computer Science, pp. 429–433. IEEE Computer Society Press, Los Alamitos (1993)

Pentus, M.: Lambek calculus is NP-complete. CUNY Ph.D. Program in Computer Science Technical Report TR–2003005. CUNY Graduate Center, New York (2003)

Savateev, Y.: Product-Free Lambek Calculus Is NP-Complete. In: Artemov, S., Nerode, A. (eds.) LFCS 2009. LNCS, vol. 5407, pp. 380–394. Springer, Heidelberg (2008)

Resumption and Island-Hood in Hausa*

Berthold Crysmann

Universität Bonn
Poppelsdorfer Allee 47, D–53115 Bonn
crysmann@uni-bonn.de

Abstract. In this paper, I shall discuss the status of Hausa resumptive pronouns. I shall argue that the asymmetry between island-sensitive wh-extraction and island-insensitive relativisation is best captured at the filler site, rather than at the gap site. The analysis proposed here builds crucially on recent HPSG work on island-insensitive rightward movement, arguing in favour of anaphoric processes within a theory of extraction.

1 Introduction

Unbounded dependency constructions in Hausa[1] make use of two different strategies: besides standard extraction as a nonlocal relation between the filler and a phonologically empty gap, the language also recognises a resumptive pronoun strategy, where a pronominal is found at the extraction site.

Resumptive elements can be free pronominals, bound pronominal affixes, or even null pronominals.[2] The choice of resumptive pronoun (free, bound, null) depends on morphological and syntactic properties of the head governing the extraction site: independent pronouns are used as objects of prepositions, bound

* A great many thanks to the anonymous reviewers, as well as to the audience of Formal Grammar 2010 for their stimulating and helpful comments, in particular Irina Agafonova and Carl Pollard. I am also gratefully indebted to Stefan Müller for his remarks on an earlier version of this paper.

[1] Hausa is an Afroasiatic language spoken mainly in Northern Nigeria and bordering areas of Niger. Both tone (high, low, falling) and length (long vs. short) are lexically and grammatically distinctive. Following common practise, I shall mark low tone with a grave accent and falling tone with a circumflex. Vowels unmarked for tone are high. Length distinctions are signalled by macron.

I shall make use of the following inventory of morphological tags in the glosses: L = "genitive linker", S = singular, P = plural, M = masculine, F = feminine, DO = direct object, IO = indirect object, IOM = indirect object marker, G = genitive, REL = relativiser, COMP = complementiser, FOC = focus marker, CPL = completive aspect, CONT = continuative aspect, HAB = habitual, FUT = future.

[2] Hausa is a null subject and null object language (Tuller, 1986): while both human and non-human subjects can be dropped with equal ease, object drop observes a restriction to non-human referents. Interpretation of null arguments is always specific, i.e. not generic (Jaggar, 2001).

P. de Groote and M.-J. Nederhof (Eds.): Formal Grammar 2010/2011, LNCS 7395, pp. 50–65, 2012.

pronominals are used for complements of nouns as well as human (animate) complements of verbs and verbal nouns, whereas null pronominals are used for subjects and inanimate objects of verbs and verbal nouns.

I shall first discuss the properties of overt resumptive pronouns, followed by a discussion of the admittedly more subtle null anaphora.

1.1 Overt Resumptive Pronouns

Complements of Prepositions. Among the constructions that make regular use of resumption, extraction from PP features prominently: in Hausa, three classes of prepositions can be distinguished, depending on their behaviour in unbounded dependency constructions. While all prepositions can be pied-piped along with their complements in focus fronting, wh-fronting and relativisation (Newman, 2000; Jaggar, 2001), for basic locative prepositions, such as *à* or *dàgà*, this seems to be the only option.

(1) a. [à Kanṑ]$_i$ akà hàifē nì $\emptyset_i$
at Kano 4.CPL give.birth 1.S.DO $\emptyset$
'It was in Kano I was born' (Jaggar, 2001)

b. * Kanṑ$_i$ akà hàifē nì à shī$_i$ / $\emptyset_i$
Kano 4.CPL give.birth 1.S.DO at 3.S.M / $\emptyset$

Non-locative basic prepositions, such as *dà* or *gàrē* do not permit extraction of their complement by way of a filler-gap dependency. Extraction is possible with these prepositions, if a resumptive pronominal is used instead of a gap.

(2) a. [dà sàndā]$_i$ sukà dṑkē shì $\emptyset_i$
with stick 3.P.CPL beat 3.S.DO
'It was a stick they beat him with.' (Jaggar, 2001)

b. sàndā$_i$ sukà dṑkē shì dà ita$_i$ / *$\emptyset_i$
stick 3.P.CPL beat 3.S.DO with 3.S.F / $\emptyset$
'It was a stick they beat him with.' (Jaggar, 2001)

The third class of prepositions, the extensive set of so-called "genitive prepositions" (Newman, 2000) or "prepositional nouns" (Wolff, 1993), feature in filler-gap dependencies, along-side pied-piping and resumptive pronoun strategies.[3]

(3) a. [ciki-n àdakā̀]$_i$ mukàn sâ kuɗi-n-mù $\emptyset_i$
inside-L box 1.P.HAB put money-L-1.P.G
'It's inside a box we usually put our money.' (Jaggar, 2001)

b. àdakā̀$_i$ mukàn sâ kuɗi-n-mù ciki-n-tà$_i$ / ciki $\emptyset_i$
box 1.P.HAB put money-L-1P inside-L-3.S.F.G / inside $\emptyset$
'It's inside a box we usually put our money.' (Jaggar, 2001)

---

[3] The presence of the genitive linker is an instance of a general property of the language, namely marking in situ direct objects of nouns, verbs, and adjectives. See, e.g., Newman (2000) and Jaggar (2001) for an extensive overview, as well as Crysmann (2005a, in press) for a unified formal analysis.

The form of the resumptive pronoun is always identical to the one normally used with a particular preposition, a generalisation which holds for resumptive pronominals across the board. Genitive prepositions take bound pronominals from the possessive set (also used with nouns), *gà/gàrē* takes bound pronominals from the accusative set (otherwise used with verbs) and all other prepositions take free pronouns from the independent set. Note, however, that basic prepositions that fail to feature in a resumptive pronoun strategy may still combine with pronominal objects, as long as this does not involve a long-distance dependency (e.g., *dàgà ita* 'from her'; Newman, 2000).

Before we proceed, I would like to briefly take stock of what we have established thus far: first, whether or not gaps or resumptive pronouns are licit, is mainly a lexical matter, depending on the governing preposition. Second, for prepositions that permit stranding, use of a resumptive pronoun is an equally viable option.

Complements of Nouns. The second major context in which resumptive pronouns surface in Hausa involves complements of nouns, including possessors.

(4) a. ['ya-r wā̀]$_i$ ka àurā ∅$_i$?
daughter-L who 2.M.CPL marry
'Whose daughter did you marry?' (Jaggar, 2001)

b. wā̀$_i$ ka àuri 'ya-r-sà$_i$ / *'yā ∅$_i$?
who 2.M.CPL marry daughter.F-L.F-3.S.M.G / daughter.F ∅
'Whose daughter did you marry?' (Jaggar, 2001)

(5) Audù$_i$ nē na ɗinkà rī̀ga-r-sà$_i$ / *rī̀gā ∅$_i$.
Audu FOC 1.S.CPL sew gown.F-L.F-3.S.M.G / gown.F ∅
'It's Audu whose gown I've sewn'

In wh-constructions, the governing nominal head can either be pied-piped along with the wh-pronoun, or a resumptive pronoun can be used at the extraction site. Extraction by filler-gap dependency, however, is impossible for complements of nouns.

The same holds for focus fronting: again, use of a resumptive pronoun is obligatory.[4]

Human Complements of Verbs. Objects of verbs, dynamic nouns, and verbal nouns can extract by way of a filler-gap dependency.[5]

(6) a. yāròn$_i$ dà sukà dṑkā ∅$_i$ yanā̀ asìbitì
boy REL 3.P.CPL beat up 3.S.M.CONT hospital
'The boy they beat up is in hospital' (Jaggar, 2001, p. 534)

[4] It is not clear to me at present, whether pied-piping would be an option here. Although it is perfectly acceptable to focus-front the entire complex NP, it remains to be clarified, whether this leads to an extension of the focus domain or not.

[5] Human direct objects cannot be pro-dropped in Hausa. Thus, whenever we encounter a zero realisation at the gap-site, we can be sure that we are not dealing with an instance of null anaphora.

b. gà yārinyàr$_i$ dà nakḕ sô $\emptyset_i$
there is girl REL 1.S.CONT want.VN
'There's the girl I love.' (Jaggar, 2001, p. 534)

c. inā littāfin$_i$ dà kakḕ màganà̄ $\emptyset_i$
where book REL 2.S.M.CONT talking
'Where is the book you're talking about?' (Jaggar, 2001, p. 534)

Jaggar (2001, p. 534) observes that "deletion is [...] the strongly preferred strategy for relativisation on direct objects." Although Jaggar does not provide any positive or negative examples for resumptive pronoun use, the wording suggests that this option is not ruled out per se, but rather highly infrequent in simple extraction contexts.

However, once we consider long extraction, we do find cases where resumption is indeed the only option: Tuller (1986) reports the following data, involving extraction across non-bridge verbs where the gap strategy leads to ungrammaticality, but resumption is fine.

(7) gà yârân dà Àli ya radà̄ minì wai ya gan-sù /
there are children REL Ali 3.S.CPL whisper 1.S.IO COMP 3.S.CPL see-3.P.DO /
*ganī $\emptyset$ gida-n giyà̄
see $\emptyset$ house-L beer
'Here are the children that Ali whispered to me that he saw in the bar.'(Tuller, 1986, p. 169; tone added)

One possible explanation for the marginal status of overt resumptives in direct object function would be to assume that the use of resumption is but a "last resort" device (Shlonsky, 1992) whose main purpose is to circumvent island violations. Yet, overt human direct object resumptive pronouns can also be found in constructions which are slightly more complex than the simple short extraction examples given above, but which nevertheless do not involve any extraction islands, as in the following examples of extraction from a sentential complement of a bridge verb:

(8) mùtumìn$_i$ dà ɗàlibai sukà san cêwā mālàma-r-sù tanà̄
man REL students 3.P.CPL know COMP teacher-L.F-3.P.GEN 3.S.F.CONT
sô-n-sà$_i$ / sô $\emptyset_i$
like.VN-L-3.S.M.GEN / like.VN
'the man that the students know that their teacher likes' (Newman, 2000, p. 539)

Similar observations can be made with across-the-board extraction from coordinate structures:

(9) [àbōkī-n-ā]$_i$ dà na zìyartà̄ $\emptyset_i$ àmmā bàn sà̄mē shì$_i$ à
friend-L-1.S.GEN REL 1.S.CPL visit but 1.S.NEG.CPL find 3.S.M.DO at home
gidā ba
NEG
'my friend that I visited but did not find at home' (Newman, 2000, p. 539)

(10) mùtumìn$_i$ dà na bā shì$_i$ aro-n bàrgō-nā àmmā
man REL 1.S.CPL give 3.S.M.DO lending-L blanket-L.1.S.G but
duk dà hakà ∅$_i$ yakè jîn sanyī
in spite of that ∅ 3.S.M.CONT feel-L cold
'the man whom I lent my blanket but who still felt cold' (Newman, 2000, p. 540)

The apparent marginality of resumption in highly local contexts observed by Hausa is highly reminiscent of similar restrictions on subject resumptives in the highest clause in Hebrew (Borer, 1984) and Irish McCloskey (1990). Note, moreover, that in all the three examples cited above, a zero gap is equally possible. Thus, a "last resort" account is anything but likely.

Turning to indirect objects, extraction using a filler-gap dependency is again the preferred option. However, in contrast to direct objects, resumptive pronouns are cited to be much more common (Jaggar, 2001; Newman, 2000).

(11) mutànên$_i$ dà sukà ƙi sayar musù$_i$ / wà ∅$_i$ dà àbinci sukà fita
men REL 3.P.CPL refuse sell 3.P.IO / IOM ∅ with food 3.P.CPL left
'the men they refused to sell food to left.' (Jaggar, 2001, p. 534)

To sum up our observations regarding objects of verbs, we find that gaps are possible in general, and that at least with indirect objects, resumption is equally possible. For direct objects, resumption appears marginal in cases of short extraction. It should be clear, however, that resumption is more than just a rescue device, given its presence in structures without any relevant extraction island.

1.2 Null Anaphora in Extraction

Hausa is a null subject language (Tuller, 1986; Jaggar, 2001): tense/aspect/mood (TAM) markers are inflected for person, number, and gender, often exhibiting fusion between agreement and TAM marking: in some paradigms, TAM categories are only expressed suprasegmentally in terms of tone and/or length distinctions. Discourse-salient subjects are typically suppressed. Pronominal subjects are either dislocated topics, or else ex situ focused constituents.

With respect to subject extraction, Tuller (1986) observes that Hausa, just like Italian, is not subject to the *that-t* effect. Given the pro-drop property this is an entirely expected pattern. In contrast to Italian, however, this does not correlate with free inversion of the subject.

(12) a. wā̀ kikḕ tsàmmānī̀ (wai/cêwā) ∅ yā tàfi Kanṑ
who 2.S.F.CONT thinking COMP 3.S.M.CPL go Kano
'Who do you think went to Kano?' (Tuller, 1986, p. 152-3; tone added)

b. wā̀ Ābù ta tàmbayā̀ kō ∅ yā tàfi Kanṑ
who Abu 3.S.F.CPL ask COMP 3.S.M.CPL go Kano
'Who did Abu ask went to Kano?' (Tuller, 1986, p. 153; tone added)

In addition to subject-drop, Hausa also features object-drop, although there is a restriction to non-human referents, as illustrated by the examples below, where the sentences in b. are answers to the questions raised in a.

(13) a. Kā ga littāfin-n Mūsa?
2.S.M.CPL see book-L Musa
'Did you see Musa's book?'

b. Ī, nā gan shì. / Ī, nā ganī
Yes 1.S.CPL see 3.S.M Yes 1.S.CPL see
'Yes, I saw it.' (Tuller, 1986, p. 61; tone added)

(14) a. Kā ga ƙanè-n Mūsa?
2.S.M.CPL see brother-L Musa
'Did you see Musa's brother?'

b. Ī, nā gan shì. / *Ī, nā ganī
Yes 1.S.CPL see 3.S.M Yes 1.S.CPL see
'Yes, I saw him.' (Tuller, 1986, p. 62; tone added)

The central observation, however, made by Tuller (1986) regarding pro-dropped subjects and objects pertains to the fact that long relativisation out of relative clauses is possible in Hausa just in those cases where the respective complement may be pro-dropped. I.e., relativisation of subjects and non-human objects is insensitive to the island nature of relative clauses, whereas relativisation of human objects is not: Since null pronominals are blocked in this case, an overt resumptive must be used instead.

(15) ? gà mātār$_i$ dà ka bā nì littāfin$_j$ dà mālàmai sukà san
here.is woman REL 2.S.M.CPL give me book REL teachers 3.P.CPL know
mùtumìn$_k$ dà $\emptyset_i$ ta rubūtā wà $\emptyset_k$ $\emptyset_j$
man REL 3.S.F.CPL write for
'Here's the woman that you gave me the book the teachers know the man she wrote it for.' (Tuller, 1986, p. 84; tone added)

(16) ? gà littāfin$_j$ dà ka gwadà minì mātār$_i$ dà mālàmai sukà
here.is book REL 2.S.M.CPL show 1.S.IO woman REL teachers 3.P.CPL
san mùtumìn$_k$ dà $\emptyset_i$ ta rubūtā wà $\emptyset_k$ $\emptyset_j$
know man REL 3.S.F.CPL write IOM
'Here's the book that you showed me the woman the teachers know the man she wrote it for.' (Tuller, 1986, p. 84; tone added)

(17) gà mùtumìn$_j$ dà ka ga yārinyàr$_i$ dà $\emptyset_i$ ta
here.is man REL 2.S.M.CPL see girl REL 3.S.F.CPL
san shì$_j$ / *sanī $\emptyset_j$
know 3.S.M.DO / know $\emptyset$
'Here's the man that you saw the girl that knows him.' (Tuller, 1986, p. 85; tone added)

Similarly, indirect objects, which do not permit pro-drop either, equally disallow long distance relativisation without an overt resumptive pronoun.

(18) gà tābōbìn$_j$ dà Àli ya san mùtumìn$_i$ dà $\emptyset_i$ zâi yī
here.is cigarettes REL Ali 3.S.M.CPL know man REL 3.S.M.FUT do
musù$_j$ / *wà $\emptyset_j$ kwālī
3.P.IO / IOM $\emptyset$ box
'Here are the cigarettes that Ali knows the man that will make a box for.' (Tuller, 1986, p. 84; tone added)

Long distance relativisation of subjects and non-human objects extends to wh-islands, again without the need for an *overt* resumptive pronoun:

(19) a. littāfin$_j$ dà ka san wā̀$_i$ $\emptyset_i$ ya rubū̀tā $\emptyset_j$
book REL 2.S.M.CPL know who 3.S.M.CPL write
'the book that you know who wrote (it)' (Tuller, 1986, p. 80; tone added)

b. mùtumìn$_i$ dà ka san mḕ$_j$ $\emptyset_i$ ya rubū̀tā $\emptyset_j$
man REL 2.S.M.CPL know what 3.S.M.CPL write
'the man that you know what (he) wrote' (Tuller, 1986, p. 80; tone added)

Tuller (1986) argues that the absence of island effects in long distance relativisation of subjects and non-human objects can be directly related to the fact that these complements can be pro-dropped. Thus, she claims that null pronominals in Hausa serve an additional function of null resumptive pronouns in relativisation.

A most important finding of Tuller's is that while long relativisation out of these islands is possible, wh-extraction is not:

(20) a. * wànè littāfì$_j$ ka san wā̀$_i$ $\emptyset_i$ ya rubū̀tā $\emptyset_j$
which book 2.S.M.CPL know who 3.S.M.CPL write
'which book do you know who wrote' (Tuller, 1986, p. 80; tone added)

b. * wànè mùtûm$_i$ ka san mḕ$_j$ $\emptyset_i$ ya rubū̀tā $\emptyset_j$
which man 2.S.M.CPL know what 3.S.M.CPL write
'which man do you know what wrote' (Tuller, 1986, p. 80; tone added)

(21) * wànè mùtûm$_i$ ka bā nì littāfin$_j$ dà $\emptyset_i$ ya rubū̀tā $\emptyset_j$
which man 2.S.M.CPL give 1.S.DO book REL 3.S.M.CPL write
'Which man did you give me the book that wrote' (Tuller, 1986, p. 81; tone added)

Interestingly enough, focus fronting patterns with wh-extraction in this respect:

(22) a. * wani mùtûm$_i$ ka bā nì littāfin$_j$ dà $\emptyset_i$ ya rubū̀tā $\emptyset_j$
a man 2.S.M.CPL give me book REL 3.S.M.CPL write
'A man, you gave me the book that wrote' (Tuller, 1986, p. 81; tone added)

b. * wani mùtûm$_i$ ka san mḕ$_j$ $\emptyset_i$ ya rubū̀tā $\emptyset_j$
a man 2.S.M.CPL know what 3.S.M.CPL write
'A man, you know what wrote' (Tuller, 1986, p. 81; tone added)

The inability of wh-phrases and focused constituents to undergo long extraction out of relative clauses does not appear to be a distinguishing property of null anaphora. Long distance wh- extraction is equally impossible with overt resumptive pronouns.

(23) a. wā̀$_j$ ka yi màganā̀ dà shī$_j$
who 2.S.M.CPL do talking with 3.S.M
'Who did you talk with?' (Tuller, 1986, p. 158; tone added)

b. * wà$_j$ ka san mātâr$_i$ dà $\emptyset_i$ ta yi màganà dà shī$_j$
who 2.S.M.CPL know woman REL 3.S.F.CPL do talking with 3.S.M
'Who do you know the woman that talked to him' (Tuller, 1986, p. 159; tone added)

(24) a. wà$_j$ ka karàntà littāfì-n-sà$_j$
who 2.S.M.CPL read book-L-3.S.M.G
'Whose book did you read?' (Tuller, 1986, p. 158; tone added)

b. * wà$_j$ ka ga yârân$_i$ dà $\emptyset_i$ sukà ƙōnè littāfì-n-sà$_j$
who 2.S.M.CPL see children REL 3.P.CPL burn book-L-3.S.M.G
'Who did you see the children that burnt his book' (Tuller, 1986, p. 159; tone added)

To summarise the empirical findings, resumptive pronouns in Hausa — be they null or overt — may license long relativisation out of relative clauses and wh-islands. Neither covert nor overt resumption, however, is capable of licensing long wh- or focus movement out of extraction islands, despite the fact that overt resumption of focused or wh-phrases is generally possible.

2 Previous Analyses

2.1 Resumption: Extraction or Anaphora?

The possibility for resumption to escape island constraint violations is an observation that has been made repeatedly in the literature. In order to model this apparent difference between filler-gap constructions and resumption, it has been repeatedly suggested to regard the latter as an anaphoric process distinct from movement (e.g. Borer, 1984; Sells, 1984). Within the framework of HPSG (Pollard and Sag, 1994), Vaillette (2001a,b) argues in favour of an analysis of Hebrew and Irish resumptives in terms of unbounded dependencies (UDCs), using a non-local feature RESUMP akin to the standard SLASH feature. A distinguishing property of the non-local RESUMP feature is, however, that it is a set of semantic indices, rather than full-fledged LOCAL values, thereby capturing the basic intuition inherent to the anaphoric binding approach. In contrast to Vaillette, Taghvaipour (2004, 2005) suggests to do away with the RESUMP feature and encode both types of UDCs as SLASH dependencies, distinguished in terms of diacritic nonlocal features indicating the type of gap (not percolated) and the type of unbounded dependency construction (percolated). The decision not to percolate information about how the foot of the dependency is realised (resumptive or gap) is motivated by the observation that ATB-extraction from coordinate structures may equate SLASH values corresponding to a gap in one conjunct, and a resumptive in the other. However, since island effects in Hausa are sensitive to the type of gap, whereas ATB-extraction in Hausa (cf. (9)) is not, it is clear that there is no trivial modification of Taghvaipour's approach to Persian that can account for both constructions at the same time.

In her GB analysis, Tuller (1986) suggests that Hausa actually has two different strategies for resumption. The first such strategy, according to Tuller, is witnessed by relative clause constructions: since subjacency restrictions are obviously not obeyed with long-distance relativisation, Tuller (1986) concludes that movement cannot be involved. Instead she claims that relative clauses in Hausa may be base-generated. The second resumptive strategy operative in Hausa is treated by Tuller (1986) as an S-structure phenomenon: since wh-extraction and focus fronting obviously obey subjacency, she concludes that these constructions must involve movement. In wh-constructions, Tuller (1986) motivates insertion of resumptive pronouns by virtue of the ECP. Following an earlier proposal by Koopman (1982) she argues that the resumptive pronouns that surface when the complement of a noun or preposition is extracted should best be analysed as "spelling out traces as pronouns". In those cases where null pronouns are independently possible, apparent surface violations of the ECP obtain, like, e.g., exemptions to the *that-t* effect. As an alternative to trace spell-out, Tuller (1986) considers Ā-binding between an operator and a resumptive, but, as suggested by footnote 31 (p. 158/217), it seems to be the less preferred analytic option.

There are several arguments that can be raised against Tuller's analysis of Hausa resumption. First, the postulation of two different resumptive processes that show a high degree in overlap regarding the resumptive elements involved is quite ad hoc: any theory of Hausa resumption that can be cast in terms of a single process must therefore seem preferable. Second, the exact conditions under which a resumptive pronoun is related to the antecedent noun remain opaque. It appears counter-intuitive at least that the more unbounded process (relativisation) should involve base-generation, whereas the one bounded at least by subjacency (wh-extraction) is regarded as a movement process. Third, if resumptive pronouns are conceived as spell-out of traces (R-expressions), it remains unexplained why the form of pronoun chosen is always the same as that of plain, non-resumptive pronouns used in the same surface-syntactic context. Fourth, in cases where no ECP violation could arise, Tuller (1986)'s analysis always assigns two distinct analyses, one involving a standard trace, the other a trace spelled out as an empty pronominal. Since the two analyses have the same semantic interpretation, we are actually facing a spurious ambiguity here.

Instead of postulating three different processes (traces, pronominal spell-out of trace, and base-generation) I shall suggest a treatment in terms of a single process, namely non-local feature percolation. Different locality restrictions on relativisation and wh-/focus fronting will be attributed to the permeability of head-filler structures towards quasi-anaphoric resumptives.

2.2 Anaphoric Processes in Rightward Movement

Apparent violations of subjacency, in particular extraction out of complex NPs and adjunct islands have also been reported for relative clause extraposition, a rightward oriented process. This unexpected behaviour with respect to Bounding Theory has notoriously been problematic for movement (Baltin, 2001) and base-generation approaches alike (Culicover and Rochemont, 1990), albeit for slightly different reasons. An alternative theory of relative clause extraposition that does not suffer from these problems regards relative clause extraposition as a clause-bound anaphoric process. Accounts along these lines have been proposed in various forms for English (Wittenburg, 1987; Kiss, 2003) and German (Kiss, 2005). What is common to all these approaches is that nouns indiscriminately percolate their semantic indices within the local clause, to be picked up (modified) by the extraposed relative clause. At sentence boundaries, percolated indices that have not yet been retrieved are simply discarded which captures the clause-boundedness of the phenomenon together with the fact that modification by a relative clause is truly optional. Furthermore, it has been shown that an anaphoric approach is also computationally much more attractive than its movement-based alternatives (Crysmann, 2005b).

Most recently, I have proposed (Crysmann, to appear) to reconcile the anaphoric approach advanced by Kiss (2005) with movement-based approaches to complement clause extraposition (e.g. Keller, 1995). Given that both these rightward-oriented processes are subject to Ross (1967)'s Right Roof Constraint, I have suggested to model them using a single non-local feature. Differences with respect to island constraints, in particular adjunct islands, are related to a difference in the status of percolated material: light semantic indices in the case of relative clause extraposition vs. full local values in the case of complement extraposition. Bounding nodes, such as the adjunct daughter in a head-adjunct structure are constrained to be permeable for "light" indices but impermeable for "heavier" full LOCAL values.

Since one of the recurring intuitions in the literature on resumption is to exploit the inherently anaphoric nature of resumptive pronouns in order to provide an explanation for the reduced sensitivity towards islands, it makes perfect sense to explore whether such a move cannot also be fruitfully applied to the case of anaphoric processes within relative clauses.

In the remainder of this paper I shall therefore develop a theory of resumption that extends the aforementioned work on extraposition towards a treatment of resumption. I shall propose that difference with respect to island-hood can be related to the nature of percolated material, i.e., a distinction between "heavier" true gaps and "lighter" resumptive elements.

3 Analysis

Before we embark on our formal analysis, which is cast within the framework of HPSG (Pollard and Sag, 1987, 1994), let me briefly summarise the desiderata of an analysis of Hausa resumption: first, resumptive use should be a modelled as a systematic property of all pronominal elements independent of their mode of realisation as null pronouns, bound affixes, or independent words. Second, spurious ambiguity between filler-gap analyses and null anaphora should be avoided. Since locality restrictions on filler-gap dependencies appear to be more strict than those operative for anaphoric processes, the problem of spurious ambiguity can only be avoided on principled grounds, if the scope of null anaphora is extended at the expense of filler-gap constructions. Third, if constraints on island-hood are best captured in terms of properties of the filler and intervening boundary nodes, such constraints should not be replicated at the extraction site in terms of local ambiguity.

In order to address our first desideratum, we need to settle on a lexical representation that does not crucially distinguish between resumptive and non-resumptive uses of any particular pronominal realisation. Vaillette (2001a) has suggested that in Hebrew, every pronoun can be subjected to a lexical rule deriving a resumptive variant for it. Since in Hausa resumptive elements are not always lexical items (they can be independent pronouns, pronominal affixes, or null anaphors), a solution along these lines would be suboptimal. Instead, I shall build on a proposal by Miller and Sag (1997) who suggest that *synsem* values can be classified according to their mode of realisation: *canonical-s(yn)s(em)*, which corresponds to in-situ lexical complements and *gap-s(yn)s(em)*, which corresponds to non-locally realised material. Orthogonal to this distinction, I shall postulate a distinction into nonpronominal and pronominal (*pron(oun)-s(yn)s(em)*) which cuts across the previous distinction between gaps and canonical *synsem* objects in a multiple-inheritance hierarchy. Following the standard HPSG theory of extraction (Sag, 1997; Ginzburg and Sag, 2001), I shall assume that *gap-ss* have their LOCAL value reentrant with the singleton member of their SLASH set.

(25) $$\textit{gap-ss} \rightarrow \begin{bmatrix} \text{LOC} & \boxed{1} \\ \text{NLOC} & \left\{ \boxed{1} \right\} \end{bmatrix}$$

In a resumptive pronoun language such as Hausa, any pronominal can, in principle, foot a non-local dependency. Building on the analogy with rightward anaphoric processes as exemplified by relative clause extraposition, I shall suggest that Hausa pronominals may percolate their INDEX value, to be picked up by an antecedent higher in the clause.

(26) $$\textit{pron(oun)-ss} \rightarrow \begin{bmatrix} \text{LOC} & \begin{bmatrix} \text{CONT} \begin{bmatrix} \text{HOOK} \begin{bmatrix} \text{INDEX} & \boxed{i} \end{bmatrix} \end{bmatrix} \end{bmatrix} \\ \text{NLOC} & \begin{bmatrix} \text{INH} \mid \text{SLASH} \left\{ \, \right\} \vee \left\{ \begin{bmatrix} \text{CONT} \begin{bmatrix} \text{HOOK} \begin{bmatrix} \text{INDEX} & \boxed{i} \end{bmatrix} \end{bmatrix} \end{bmatrix} \right\} \end{bmatrix} \end{bmatrix}$$

Building on my earlier work on rightward movement, I will also incorporate a distinction of LOCAL values according to weight: more specifically, I shall distinguish two subtypes of *local*, namely *full-local*, which has both CAT and CONT as appropriate attributes, and an impoverished *index-local* which crucially fails to incorporate syntactic information (CAT) and where the semantic contribution is limited to index features (empty RELS list).[6] While *synsem* objects will always

[6] I shall assume Minimal Recursion Semantics (Copestake et al., 2005) as meaning representation language.

select *full-local* as the type of the LOCAL attribute they introduce, SLASH elements (just like the EXTRA elements of Crysmann to appear) are underspecified as to the type of *local* objects they can contain.

(27)

$$\begin{bmatrix} \textit{local} \\ \text{CONT} \quad \textit{mrs} \end{bmatrix} \qquad \textit{synsem} \rightarrow \begin{bmatrix} \text{LOC} \quad \textit{full-local} \end{bmatrix}$$

$$\begin{bmatrix} \textit{full-local} \\ \text{CAT} \quad \textit{cat} \end{bmatrix} \quad \begin{bmatrix} \textit{index-local} \\ \text{CONT} \quad \begin{bmatrix} \text{RELS} \ \langle\rangle \end{bmatrix} \end{bmatrix}$$

For lexical pronouns, the reentrancy between the pronoun's INDEX with that of its SLASH element will be introduced by the lexical entry of the pronoun directly (requiring the pronoun's SYNSEM value to be of type *pron-ss*), whereas for pronominal affixes and null pronominals it will be introduced by virtue of a valence-reducing lexical rule, along the lines given below. Of course, this general rule type will be further differentiated to introduce appropriate exponents for bound pronominals, depending on the category of the host (noun or verb) and the index features of the pronominal complement. For null object pronominals, application will be restricted to the sort of non-human referents.

(28)

$$\begin{bmatrix} \text{ARG-ST} \quad \left\langle \ldots, \boxed{0}, \ldots \right\rangle \\ \text{SYNSEM} \quad \begin{bmatrix} \text{LOC} \mid \text{CAT} \mid \text{COMPS} \quad \boxed{1} \oplus \left\langle \boxed{0} \ \textit{pron-ss} \right\rangle \oplus \boxed{2} \end{bmatrix} \end{bmatrix}$$

$$\mapsto \begin{bmatrix} \text{SYNSEM} \quad \begin{bmatrix} \text{LOC} \mid \text{CAT} \mid \text{COMPS} \quad \boxed{1} \oplus \boxed{2} \end{bmatrix} \end{bmatrix}$$

Following the head-driven theory of extraction advanced by Sag (1997) and Ginzburg and Sag (2001), the (anaphoric) slashes introduced by lexical pronouns or (null) pronominal lexical rules will be amalgamated from the ARG-ST list onto the SLASH set of the head. Further percolation will be effected by the Generalised Head Feature Principle.

Alongside introduction of the indices of pronominals, I shall also postulate at least one lexical slash insertion rule dedicated to human direct object complements.[7] Since human direct objects cannot enter into the same kinds of anaphoric relations as overt pronouns and null subjects and non-human objects do, this difference will be signalled by a sortal feature. Note that such a sortal distinction is independently required to ensure proper selection of wh-pronouns in Hausa, i.e. *wà̃* 'who' vs. *mè̃* 'what'.

[7] In addition to this lexical SLASH introduction rule, I shall also postulate a syntactic rule for adjunct extraction (Levine, 2003), in order to cover, inter alia, PP pied-piping.

(29)
$$\begin{bmatrix} \text{ARG-ST} \left\langle \ldots, \boxed{0}, \ldots \right\rangle \\ \text{SS} \begin{bmatrix} \text{L} \mid \text{CAT} \begin{bmatrix} \text{HEAD} & \textit{noun} \vee \textit{verb} \\ \text{COMPS} & \boxed{1} \oplus \left\langle \boxed{0} \begin{bmatrix} \textit{gap-ss} \\ \text{LOC} \mid \text{CONT} \mid \text{HOOK} \mid \text{IND} \mid \text{SORT} \; \textit{hum} \end{bmatrix} \right\rangle \oplus \boxed{2} \end{bmatrix} \end{bmatrix} \end{bmatrix}$$

$$\mapsto \begin{bmatrix} \text{SS} \begin{bmatrix} \text{L} \mid \text{CAT} \mid \text{COMPS} & \boxed{1} \oplus \boxed{2} \end{bmatrix} \end{bmatrix}$$

Since we allow both full sharing of local values and mere index sharing, we need to make sure that fillers binding *full-local* SLASH elements are properly terminated at the bottom of the dependency. In order to enforce reentrancy with a local value, I shall introduce the following constraint:

(30)
$$\begin{bmatrix} \text{LOC} & \begin{bmatrix} \text{CONT} \mid \text{HOOK} \mid \text{INDEX} \; \boxed{i} \end{bmatrix} \\ \text{NLOC} & \begin{bmatrix} \text{INH} \mid \text{SL} \left\{ \begin{bmatrix} \textit{full-local} \\ \text{CONT} \mid \text{HOOK} \mid \text{INDEX} \; \boxed{i} \end{bmatrix} \right\} \end{bmatrix} \end{bmatrix} \rightarrow \begin{bmatrix} \text{LOC} & \boxed{l} \\ \text{NLOC} & \begin{bmatrix} \text{INH} \mid \text{SL} \left\{ \boxed{l} \right\} \end{bmatrix} \end{bmatrix}$$

Having discussed how slash dependencies are introduced and percolated, I shall now turn to the representation of filler-head structures, i.e., those structures where filler-gap dependencies, be they resumptive or not, will be bound off. In the most general case, a filler must at least bind an index contributed by a gap, a situation that holds both for (anaphoric) resumption and extraction proper.

(31) *filler-head-struc* →
$$\begin{bmatrix} \text{SS} & \begin{bmatrix} \text{NLOC} \mid \text{INH} \mid \text{SLASH} & \textit{set(index-local)} \end{bmatrix} \\ \text{DTRS} & \begin{bmatrix} \text{FILLER-DTR} & \begin{bmatrix} \text{L} \mid \text{CONT} \mid \text{HOOK} \mid \text{INDEX} & \boxed{i} \end{bmatrix} \\ \text{HD-DTR} & \begin{bmatrix} \text{SS} \begin{bmatrix} \text{L} \mid \text{CAT} \begin{bmatrix} \text{HD} & \begin{bmatrix} \text{PRD} & + \end{bmatrix} \\ \text{SUBJ} & \langle\rangle \\ \text{COMPS} & \langle\rangle \end{bmatrix} \\ \text{NLOC} \mid \text{TO-BIND} \mid \text{SLASH} \left\{ \begin{bmatrix} \text{CONT} \mid \text{HOOK} \mid \text{INDEX} \; \boxed{i} \end{bmatrix} \right\} \end{bmatrix} \end{bmatrix} \end{bmatrix} \end{bmatrix}$$

Recall from our discussion of the data that we observed a fundamental asymmetry between true gaps (zero human direct objects) and overt or covert resumption: while UDCs involving non-resumptive gaps cannot escape any extraction islands, those involving resumption can. Given our distinction of local values into *full-local* and impoverished *index-local*, the selective permeability of filler-head structures towards light indices can be modelled straightforwardly as a constraint on the mother's SLASH value.

The first subtype of filler-head structures to be considered are relative-head structures:

(32) *rel-head-struc* →

$$\left[\begin{array}{l} \textit{filler-head-struc} \\ \text{DTRS} \left[\text{FILLER-DTR} \left[\begin{array}{ll} \text{L} \mid \text{CONT} \mid \text{HOOK} \mid \text{INDEX} & \boxed{i}\ \textit{index} \\ \text{NLOC} \mid \text{INH} \mid \text{REL} & \left\{\boxed{i}\right\} \end{array}\right]\right] \end{array}\right]$$

As depicted above, relative filler-head structures inherit most of their constraints from *head-filler-struc*, mainly adding a restriction for the filler to contribute a non-eventual index. Since relative filler-head structures do not equate the filler's local value with the head-daughter's TO-BIND|SLASH, there is no coercion to *full-local*, and, therefore, relative fillers can easily bind either *full-local* or *index-local* SLASH elements, i.e., long-distance percolated indices originating inside an extraction island. In sum, this underspecification will enable the relative pronoun to bind true gaps (e.g. human direct objects), as well as resumptive elements.

Wh-extraction and focus fronting, by contrast, cannot apply long distance out of wh-islands or relative clauses. If our above characterisation of the permeability of filler-head structures is correct, we can represent the island-sensitivity of these latter processes by requiring these fillers to have their LOCAL value reentrant with the head daughter's TO-BIND|SLASH. Taking the case of wh-fillers as an example[8], we can model this selectivity using the standard constraint on filler-head structures familiar from English (Pollard and Sag, 1994):

(33) *wh-head-struc* →

$$\left[\begin{array}{l} \textit{filler-head-struc} \\ \text{DTRS} \left[\begin{array}{l} \text{FILLER-DTR} \left[\begin{array}{ll} \text{L} & \boxed{l} \\ \text{NLOC} \mid \text{INH} \mid \text{QUE} & \left\{\boxed{\ }\right\} \end{array}\right] \\ \text{HD-DTR} \left[\text{SS} \mid \text{NLOC} \mid \text{TO-BIND} \mid \text{SLASH} \left\{\boxed{l}\right\}\right] \end{array}\right] \end{array}\right]$$

Since gaps introduced by pronominal synsems are underspecified for their LOCAL value, they can also function as gaps for wh-fillers. Needless to say that proper gaps, being of type *full-local* by means of rentrancy with a sign's LOCAL value, can also be bound by a wh-phrase. Once percolation of non-local features crosses a filler-head structure, all elements in SLASH are forced to be interpreted as light *index-local* elements, thereby deriving the island-sensitivity of wh-expressions.

The underspecification approach advocated here, where SLASH values lexically contributed by pronominals subsume the SLASH value characteristic of gaps, readily accounts for ATB extraction: as witnessed by (9), Hausa, just like Hebrew (Sells, 1984; Vaillette, 2001a) or Persian (Taghvaipour, 2005), allows a gap in one conjunct with a resumptive in the other. In contrast to Vaillette (2001a), the present approach can capture this directly using the standard Coordination Principle of Pollard and Sag (1994) without any disjunctive specifications.

---

8 A more general version covering both wh-expressions and focused fillers can probably be formulated quite easily by means of reference to information structure, given that wh-expression are inherently focused, but relative phrases are most likely not.

4 Conclusion

In this paper we have argued in favour of an approach to resumption in Hausa where pronominal elements are typically underspecified with respect to their status as pronominals, light "anaphoric" elements and full-fledged resumptive gaps, by means of optionally introducing a SLASH dependency which minimally contains the pronoun's index, but which can be further restricted to involve full sharing of local values. Given the underspecification at the gap site, differences with respect to island-hood are defined instead as properties of fillers: while wh-fillers and focus fronted constituents coerce the gaps they bind to be full local values, relative pronouns do not do so. The observed asymmetry between island-sensitive wh-extraction and island-insensitive relativisation was easily captured then by a single constraint on head-filler-structures to be permeable for light indices, yet opaque for full local values.

The analysis advanced here builds crucially on the notion of percolation of "anaphoric" indices. This approach not only paves the way for an analysis which is entirely free of any spurious ambiguity between gap and resumptive analyses of null pronominals, but it also connects the anaphoric nature of resumptive processes to similar phenomena in the area of rightward movement, namely relative clause extraposition, which happens to be equally insensitive to island constraints.

References

Baltin, M.: Extraposition, the right roof constraint, result clauses, relative clause extraposition, and pp extraposition. ms., New York University (2001)

Borer, H.: Restrictive relatives in Modern Hebrew. NLLT 2, 219–260 (1984)

Copestake, A., Flickinger, D., Pollard, C., Sag, I.: Minimal recursion semantics: an introduction. Research on Language and Computation 3(4), 281–332 (2005)

Crysmann, B.: An inflectional approach to Hausa final vowel shortening. In: Booij, G., van Marle, J. (eds.) Yearbook of Morphology 2004, pp. 73–112. Kluwer (2005a)

Crysmann, B.: Relative clause extraposition in German: An efficient and portable implementation. Research on Language and Computation 3(1), 61–82 (2005b)

Crysmann, B.: A Unified Account of Hausa Genitive Constructions. In: de Groote, P., Egg, M., Kallmeyer, L. (eds.) Formal Grammar. LNCS, vol. 5591, pp. 102–117. Springer, Heidelberg (2011)

Crysmann, B.: On the locality of complement clause and relative clause extraposition. In: Webelhuth, G., Sailer, M., Walker, H. (eds.) Rightward Movement in a Comparative Perspective. John Benjamins, Amsterdam (to appear)

Culicover, P., Rochemont, M.: Extraposition and the complement principle. Linguistic Inquiry 21, 23–47 (1990)

Ginzburg, J., Sag, I.: Interrogative Investigations: the Form, Meaning and Use of English Interrogatives. CSLI publications, Stanford (2001)

Jaggar, P.: Hausa. John Benjamins, Amsterdam (2001)

Keller, F.: Towards an account of extraposition in HPSG. In: Proceedings of the Ninth Meeting of the European ACL, pp. 301–306. Association for Computational Linguistics, Dublin (1995)

Kiss, T.: Phrasal typology and the interaction of topicalization, wh-movement, and extraposition. In: Kim, J.-B., Wechsler, S. (eds.) Proceedings of the 9th International Conference on Head-Driven Phrase Structure Grammar, Kyung Hee University, Seoul, August 5-7, CSLI Publications, Stanford (2002)

Kiss, T.: Semantic constraints on relative clause extraposition. Natural Language and Linguistic Theory 23, 281–334 (2005)

Koopman, H.: Control from COMP and comparative syntax. The Linguistic Review 2, 365–391 (1982)

Levine, R.D.: Adjunct valents: cumulative scoping adverbial constructions and impossible descriptions. In: Kim, J., Wechsler, S. (eds.) The Proceedings of the 9th International Conference on Head-Driven Phrase Structure Grammar, pp. 209–232. CSLI Publications, Stanford (2003)

McCloskey, J.: Resumptive pronouns, a'-binding and levels of representation in Irish. In: Hendrick, R. (ed.) The Syntax of the Modern Celtic Languages. Syntax and Semantics, vol. 23. Academic Press, London (1990)

Miller, P., Sag, I.: French clitic movement without clitics or movement. Natural Language and Linguistic Theory 15(3), 573–639 (1997)

Newman, P.: The Hausa Language. An Encyclopedic Reference Grammar. Yale University Press, New Haven (2000)

Pollard, C., Sag, I.: Information–Based Syntax and Semantics, vol. 1. CSLI, Stanford (1987)

Pollard, C., Sag, I.: Head–Driven Phrase Structure Grammar. CSLI and University of Chicago Press, Stanford (1994)

Ross, J.R.: Constraints on Variables in Syntax. PhD thesis, MIT (1967)

Sag, I.: English relative clause constructions. Journal of Linguistics 33(2), 431–484 (1997)

Sells, P.: Syntax and Semantics of Resumptive Pronouns. PhD thesis. University of Massachusetts at Amherst (1984)

Shlonsky, U.: Resumptive pronouns as a last resort. Linguistic Inquiry 23, 443–468 (1992)

Taghvaipour, M.: An HPSG analysis of Persian relative clauses. In: Müller, S. (ed.) Proceedings of the HPSG 2004 Conference, Center for Computational Linguistics, Katholieke Universiteit Leuven, pp. 274–293. CSLI Publications, Stanford (2004)

Taghvaipour, M.A.: Persian free relatives. In: Müller, S. (ed.) The Proceedings of the 12th International Conference on Head-Driven Phrase Structure Grammar, Department of Informatics, University of Lisbon, pp. 364–374. CSLI Publications, Stanford (2005)

Tuller, L.A.: Bijective Relations in Universal Grammar and the Syntax of Hausa. PhD thesis. UCLA, Ann Arbor (1986)

Vaillette, N.: Hebrew relative clauses in HPSG. In: Flickinger, D., Kathol, A. (eds.) The Proceedings of the 7th International Conference on Head-Driven Phrase Structure Grammar, pp. 305–324. CSLI Publications, Stanford (2001a)

Vaillette, N.: Irish gaps and resumptive pronouns in HPSG. In: Van Eynde, F., Beermann, D., Hellan, L. (eds.) The Proceedings of the 8th International Conference on Head-Driven Phrase Structure Grammar, pp. 284–299. CSLI Publications, Stanford (2001b)

Wittenburg, K.: Extraposition from NP as anaphora. In: Huck, G., Ojeda, A. (eds.) Discontinuous Constituency. Syntax and Semantics, vol. 20, pp. 428–445. Academic Press, New York (1987)

Wolff, E.: Referenzgrammatik des Hausa. LIT, Münster (1993)

Iterated Dependencies and Kleene Iteration

Michael Dekhtyar[1], Alexander Dikovsky[2], and Boris Karlov[1,*]

[1] Dept. of Computer Science, Tver State University, Tver, Russia, 170000
Michael.Dekhtyar@tversu.ru, bnkarlov@gmail.com
[2] LINA CNRS UMR 6241, Université de Nantes
Alexandre.Dikovsky@univ-nantes.fr

Abstract. Categorial Dependency Grammars (CDG) is a class of simple and expressive categorial grammars defining projective and discontinuous dependency structures in a strongly compositional way. They are more expressive than CF-grammars, are polynomial time recognizable and different from the mildly context sensitive grammars. CDG languages are proved to be closed under all AFL operations, but `iteration`. In this paper, we explain the connection between the iteration closure and the iterated dependencies (optional repeatable dependencies, inevitable in dependency syntax) and show that the CDG extended by a natural multimodal rule define an AFL, but the membership problem in this extended family is NP-complete.

Keywords: Dependency Grammar, Categorial Dependency Grammar, Iterated Dependency, Iteration.

1 Introduction

In this paper are studied `iterated`, i.e. optional repeatable dependencies, such as `modifier` dependencies of nouns (e.g. *optional* $\overset{modif}{\longleftarrow}$ *dependencies* and `circumstantial` dependencies of verbs (e.g. $fits \overset{circ}{\longrightarrow}$ *well*). The main question is whether the dependency grammars expressing such dependencies generate languages closed under Kleene iteration, or more generally, is there a direct connection between the former and the latter. It should be made clear what do we mean by "express iterated dependencies". In fact, in the traditional linguistics it is generally accepted that the ultimately repeatable modifiers / circumstantials share the same governor (e.g. see [12]). I.e., the adequate dependency structure for *a tall blond young girl* is that in Fig. 1(a) and not that in Fig. 1(b).[1]

* This work was sponsored by the Russian Fundamental Studies Foundation (Grants No. 10-01-00532-a and 08-01-00241-a).

[1] By indirection, this means that the dependency structures corresponding to the traditional recursive types of modifiers / circumstantials used in the categorial grammars [1,16], in Lambek grammar [11] and, more generally, in the type logical grammars interfacing the semantics of Montague [14] are not adequate from the traditional linguistic point of view. In [13] this structural defect is amended using types extended with modalities, but the resulting calculus is computationally untractable.

P. de Groote and M.-J. Nederhof (Eds.): Formal Grammar 2010/2011, LNCS 7395, pp. 66–81, 2012.

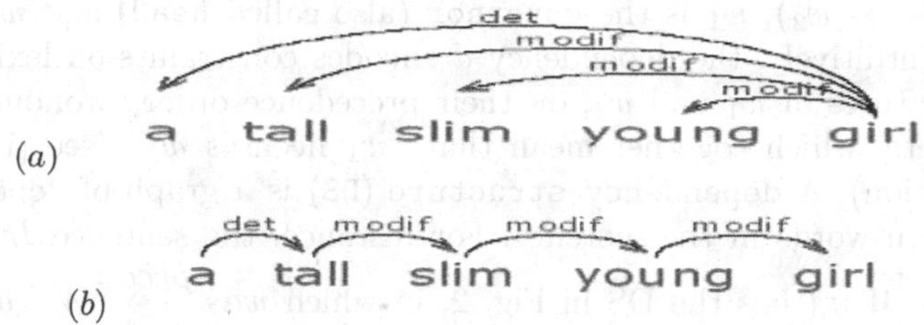

Fig. 1. Dependency structures: iterative vs. recursive

As it concerns the grammars not expressing discontinuous (crossing) dependencies and generating exactly the context-free languages, such as for instance the link grammars [15], the question is trivially answered in the positive sense. For the (rare) grammars expressing discontinuous dependencies, there is no general solution. E.g., for constraint grammars with NP-hard membership problem[2], such as topological dependency grammars [7], the answer is positive. But for polynomially analyzed dependency grammars expressing discontinuous dependencies the problem needs a specific solution for every particular class.

In this paper, we try to establish a connection between the iterated dependencies and the Kleene star closure in the class of categorial dependency grammars (CDG) [6]. CDG are categorial grammars based on a simple calculus of dependency types. They express unlimited discontinuous dependencies using simple polarized valencies' pairing rules and are polynomially parsed. CDG express iterated dependencies explicitly through iterated types of the form $t*$. For instance, in Fig. 1(a), for the word *girl* is used the type $[modif * \backslash det \backslash S]$, S being the axiom. It may seem that in the presence of such iterated types the generated languages are immediately closed under the Kleene iteration. This illusion has let down the authors of [3] who stated that the CDG languages constitute an AFL, and in particular are closed under iteration.[3] As we show below, in general, the direct closure construction doesn't work because of the CDG's valency pairing rule. We arrive to find rather a minimal modality extension of the basic CDG for which we finally prove the iteration closure property. However, the resulting extended CDG turn out to be NP-complete.

2 Basics of Dependency Structures

Tesnière [17] was the first who systematically described the sentence structure in terms of named binary relations between words (`dependencies`). When two words w_1 and w_2 are related in a sentence through dependency relation

[2] Membership problem in a family $\mathcal{F}$ is the problem $w \in L(G)$ for a given grammar $G \in \mathcal{F}$ (not to confuse with the "uniform membership problem" $\{< w, G > | w \in L(G), G \in \mathcal{F}\}$).

[3] This is the only assertion in [3] stated without proof because of its "evidence".

d (denoted $w_1 \xrightarrow{d} w_2$), w_1 is the **governor** (also called **head**) and w_2 is the **subordinate**. Intuitively, the dependency d encodes constraints on lexical and grammatical features of w_1 and w_2, on their precedence order, pronominalization, context, etc. which together mean that "w_1 licenses w_2" (see [12] for a detailed exposition). A **dependency structure** (**DS**) is a graph of dependency relations between words in the sentence. For instance, the sentence *In the beginning was the Word* has the DS in Fig. 2, in which *was* $\xrightarrow{pred}$ *Word* stands for the predicative dependency between the copula *was* and the subject *Word*. There is no general agreement on the notion of DS: sometimes it is separated from the precedence order in the sentence, sometimes it is linearly ordered by the precedence, most people require it be a tree (the tradition going back to [17]), some others do not (cf. [8]) because without this constraint one can define mixed structures (e.g. taking in account the co-reference). When a DS is a tree it is called **dependency tree** (**DT**).

CDG use DS linearly ordered by the precedence order in the sentence. Without this order some fundamental properties of DS cannot be expressed. This is the case of one of the most important properties of DS, called **projectivity**. This property is expressed in terms of the **immediate dominance** relation: $w_1 \Rightarrow w_2 \equiv \exists d\ (w_1 \xrightarrow{d} w_2)$ and of its reflexive-transitive closure $\Rightarrow^*$ called **dominance**. A DT of a sentence x is **projective** if, for every word w in x, the set of all words dominated by w: $proj(w) = \{w' \mid w \Rightarrow^* w'\}$ (called **projection** of w) is an **interval** of x with respect to the precedence order. In all languages, the majority of DS are projective DT (an example is given in Fig. 2). But even if this is true, the projectivity is not a norm. Non-projective dependencies are often due to discontinuous constructions such as comparatives (cf. *more..than* in English or negation *ne..pas* in French). They are also caused by verb complements' dislocation, clefting, scrambling, conversion of complements into clitics and other regular constructions marking for a communicative structure of sentences. We show two examples of non-projective DT in Figs. 3, 4.

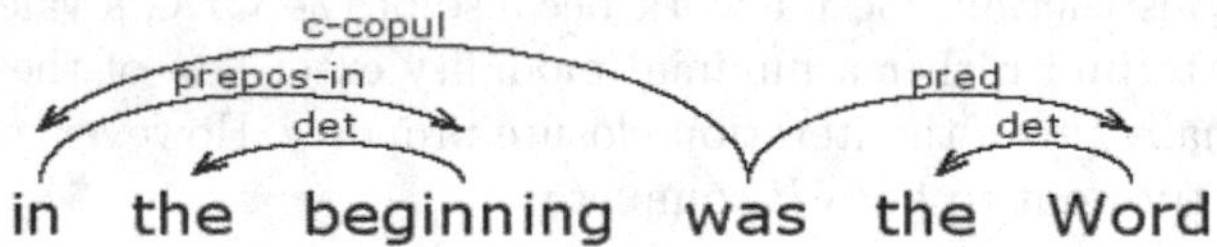

Fig. 2. A projective DT

Definition 1. *Let $w = a_1 \ldots a_n$ be a string, W be the set of all occurrences of symbols in w and $C = \{d_1, \ldots, d_m\}$ be a set of* **dependency names**. *A graph $D = (W, E)$ with labeled arcs is a* **DS** *of w if it has a* **root**, *i.e. a node $a_0 \in W$ such that (i) for any node $a \in W$, $a \neq a_0$, there is a path from a_0 to a and (ii) there is no arc (a', d, a_0).*[4] *An arc $(a_1, d, a_2) \in E$ is called* **dependency** *d from a_1 to a_2. The linear order on W induced by w is the* **precedence order** *on D.*

[4] Evidently, every DS is connected and has a unique root.

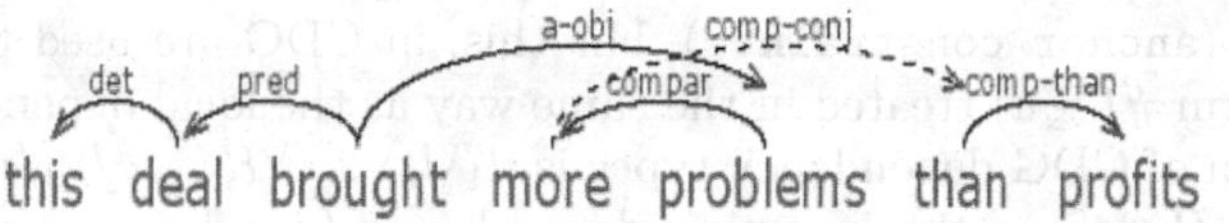

Fig. 3. A non-projective DT in English

Fig. 4. A non-projective DT in French (**she it$_{\mathrm{FEM}}$ to-him has given*)

3 Categorial Dependency Grammars

As all categorial grammars, the **categorial dependency grammars** (CDG) may be seen as assignments of dependency types to words. Every dependency type assigned to a word w defines its possible **local neighborhood** in grammatically correct DS. The neighborhood of w consists of the **incoming** dependency, i.e. the dependency relation d through which w is subordinate to a word G, its **governor**, and also of a sequence of **outgoing** dependencies, i.e. the dependency relations d_i through which w governs a subordinate word w_i. For instance, the type assignment:

$in \mapsto [c\!-\!copul/prepos\!-\!in]$, $the \mapsto [det]$, $beginning \mapsto [det\backslash prepos\!-\!in]$,
$was \mapsto [c\!-\!copul\backslash S/pred]$, $Word \mapsto [det\backslash pred]$

determines the DS in Fig. 2. In particular, the type $[c\!-\!copul/\ prepos\!-\!in]$ of *in* defines its local neighborhood, where $c\!-\!copul$ is the incoming dependency and $prepos\!-\!in$ is the right outgoing dependency. The verb *was* in the root has the head type S which serves as the grammar's **axiom**. CDG use iteration to express all kinds of repetitive dependencies and in particular the coordination relations. This provides more adequate DS than those traditionally defined in categorial grammars through recursive types $[X/X]$. For instance, the type assignment: $a \mapsto [det]$, $tall$, $slim$, $young \mapsto [modif]$, $girl \mapsto [modif*\backslash det\ S]$ determines the DT in Fig. 1(a). Remark that in the canonical categorial grammars the types of articles and adjectives are functional.

In CDG, the non-projective dependencies are expressed using so called polarized valencies. Namely, in order that a word G may govern through a discontinous dependency d a word D situated somewhere on its right, D should have a type declaring the positive valency $\nearrow d$, whereas its subordinate D should have a type declaring the **negative** valency $\searrow d$. Together these **dual** valencies define the discontinous dependency d. By the way, the pairing itself of dual valencies is not enough to express the constraints of adjacency of the distant subordinate to a host word (anchor constraints). For this, in CDG are used the anchor types of the form $\#(\searrow d)$ treated in the same way as the local dependencies. So the general form of CDG dependency types is $[l_1\backslash l_2\backslash \ldots \backslash H/\ldots/r_2/r_1]^P$, where the head type H defines the incoming dependency, l_i and r_i are respectively the left and the right outgoing dependencies or anchors and P is the potential, i.e. a string of polarized valencies defining incoming and outgoing discontinuous long distance dependencies.[5] We show two examples of non-projective DT in Figs. 3, 4. E.g., the non projective DT in Fig. 4 is defined by the assignment:

$elle \mapsto [pred]$
$la \mapsto [\#(\swarrow clit-a-obj)]^{\swarrow clit\text{-}a\text{-}obj}$
$lui \mapsto [\#(\swarrow clit-3d-obj)]^{\swarrow clit\text{-}3d\text{-}obj}$
$a \mapsto [\#(\swarrow clit-3d-obj)\backslash \#(\swarrow clit-a-obj)\backslash pred\backslash S/aux]$
$donn\acute{e}e \mapsto [aux]^{\nwarrow clit\text{-}3d\text{-}obj \nwarrow clit\text{-}a\text{-}obj}$.

Definition 2. *Let* **C** *be a set of* local dependency names *and* **V** *be a set of* valency names.

The expressions of the form $\swarrow v$, $\nwarrow v$, $\searrow v$, $\nearrow v$, *where* $v \in \mathbf{V}$, *are called* polarized valencies. $\nwarrow v$ *and* $\nearrow v$ *are* positive, $\swarrow v$ *and* $\searrow v$ *are* negative; $\nwarrow v$ *and* $\swarrow v$ *are* left, $\nearrow v$ *and* $\searrow v$ *are* right. *Two polarized valencies with the same valency name and orientation, but with the opposite signs are* dual.

An expression of one of the forms $\#(\swarrow v)$, $\#(\searrow v)$, $v \in \mathbf{V}$, *is called* anchor type *or just* anchor. *An expression of the form* d^* *where* $d \in \mathbf{C}$, *is called* iterated dependency type.

Local dependency names, iterated dependency types and anchor types are primitive types.

An expression of the form $t = [l_m\backslash \ldots \backslash l_1\backslash H/\ldots/r_1\ldots/r_n]$ *in which* $m, n \geq 0$, $l_1, \ldots, l_m, r_1, \ldots, r_n$ *are primitive types and* H *is either a local dependency name or an anchor type, or empty, is called* basic dependency type. $l_1, \ldots, l_m$ *and* $r_1, \ldots, r_n$ *are respectively left and right* argument subtypes *of* t. *If nonempty,* H *is called* head subtype *of* t *(or* head type *for short).*

A (possibly empty) string P *of polarized valencies is called* potential.

A dependency type *is an expression* B^P *in which* B *is a basic dependency type and* P *is a potential.* $\mathbf{CAT}(\mathbf{C}, \mathbf{V})$ *and* $\mathbf{B}(\mathbf{C})$ *will denote respectively the set of all dependency types over* **C** *and* **V** *and the set of all basic dependency types over* **C**.

CDG are defined using the following calculus of dependency types [6]

$\mathbf{L^l}$. $C^{P_1}[C\backslash\beta]^{P_2} \vdash [\beta]^{P_1P_2}$
$\mathbf{I^l}$. $C^{P_1}[C^*\backslash\beta]^{P_2} \vdash [C^*\backslash\beta]^{P_1P_2}$
$\mathbf{\Omega^l}$. $[C^*\backslash\beta]^P \vdash [\beta]^P$

[5] All subtypes being dependency names or iterated, this means that the CDG types are first order.

[6] We show left-oriented rules. The right-oriented are symmetrical.

D^1. $\alpha^{P_1(\swarrow C)P(\nwarrow C)P_2} \vdash \alpha^{P_1PP_2}$, if the potential $(\swarrow C)P(\nwarrow C)$ satisfies the following pairing rule **FA** (`first available`):

$$\textbf{FA}: \quad P \text{ has no occurrences of } \swarrow C, \nwarrow C.$$

L^1 is the classical elimination rule. Eliminating the argument subtype $C \neq \#(\alpha)$ it constructs the (`projective`) dependency C and concatenates the potentials. $C = \#(\alpha)$ creates the `anchor dependency`. **I**^{1} derives $k > 0$ instances of C. $\boldsymbol{\Omega}^1$ serves for the case $k = 0$. **D**^{1} creates `discontinuous dependencies`. It pairs and eliminates dual valencies with name C satisfying the rule **FA** to create the discontinuous dependency C.

Definition 3. *A* categorial dependency grammar (CDG)[7] *is a system* $G = (W, \mathbf{C}, S, \lambda)$, *where* W *is a finite set of words,* $\mathbf{C}$ *is a finite set of local dependency names containing the selected name* S *(an axiom), and* λ, *called* lexicon, *is a finite substitution on* W *such that* $\lambda(a) \subset \mathbf{CAT}(\mathbf{C}, \mathbf{V})$ *for each word* $a \in W$.
For a DS D *and a string* $x \in W^*$, *let* $G(D, x)$ *denote the relation:* D *is constructed in a proof* $\Gamma \vdash S$ *for some* $\Gamma \in \lambda(x)$. *Then the* language *generated by* G *is the set* $L(G)=_{df} \{w \mid \exists D\ G(D, w)\}$ *and the* DS-language *generated by* G *is the set* $\Delta(G)=_{df} \{D \mid \exists w\ G(D, w)\}$. $\mathcal{D}(CDG)$ *and* $\mathcal{L}(CDG)$ *will denote the families of DS-languages and languages generated by these grammars.*

Below we cite from [4,3] several examples and facts showing that CDG are very expressive. Evidently, they generate all context-free languages. They can also generate non-CF languages.

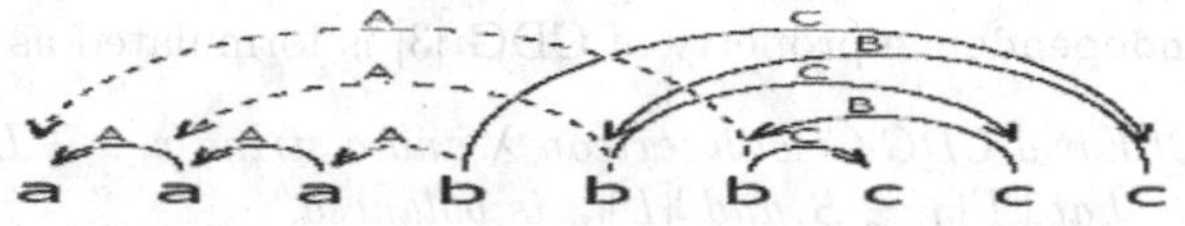

Fig. 5. DS for $a^3b^3c^3$

Example 1. The CDG $G_{abc}: a \mapsto A^{\swarrow A}, [A\backslash A]^{\swarrow A}, b \mapsto [B/C]^{\nwarrow A}, [A\backslash S/C]^{\nwarrow A}, c \mapsto C, [B\backslash C]$ generates the language $\{a^n b^n c^n \mid n > 0\}$. E.g., $G_{abc}(D^{(3)}, a^3b^3c^3)$ holds for the DS in Fig 5 and the string $a^3b^3c^3$ due to the proof in Fig. 6.

Seemingly, $\mathcal{L}(CDG)$ is different from mildly context-sensitive languages [9,18] generated by multi-component TAG, linear CF rewrite systems and some other grammars. $\mathcal{L}(CDG)$ contains non-TAG languages, e.g. $L^{(m)} = \{a_1^n a_2^n ... a_m^n \mid n \geq 1\}$ for all $m > 0$. In particular, it contains the language $MIX = \{w \in \{a, b, c\}^+ \mid |w|_a = |w|_b = |w|_c\}$ [2], for which E. Bach has conjecture that it is not mildly CS. On the other hand, in [3] it is conjectured that this family does not contain the copy language $L_{copy} = \{xx \mid x \in \{a, b\}^*\}$, which is TAG. This comparison

[7] They are called `generalized CDG` in [3] in order to distinguish them from CDG generating DT, which we do not consider here.

$$\frac{[B/C]^{\nwarrow A}\ C}{B^{\nwarrow A}}\ (\mathbf{L}^r)\qquad \frac{B^{\nwarrow A}\ \ [B\backslash C]}{C^{\nwarrow A}}\ (\mathbf{L}^l)\qquad \frac{[B/C]^{\nwarrow A}\ \ C^{\nwarrow A}}{B^{\nwarrow A\nwarrow A}}\ (\mathbf{L}^r)\qquad \frac{B^{\nwarrow A\nwarrow A}\ \ [B\backslash C]}{C^{\nwarrow A\nwarrow A}}\ (\mathbf{L}^l)$$

$$\frac{[A]^{\swarrow A}\ \ [A\backslash A]^{\swarrow A}}{[A]^{\swarrow A\swarrow A}}\ (\mathbf{L}^l)\qquad \frac{[A]^{\swarrow A\swarrow A}\ \ [A\backslash A]^{\swarrow A}}{[A]^{\swarrow A\swarrow A\swarrow A}}\ (\mathbf{L}^l)\qquad \frac{[A\backslash S/C]^{\nwarrow A}\ \ C^{\nwarrow A\nwarrow A}}{[A\backslash S]^{\nwarrow A\nwarrow A\nwarrow A}}\ (\mathbf{L}^l)$$

$$\frac{[A]^{\swarrow A\swarrow A\swarrow A}\ \ [A\backslash S]^{\nwarrow A\nwarrow A\nwarrow A}}{[S]^{\swarrow A\swarrow A\swarrow A\nwarrow A\nwarrow A\nwarrow A}}\ (\mathbf{L}^l)\qquad \frac{[S]^{\swarrow A\swarrow A\swarrow A\nwarrow A\nwarrow A\nwarrow A}}{S}\ (\mathbf{D}^l\times 3)$$

Fig. 6. DS correctness proof

shows a specific nature of the valencies' pairing rule **FA**. This rule implies an important property of independence of basic types and of polarized valencies expressed in terms of `projections` of types and `‘‘well-bracketing’’` criteria for potentials.

For every type α and every sequence of types γ the `local projection` $\|\gamma\|_l$ and the `valency projection` $\|\gamma\|_v$ are defined as follows:

1. $\|\varepsilon\|_l = \|\varepsilon\|_v = \varepsilon$; $\|\alpha\gamma\|_l = \|\alpha\|_l\|\gamma\|_l$ and $\|\alpha\gamma\|_v = \|\alpha\|_v\|\gamma\|_v$.
2. $\|C^P\|_l = C$ et $\|C^P\|_v = P$ for every type C^P.

To speak about "well-bracketing" of potentials, it is useful to interpret $\swarrow d$ and $\nearrow d$ as `left brackets` and $\nwarrow d$ and $\searrow d$ as `right brackets`. Then a potential is `balanced` if it is well bracketed in the usual sense.

Let **c** be the projective core of the dependency calculus, consisting of the rules **L**, **I** and $\mathbf{\Omega}$ and $\vdash_{\mathbf{c}}$ denote the provability relation in this sub-calculus. Then the `projections independence` property of CDG [3] is formulated as follows.

Theorem 1. *[3] For a CDG G with lexicon λ and a string x, $x \in L(G)$ iff there is $\Gamma \in \lambda(x)$ such that $\|\Gamma\|_l \vdash_{\mathbf{c}}^* S$ and $\|\Gamma\|_v$ is balanced.*

On this property resides a polynomial time parsing algorithm for CDG [3].

4 Problem of Iteration and a Multimodal Solution

As we saw, the types of CDG admit iterated subtypes. So it may seem that the family of CDG languages is trivially closed under Kleene iteration. To see why the straightforward construction does not work, let us consider the CDG:

$a \mapsto A^{\swarrow A}$, $[A\backslash A]^{\swarrow A}$,
$b \mapsto [B/C]^{\nwarrow A}$, $[A\backslash S1/C]^{\nwarrow A}$, $[A\backslash S/S1/C]^{\nwarrow A}$,
$c \mapsto C$, $[B\backslash C]$

It may seem that $L(G) = L(G_{abc})L(G_{abc})$, but it is not so because it contains, for example, the string *aaabbccabbcc* which has the DS in Fig. 7. We see that this effect is due to the fact that dual valencies may sometimes be paired across the limits of concatenated / iterated strings and not within the limits, as needed. Of

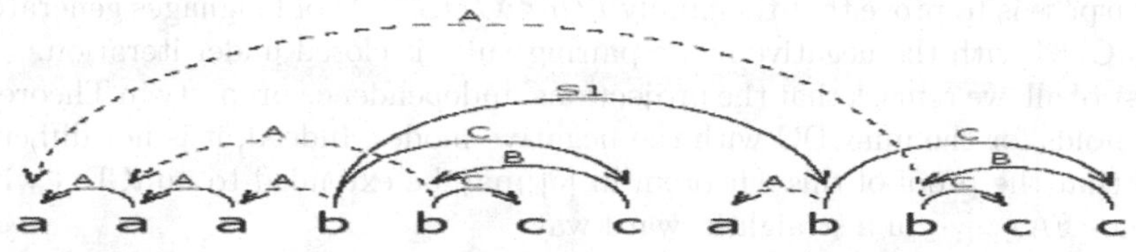

Fig. 7. DS of *aaabbccabbcc*

course, one can easily avoid this effect by renaming the valencies $\swarrow A$ and $\nwarrow A$. Indeed, this may work for concatenation and for any finite power $L(G_{abc})^k$, but this won't work for $L(G_{abc})^*$.

Now that the source of the problem is found, we will try to use possibly economical means to express the constraint of a "limit impenetrable for discontinuous dependencies". For that, we will follow the proposal of [5] where are introduced the so called multimodal CDG in which it is possible that every polarized valency has its own pairing rule (`pairing mode`).

Definition 4. *$G=(W,\mathbf{C},S,\lambda,\mu)$ is a* multimodal CDG *(mmCDG) if $(W,\mathbf{C},S,\lambda)$ is a CDG, in which are admitted empty head types ε, and μ is a function assigning to each polarized valency α a pairing principle $\mathbf{M}_\alpha$. There are rules $\mathbf{D}^l_{\mathbf{M}_\alpha}$ and $\mathbf{D}^r_{\mathbf{M}_\alpha}$ in the multimodal dependency calculus $\vdash_\mu$ for every valency α used in μ. The language (DS-language) generated by G using a set of modes M is denoted $L^M(G)$ $(\Delta^M(G))$. $mmCDG^M$ is the family of all such mmCDG.*

For instance, in [5] the calculus rule $\mathbf{D}^l$ is replaced by a new rule $\mathbf{D}_{\mathbf{FC}^l}$ in which in the place of the pairing rule **FA** is used the following pairing rule $\mathbf{FC}^l$ (`first cross`):

$$\mathbf{D}_{\mathbf{FC}^l}.\quad \alpha^{P_1(\swarrow C)P(\nwarrow C)P_2} \vdash \alpha^{P_1PP_2},$$

if $P_1(\swarrow C)P(\nwarrow C)$ satisfies the pairing rule

$\mathbf{FC}^l$: $\quad P_1$ has no occurrences of $\swarrow C$ and P has no occurrences of $\nwarrow C$.

This rule was used to show that the so called `unlimited cross-serial dependencies` in Dutch are naturally expressed in mmCDG.

In this section we will show how the iteration problem can be resolved using multimodal CDG. For this, we will use `negative mode` pairing rules $\mathbf{FA}_{C:\pi(C)}$ which pair dual valencies C under the negative condition that the resulting discontinous dependency C `does not cross` the discontinuous dependencies belonging to a fixed list $\pi(C)$. More precisely, in the discontinuous dependency rule

$$\mathbf{D}_{\mathbf{FA}^l_{C:\pi(C)}}\quad \alpha^{P_1(\swarrow C)P(\nwarrow C)P_2} \vdash \alpha^{P_1PP_2},$$

$(\swarrow C)P(\nwarrow C)$ satisfies the pairing rule $\mathbf{FA}_{C:\pi(C)}$:

P has no occurrences of $\swarrow C, \nwarrow C$ and also of $\swarrow A, \nwarrow A, \nearrow A, \searrow A$ for all $A \in \pi(C)$.

The mmCDG with this pairing rule will be denoted $(W,\mathbf{C},S,\lambda,\mu,\pi)$.

Our purpose is to prove that the family $\mathcal{L}(mmCDG^{-FA})$ of languages generated by mmCDG with the negative mode pairing rules is closed under iteration.

First of all, we remark that the projections' independence property in Theorem 1 also holds for the mmCDG with the negative modes. Indeed, it is not difficult to see that the proof of this Theorem in [5] may be extended to mmCDG with the rules $\mathbf{FA}_{C:\pi(C)}$ in a straightforward way.

Then, as it shows the following Lemma, one can consider without loss of generality only *mmCDG* in Greibach normal form.

Lemma 1. *[2,10] For every CDG G there is an equivalent CDG*[8] *G′ such that every type has one of the forms:* $[A]^P$, $[A/B]^P$ *or* $[A/B/C]^P$ *where* B, C *are primitive types different from* S.

Theorem 2. $\mathcal{L}(mmCDG^{-FA})$ *is closed under iteration.*

Proof. Let us suppose that $G = (W, \mathbf{C}, S, \lambda, \mu, \pi)$ is an mmCDG in the normal form. We will define from G a sequence of mmCDG $G_i = (W, \mathbf{C}_i, S, \lambda_i, \mu, \pi)$ by the following induction on i.
I. $i = 1$. $\lambda_1 = \lambda \cup \{w \mapsto [S/A'/\alpha]^P \mid (w \mapsto [S/A/\alpha]^P) \in \lambda\}$ for a new local dependency name $A' \in \mathbf{C}_1$.
II. $i > 1$. Let $A'_1, \ldots, A'_k \in \mathbf{C}_i$ be all new local dependency names in the grammar G_i. Let us first consider the auxiliary extended lexicon

$\lambda''_{i+1} = \lambda_i \ \cup \ \{w \mapsto [A'_j/B]^P \mid (w \mapsto [A_j/B]^P) \in \lambda_i, 1 \le j \le k\} \cup$
$\{w \mapsto [A'_j/B/C]^P \mid (w \mapsto [A_j/B/C]^P) \in \ \lambda_i, 1 \le j \le k\}$. Then let us set

$$\lambda_{i+1} = \lambda''_{i+1} \cup \{w \mapsto [A'_j/A'/\alpha]^P \mid (w \mapsto [A'_j/A/\alpha]^P) \in \lambda''_{i+1}\}.$$

New types A'_j, A' are added to $\mathbf{C}_{i+1}$. This construction converges to a mmCDG $G_m = (W, \mathbf{C}_m, S, \lambda_m, \mu, \pi)$, $m \le |\mathbf{C}|$. Let $b \notin \mathbf{C}_m$ be a new local dependency name. Let us consider an auxiliary mmCDG $G'_m = (W, \mathbf{C}_m \cup \{b\}, S, \lambda'_m, \mu', \pi')$ constructed from G_m as follows. In λ_m, every type $[A']^P$ is replaced by $[A']^{P\searrow b}$, every type $[S/A'/\alpha]^P$ is replaced by $[S/A'/\alpha]^{\nearrow bP}$ and every type $[S]^P$ is replaced by $[S]^{\nearrow bP \searrow b}$. Let also $\pi'(A) = \pi(A) \cup \{b\}$ for all $A \in \mathbf{C}_m \cup \{b\}$. Now, the mmCDG G' defining the iteration of $L(G)$ can be defined as follows. $G' = (W, \mathbf{C}_m \cup \{b, S_0\}, S_0, \lambda', \mu', \pi')$, where $\lambda' = \lambda'_m \cup \{w \mapsto [S_0/S*/\alpha]^P \mid (w \mapsto [S/\alpha]^P) \in \lambda'_m\}$.

Let us prove that $L(G') = L(G)^*$.

Lemma 2. *If in* G'_m $\gamma_1 \ldots \gamma_n \vdash^* [A]^P$, *where* $A \in \mathbf{C}$, *then there are no types* A' *in* $\gamma_1 \ldots \gamma_n$.

Proof. By straightforward induction on n. E.g., if $\gamma_1 = [A/B]^{P_1}$, then $\gamma_2 \ldots \gamma_n \vdash^* [B]^{P_1}$, $B \in \mathbf{C}$ and therefore, there are no types A' in $\gamma_2 \ldots \gamma_n$.

Lemma 3. *Let* $\gamma_1 \ldots \gamma_n \vdash^* [A']^P$ *in* G'_m. *Then* $\gamma_n = [X']^{P'\searrow b}$ *and there are no occurrences of* b *in the potentials of types* $\gamma_1 \ldots \gamma_{n-1}$.

[8] Here **equivalent** corresponds to **weakly equivalent**, i.e. generating the same language (possibly not the same DS-language).

Proof. By induction on n.
When $n=1$, $\gamma_1=[X']^{P'\searrow b}$ and $P=P'\searrow b$ by construction.
Let $\gamma_1\dots\gamma_{n+1}\vdash^*[A']^P$ in G'_m.
1) If $\gamma_1=[A'/B']^{P_1}$, then $\gamma_2\dots\gamma_{n+1}\vdash^*[B']^{P_2}$. By hypothesis, $\gamma_{n+1}=[X']^{P'\searrow b}$ and potentials of $\gamma_2\dots\gamma_{n+1}$ do not contain b. P_1 does contain b by construction.
2) If $\gamma_1=[A'/B'/C]^{P_1}$, then $\gamma_2\dots\gamma_r\vdash^*[C]^{P_2}$ and $\gamma_{r+1}\dots\gamma_{n+1}\vdash^*[B']^{P_3}$ for some r. By hypothesis, $\gamma_{n+1}=[X']^{P'\searrow b}$ and potentials of $\gamma_{r+1}\dots\gamma_n$ do not contain b. By Lemma 2, $\gamma_2\dots\gamma_r$ do not contain subtypes Y' so their potentials do not contain b by construction. P_1 also does not contain b by construction.
3) Other cases are impossible when the derived type has the form $[A']^P$.

Lemma 4. *Let $\Gamma=\gamma_1\dots\gamma_n\in\lambda'_m(w)$ and $\Gamma\vdash^*_{G'_m}[A]^P$ or $\Gamma\vdash^*_{G'_m}[A']^{P\searrow b}$ for some $A\neq S$. Then there is $\Gamma'=\gamma'_1\dots\gamma'_n\in\lambda(w)$ such that $\Gamma'\vdash^*_G[A]^P$.*

Proof. If $\Gamma\vdash^*_{G'_m}[A]^P$, then by Lemma 2 the types $\gamma_1,\dots,\gamma_n$ may also be assigned in G. So $\Gamma'=\Gamma$ in this case.
When $\Gamma\vdash^*_{G'_m}[A']^{P\searrow b}$, the proof proceeds by induction on n.
For $n=1$, $\gamma_1=[A']^{P\searrow b}$ and $\gamma'_1=[A]^P$.
Let $\Gamma=\gamma_1\dots\gamma_{n+1}\vdash^*_{G'_m}[A']^{P\searrow b}$ for some $A\neq S$.
1) $\gamma_1=[A'/B']^{P_1}$. Then $\gamma_2\dots\gamma_{n+1}\vdash^*_{G'_m}[B']^{P_2\searrow b}$. By hypothesis, there is a proof $\gamma'_2\dots\gamma'_{n+1}\vdash^*_G[B]^{P_2}$. Let us set $\gamma'_1=[A/B]^{P_1}$. Then $\gamma'_1\dots\gamma'_{n+1}\vdash^*_G[A/B]^{P_1}[B]^{P_2}\vdash_G[A]^{P_1P_2}=[A]^P$.
2) Case $\gamma_1=[A'/B'/C]^{P_1}$ is similar.

Lemma 5. *Let $\Gamma=\gamma_1\dots\gamma_n\in\lambda(w)$, $\Gamma\vdash^*_G[A]^P$ and $A'\in\mathbf{C}'_m$. Then there is $\Gamma'=\gamma'_1\dots\gamma'_n\in\lambda'_m(w)$ such that $\Gamma'\vdash^*_{G'_m}[A']^{P\searrow b}$.*

Proof. Induction on n.
For $n=1$, $\gamma_1=[A]^P$. So we can set $\gamma'_1=[A']^{P\searrow b}$.
Let $\Gamma=\gamma_1\dots\gamma_{n+1}$.
1) $\gamma_1=[A/B]^{P_1}$. Then $\gamma_2\dots\gamma_{n+1}\vdash^*_G[B]^{P_2}$, where $P_1P_2=P$. As $A'\in\mathbf{C}'_m$, B' is also added by construction: $B'\in\mathbf{C}'_m$. By hypothesis, there is $\gamma'_2\dots\gamma'_{n+1}$ such that $\gamma_2\dots\gamma_{n+1}\vdash^*_{G'_m}[B']^{P_2\searrow b}$. Let us set $\gamma'_1=[A'/B']^{P_1}$. Then $\gamma'_1\dots\gamma'_{n+1}\vdash^*_{G'_m}[A'/B']^{P_1}[B']^{P_2\searrow b}\vdash_{G'_m}[A']^{P_1P_2\searrow b}=[A]^{P\searrow b}$.
2) $\gamma_1=[A/B/C]^{P_1}$. In this case, $\gamma_2\dots\gamma_r\vdash^*_G[C]^{P_2}$, $\gamma_{r+1}\dots\gamma_{n+1}\vdash^*_G[B]^{P_3}$ and $P_1P_2P_3=P$. All types of G not containing S remain in G'_m. Besides this, $B'\in\mathbf{C}'_m$. Therefore, there is $\gamma'_2\dots\gamma'_r$ such that $\gamma'_2\dots\gamma'_r\vdash^*_{G'_m}[C]^{P_2}$ ($\gamma'_i=\gamma_i$) and $\gamma'_{r+1}\dots\gamma'_{n+1}$ such that $\gamma'_{r+1}\dots\gamma'_{n+1}\vdash^*_{G'_m}[B']^{P_3\searrow b}$ (induction hypothesis). So $\gamma'_1\dots\gamma'_{n+1}\vdash^*_{G'_m}[A'/B'/C]^{P_1}[C]^{P_2}[B']^{P_3\searrow b}\vdash^*_{G'_m}[A']^{P_1P_2P_3\searrow b}=[A']^{P\searrow b}$.

Lemma 6. $L(G'_m)=L(G)$.

Proof. [$\Rightarrow$]. Let $w=w_1\dots w_n\in L(G'_m)$. Then there is $\gamma_1\dots\gamma_n\in\lambda'_m(w)$ such that $\gamma_1\dots\gamma_n\vdash^*_{G'_m}[S]$.
1) If $\gamma_1=[S]^{\nearrow bP\searrow b}$, then we should just replace it by $[S]^P$.

2) Let $\gamma_1 = [S/A']^{\nearrow bP_1}$. Then $\gamma_2 \dots \gamma_n \vdash^*_{G'_m} [A']^{P_2 \searrow b}$ and the potential P_1P_2 is balanced. By Lemma 4, there is $\gamma'_2 \dots \gamma'_n \in \lambda(w_2 \dots w_n)$ such that $\gamma'_2 \dots \gamma'_n \vdash^*_G [A]^{P_2}$. Let us set $\gamma'_1 = [S/A]^{P_1} \in \lambda(w_1)$. Then $\gamma'_1 \dots \gamma'_n \vdash^*_G [S/A]^{P_1}[A]^{P_2} \vdash_G [S]^{P_1P_2} \vdash^*_G [S]$.
3) The case $\gamma_1 = [S/A'/B]^{\nearrow bP_1}$ is similar. So $w \in L(G)$.
[$\Leftarrow$]. Let $w = w_1 \dots w_n \in L(G)$. Then there is $\gamma_1 \dots \gamma_n \in \lambda(w)$ such that $\gamma_1 \dots \gamma_n \vdash^*_G [S]$.
1) If $\gamma_1 = [S]^P$, we can just replace it by $[S]^{\nearrow bP \searrow b}$.
Let us prove the case 3) $\gamma_1 = [S/A/B]^{P_1}$ (case 2) is similar). In this case, $\gamma_2 \dots \gamma_r \vdash^*_G [B]^{P_2}$ and $\gamma_{r+1} \dots \gamma_n \vdash^*_G [A]^{P_3}$ for some r and the potential $P_1P_2P_3$ is balanced. In G'_m w_1 has the type $\gamma'_1 = [S/A'/B]^{\nearrow bP_1} \in \lambda'_m(w_1)$. As all types without S are kept in G'_m, we have $\gamma_2 \dots \gamma_r \in \lambda'_m(w_2 \dots w_r)$ and $\gamma_2 \dots \gamma_r \vdash^*_{G'_m} [B]^{P_2}$. So we set $\gamma'_2 = \gamma_2, \dots, \gamma'_r = \gamma_r$. Besides this, by Lemma 5, there is $\gamma'_{r+1} \dots \gamma'_n \in \lambda'_m(w_{r+1} \dots w_n)$ such that $\gamma'_{r+1} \dots \gamma'_n \vdash^*_{G'_m} [A']^{P_3 \searrow b}$. Then $\gamma'_1 \dots \gamma'_n \vdash^*_{G'_m} [S/A'/B]^{\nearrow bP_1}[B]^{P_2}[A']^{P_3 \searrow b} \vdash^*_{G'_m} [S]^{\nearrow bP_1P_2P_3 \searrow b} \vdash^*_{G'_m} [S]$. Therefore, $w \in L(G'_m)$.

By this Lemma, it is now sufficient to prove that $L(G'_m)^* = L(G')$.
[$\Rightarrow$] $L(G'_m)^* \subseteq L(G')$. This inclusion is rather evident. If $x = x_1 \dots x_n \in L(G'_m)^*$ and $x_1, \dots, x_n \in L(G'_m)$, then there are type assignments $\Gamma_i \in \lambda'_m(x_i)$ such that $\Gamma_i \vdash^*_{G'_m} [S]$, $1 \le i \le n$. The first type in Γ_1 has the form $[S/\alpha]^{\nearrow bP_1}$. We will replace this type by $[S_0/S*/\alpha]^{\nearrow bP_1}$ obtaining a new sequence $\Gamma'_1 \in \lambda'(x_1)$ such that $\Gamma'_1 \vdash_{G'} [S_0/S*]$. Now we can assign to x the sequence $\Gamma'_1\Gamma_2 \dots \Gamma_n \in \lambda'(x)$ such that $\Gamma'_1\Gamma_2 \dots \Gamma_n \vdash^*_{G'} [S_0/S*][S] \dots [S] \vdash^*_{G'} [S_0]$.

[$\Leftarrow$] $L(G') \subseteq L(G'_m)^*$. $x \in L(G')$ means that there is $\Gamma \in \lambda'(x)$ such that $\Gamma \vdash^*_{G'} [S_0]$. We can decompose this proof into the subproofs which eliminate consecutive iterated S: $\Gamma = \Gamma_1 \dots \Gamma_n$, where $\Gamma_i \in \lambda'(x_i)$ $1 \le i \le n$, and $\Gamma_1 \vdash^*_{G'} [S_0/S*]^{P_1}$, $\Gamma_j \vdash^*_{G'} [S]^{P_j}$, $2 \le j \le n$. For all $i, 1 \le i \le n$, there also exist proofs $\Gamma_i \vdash^*_{G'_m} [S]^{P_i}$ in which the potentials P_i are balanced. Indeed, if $|x_i| > 1$, then $\Gamma_i = [S/\alpha_1]^{\nearrow bP_i^1}[\alpha_2]^{P_i^2} \dots [\alpha_{k-i}]^{P_i^{k-1}}[A']^{P_i^k \searrow b}$. The last type has the form $[A']^{P_i^k \searrow b}$ due to Lemma 3. By Lemmas 2,3 the potential $P_i^1 \dots P_i^k$ does not contain occurrences of b. So by definition of the rule $\mathbf{FA}_{C:\pi(C)}$ this potential is balanced. Therefore, P_i is balanced too. If otherwise $|x_i| = 1$, then $P_i = \nearrow bP'_i \searrow b$ is also balanced. As a result, $\Gamma_i \vdash^*_{G'_m} [S]$, i.e. $x_i \in L(G'_m)$ for all $i, 1 \le i \le n$, and so $x \in L(G'_m)^*$.

Corollary 1. *The family of $mmCDG^{-FA}$-languages is an AFL.*

Proof. By Theorem 2 and Theorem 4 in [3].

5 Expressiveness of mmCDG with Negative Modes

The negative constraints used for the iteration closure turn out to be rather expressive. E.g., using such constraints one can generate an exponential length strings language.

Let $L_{exp} = \{1010^21 \ldots 10^{2^n}1 \mid n > 1\}$ and G_{exp} be the following grammar:

$$1 \mapsto [S/A_1]^{\nearrow X}, [D_1/C]^{\nearrow Y}, [B/A]^{\searrow X \nearrow X}, [D/C]^{\searrow Y \nearrow Y}, [B/A_2]^{\searrow X \nearrow X}, [D/C_2]^{\searrow Y \nearrow Y}, [A_2]^{\searrow X \searrow Y}, [C_2]^{\searrow X \searrow Y}$$

$$0 \mapsto [A_1/D_1]^{\nearrow B \nearrow B}, [A/A]^{\searrow A \nearrow B \nearrow B}, [A/D]^{\searrow A \nearrow B \nearrow B}, [C/C]^{\searrow B \nearrow A \nearrow A}, [C/B]^{\searrow B \nearrow A \nearrow A}, [A_2/A_2]^{\searrow A}, [C_2/C_2]^{\searrow B}$$

In this mmCDG is used the **FA** pairing rule with two negative modalities: $\pi(X) = \{A\}, \pi(Y) = \{B\}$. The intuitive idea is that every 0 takes one negative valency ($\searrow A$ or $\searrow B$) and puts out two positive valencies ($\nearrow B$ or $\nearrow A$). A and B are alternated in order that zeros in the same block couldn't be linked by a dependency. In order that a zero were linked with another zero in the next block, the consecutive symbols 1 are linked by discontinuous dependencies X (for even blocks) or Y (for odd blocks). Due to the negative modalities π, X does not let pass dependency A and Y does not let pass B. As a result, $L(G_{exp}) = L_{exp}$. A formal proof of this equality is a consequence of the following fact.

Lemma 7. *Let $w \in L(G_{exp})$, $w = w'w''$ and $w' = 10^{i_1}1 \ldots 10^{i_k}1$, where $i_j \geq 0$. Let $\Gamma \in \lambda_{G_{exp}}(w)$, $\Gamma \vdash^*_{G_{exp}} [S]$ and $\Gamma = \Gamma'\Gamma''$, where $\Gamma' = \lambda_{G_{exp}}(w')$ and $\Gamma'' = \lambda_{G_{exp}}(w'')$. Then:*
1) $i_j = 2^{j-1}$ for $1 \leq j \leq k$. 2) If $w'' \neq \varepsilon$ and k is odd, then $\|\Gamma'\|_v \vdash^* \nearrow X(\nearrow B)^{2^k} \nearrow Y$[9] and to the last 1 in w' is assigned in Γ' one of types: $[D_1/C]^{\nearrow Y}$, $[D/C]^{\searrow Y \nearrow Y}$ or $[D/C_2]^{\searrow Y \nearrow Y}$. 3) If $w'' \neq \varepsilon$ and k is even, then $\|\Gamma'\|_v \vdash^* \nearrow Y(\nearrow A)^{2^k} \nearrow X$ and to the last 1 in w' is assigned in Γ' one of types: $[B/A]^{\searrow X \nearrow X}$ or $[B/A_2]^{\searrow X \nearrow X}$.

Proof. By induction on k.

Corollary 2. *Languages in $\mathcal{L}(mmCDG^{-FA})$ may be not semilinear.*

It should be remarked that the problem of semilinearity is still open for CDG.

The example of L_{exp} suggests that languages in $\mathcal{L}(mmCDG^{-FA})$ may be rather complex. Indeed, we prove that the membership problem in this family is NP-complete.

Theorem 3. *Membership problem for $mmCDG^{-FA}$ is NP-complete.*

Proof. [NP – $hardness$]. We will reduce the problem of satisfiability of $3-CNF$ to the membership problem for $mmCDG^{-FA}$. For this, we will define an $mmCDG^{-FA}$-grammar $G^{(3)}$ and, for every 3–CNF Φ, we will construct a string $w(\Phi)$ such that $w(\Phi) \in L(G^{(3)})$ iff $\Phi \in SAT$. This is a definition of $G^{(3)}$:

Dictionary: $W = \{*, x, \bar{x}, y, b, f, F\}$.
x corresponds to occurrences of the propositional letters in the clauses of 3–CNF. $\bar{x}$ corresponds to occurrences of the negated propositional letters. y corresponds to the propositional letters which have no occurrences in a clause. The following

[9] I.e. the valency projection is reducible to this string of valencies.

example explains how the clauses are coded using these symbols. Supposing that there are only seven letters $x_1, \dots, x_7$, the clause $x_2 \vee \neg x_4 \vee x_7$ is represented by the string $yxy\bar{x}yyx$. For a clause C, $g(C)$ will denote the string representing C.

Elementary types:
$\mathbf{C} = \{S, 0, 1, 0', 1', A, B, T\}$.

Modes:
$\pi(0) = \{1\}$, $\pi(1) = \{0\}$, $\pi(0') = \{1'\}$, $\pi(1') = \{0'\}$. Intuitively, these negative modes mean that the dependencies 0 and 1 (respectively, $0'$ and $1'$) cannot cross.

Lexicon λ:

$$F \mapsto [S],$$
$$b \mapsto \{[\varepsilon]^{\nearrow 0},\ [\varepsilon]^{\nearrow 1}\},$$
$$f \mapsto \left\{\begin{array}{l} [\varepsilon]^{\searrow 0}, \\ [\varepsilon]^{\searrow 0'}, \\ [\varepsilon]^{\searrow 1}, \\ [\varepsilon]^{\searrow 1'} \end{array}\right\} \qquad y \mapsto \left\{\begin{array}{ll} [A\backslash A]^{\searrow 1 \nearrow 1'}, & [B\backslash B]^{\searrow 1' \nearrow 1}, \\ [A\backslash A]^{\searrow 0 \nearrow 0'}, & [B\backslash B]^{\searrow 0' \nearrow 0}, \\ [A/A]^{\searrow 1 \nearrow 1'}, & [B/B]^{\searrow 1' \nearrow 1}, \\ [A/A]^{\searrow 0 \nearrow 0'}, & [B/B]^{\searrow 0' \nearrow 0}, \\ [A]^{\searrow 1 \nearrow 1'}, & [B]^{\searrow 1' \nearrow 1}, \\ [A]^{\searrow 0 \nearrow 0'}, & [B]^{\searrow 0' \nearrow 0} \end{array}\right\}$$
$$* \mapsto [T\backslash\varepsilon]$$
$$x \mapsto \left\{\begin{array}{ll} [A\backslash T/A]^{\searrow 1 \nearrow 1'}, & [B\backslash T/B]^{\searrow 1' \nearrow 1}, \\ [A\backslash A]^{\searrow 1 \nearrow 1'}, & [B\backslash B]^{\searrow 1' \nearrow 1}, \\ [A\backslash A]^{\searrow 0 \nearrow 0'}, & [B\backslash B]^{\searrow 0' \nearrow 0}, \\ [A/A]^{\searrow 1 \nearrow 1'}, & [B/B]^{\searrow 1' \nearrow 1}, \\ [A/A]^{\searrow 0 \nearrow 0'}, & [B/B]^{\searrow 0' \nearrow 0}, \\ [A\backslash T]^{\searrow 1 \nearrow 1'}, & [B\backslash T]^{\searrow 1' \nearrow 1}, \\ [A]^{\searrow 1 \nearrow 1'}, & [B]^{\searrow 1' \nearrow 1}, \\ [A]^{\searrow 0 \nearrow 0'}, & [B]^{\searrow 0' \nearrow 0}, \\ [T/A]^{\searrow 1 \nearrow 1'}, & [T/B]^{\searrow 1' \nearrow 1} \end{array}\right\} \quad \bar{x} \mapsto \left\{\begin{array}{ll} [A\backslash T/A]^{\searrow 0 \nearrow 0'}, & [B\backslash T/B]^{\searrow 0' \nearrow 0}, \\ [A\backslash A]^{\searrow 1 \nearrow 1'}, & [B\backslash B]^{\searrow 1' \nearrow 1}, \\ [A\backslash A]^{\searrow 0 \nearrow 0'}, & [B\backslash B]^{\searrow 0' \nearrow 0}, \\ [A/A]^{\searrow 1 \nearrow 1'}, & [B/B]^{\searrow 1' \nearrow 1}, \\ [A/A]^{\searrow 0 \nearrow 0'}, & [B/B]^{\searrow 0' \nearrow 0}, \\ [A\backslash T]^{\searrow 0 \nearrow 0'}, & [B\backslash T]^{\searrow 0' \nearrow 0}, \\ [A]^{\searrow 1 \nearrow 1'}, & [B]^{\searrow 1' \nearrow 1}, \\ [A]^{\searrow 0 \nearrow 0'}, & [B]^{\searrow 0' \nearrow 0}, \\ [T/A]^{\searrow 0 \nearrow 0'}, & [T/B]^{\searrow 0' \nearrow 0} \end{array}\right\}$$

Let $\Phi = C_1 \wedge \dots \wedge C_m$ be a 3–CNF with propositional letters in $X = \{x_1, \dots, x_n\}$.

String $w(\Phi)$ encoding Φ:
$w(\Phi) = b^n F g(C_1)^R * g(C_2) * \cdots * g(C_m)^{(R)} * f^n$, where every even member i is $g(C_i)$ and every odd member i is the mirror image of $g(C_i)$.

Lemma 8. *Φ is satisfiable iff $w(\Phi) \in L(G^{(3)})$.*

Proof. 1. If Φ is satisfied by values $x = v_1, \dots, x_n = v_n$, we assign to every occurrence i of b in $w(\Phi)$ the type $[\varepsilon]^{\nearrow v_i}$. The choice of types for the symbols x, $\bar{x}$ and y depends on the number j of the block $g(C_j)$ in $w(\Phi)$ and on the position of their occurrence in the block. If j is odd, then the assigned type has argument subtypes A, otherwise it has argument subtypes B. We show the exact choice of types trough an example. Let us suppose that j is odd (the other case is similar). The clause C_j is made true by some literal x_i or $\neg x_i$. In the first case, for the occurrence i of x in $g(C_j)$ is selected the type $[T/A]^{\searrow 1 \nearrow 1'}$ when $i = 1$, the type $[A\backslash T]^{\searrow 1 \nearrow 1'}$ when $i = n$ and $[A\backslash T/A]^{\searrow 1 \nearrow 1'}$ otherwise. In the second case, the corresponding types will be $[T/A]^{\searrow 0 \nearrow 0'}$, $[A\backslash T]^{\searrow 0 \nearrow 0'}$ and $[A\backslash T/A]^{\searrow 0 \nearrow 0'}$. Every other symbol in a position $k \neq i$ will have a type $t^{\searrow v_k \nearrow v'_k}$, where t is one of the basic types $[A]$, $[A\backslash A]$, $[A/A]$. For instance, if $x_1 = 0, x_2 = 1, x_3 = 0$ and $g(C_j) = xy\bar{x}$, then the first x has the type $[A]^{\searrow 0 \nearrow 0'}$, y has the type $[A\backslash A]^{\searrow 1 \nearrow 1'}$ and the second x has the type $[A\backslash T]^{\searrow 0 \nearrow 0'}$. Finally, $[S]$ is assigned to F, $[T\backslash\varepsilon]$ is assigned to $*$ and for every $1 \leq i \leq n$, if m is odd, then for the occurrence i of f we choose the type $[\varepsilon]^{\searrow v'_i}$, otherwise we choose the type $[\varepsilon]^{\searrow v_{n-i+1}}$. Let us denote by $\Gamma(\Phi, v_1, \ldots, v_n)$ the string of types assigned in this manner to the string $w(\Phi)$ and by $\Gamma(C_j, v_1, \ldots, v_n)$ the substring of types assigned to $g(C_i)$. It is not difficult to see that $\Gamma(\Phi, v_1, \ldots, v_n) \vdash S$. Indeed, the potentials of the types assigned to n consecutive occurrences of b send the positive valencies encoding the values v_i of the corresponding propositional letters x_i. These valencies are intercepted by the dual negative valencies of the types chosen for the letters x, $\bar{x}$ and y in the closest block $g(C_1)$. Due to the definition of the modes π, the discontinuous dependencies 0 do not cross the discontinuous dependencies 1. This means that the valencies will be intercepted in the inverse order. Then every letter in $g(C_1)$ receiving the valency v will send on the positive valency v' and again, because the discontinuous dependencies $0'$ do not cross the discontinuous dependencies $1'$, the valencies will be intercepted in the inverse order (i.e. in the original order of the symbols b), etc. till the symbols f. This means that all valencies will be paired and eliminated. Besides this, $\Gamma(C_j, v_1, \ldots, v_n) \vdash [T]^{P_j}$ for some P_j and for every $1 \leq j \leq m$. Finally, every type $[T]$ is eliminated by the type assigned to the corresponding occurrence of $*$. So by Theorem 1, $\Gamma(\Phi, v_1, \ldots, v_n) \vdash S$.

2. Let Φ be not satisfiable. Let us assume that $w(\Phi) \in L(G^{(3)})$. Then the types assigned to the occurrences of symbols b correspond to the assignments of the coded values to the corresponding propositional letters. Let us suppose that C_{j_0} is the first false clause in Φ. Without loss of generality, we can suppose that j_0 is odd. Let Γ be a string of types assigned to $w(\Phi)$ in $G^{(3)}$ and Γ_{j_0} be its substring of types assigned to $g(C_{j_0})^R *$. Then Γ_{j_0} must reduce to $[\varepsilon]^{P_{j_0}}$ for some potential P_{j_0}. Therefore, there should be an occurrence i of one of the symbols x or $\bar{x}$ to which is assigned a type with the head dependency T, for instance the type $[A\backslash T/A]^{\searrow 1 \nearrow 1'}$ (or respectively $[A\backslash T/A]^{\searrow 0 \nearrow 0'}$). But, to be eliminated, this type needs paring the valency $\searrow 1$ (respectively $\searrow 0$), which is impossible, because C_{j_0} is false, so the value of x_i is dual to the choice of valencies. This means that the potential in Γ cannot be balanced. Hence, $w(\Phi) \notin L(G(\phi))$.

[*NP − completeness*]. Let us define the following relation $\prec$ on the set of discontinuous dependencies (i.e. pairs of the form $a(d) = \nearrow d \searrow d$ or $a(d) = \swarrow d \nwarrow d$): $a(d_1) \prec_\pi a(d_2)$ if the two dependencies cross and $d_1 \in \pi(d_2)$. With respect to the pairing rule **FA**, this relation means that $a(d_1)$ must be paired before $a(d_2)$. More precisely, the following proposition holds:

Lemma 9. *A potential P is balanced with respect to $\mathbf{FA}^l_{C:\pi(C)}$ with negative modalities π iff there is such pairing of valencies in P that $\prec_\pi$ has no circles on crossing dependencies.*

So the nondeterministic polynomial algorithm for membership $w \in L(G)$ is as follows:

1) guess $\Gamma \in \lambda(w)$,
2) check $\|\Gamma\|_l \vdash_c [S]$,
3) guess a pairing of valencies in $\|\Gamma\|_v^*$,
4) check that it is balanced and
5) check that $\prec_\pi$ has no circles on crossing dependencies.

By the way, using these techniques we can define a $mmCDG^{-FA}$ generating the language $\{ww^Rw \mid w \in \{a,b\}^+\}$ and, for every Turing machine, define a $mmCDG^{-FA}$ generating the "protocols" of its computations. We can also generate the copy language L_{copy} using the pairing rule $\mathbf{FC}_\pi$ with negative modes.

6 Conclusion

This study shows that the class of $mmCDG^{-FA}$-grammars may serve as a general theoretical frame for categorial dependency grammars, in which the languages form an AFL, may be not semilinear and the membership problem is NP-complete. Meanwhile, its subset of polynomially parsed $mmCDG$ without negative modes is perfectly adequate and sufficient for practical use because in the text corpora and more generally, in the written speech the sentences are explicitly separated by punctuation markers. So they are analyzed independently. As to the iterated constructions in the sentences, they are immediately definable through the primitive iterated types. On the other hand, for this subfamily $mmCDG$, the problems of semilinearity, of closure under iteration and of inclusion $L_{copy} \in \mathcal{L}(mmCDG)$ still rest open.

The two main lessons of this study are that:

(a) checking individual long distance discontinuous dependencies is a polynomial time task, whereas checking interaction of at least four of them may be untractable;
(b) for categorial dependency grammars, their closure under Kleene iteration may be obtained by prohibition of crossing one selected discontinuous dependency.

Acknowledgements. We are grateful to an anonymous reviewer who checked the proofs and pointed out several imprecisions and misprints.

References

1. Bar-Hillel, Y., Gaifman, H., Shamir, E.: On categorial and phrase structure grammars. Bull. Res. Council Israel 9F, 1–16 (1960)
2. Béchet, D., Dikovsky, A.J., Foret, A.: Dependency Structure Grammars. In: Blache, P., Stabler, E.P., Busquets, J.V., Moot, R. (eds.) LACL 2005. LNCS (LNAI), vol. 3492, pp. 18–34. Springer, Heidelberg (2005)
3. Dekhtyar, M., Dikovsky, A.: Generalized Categorial Dependency Grammars. In: Avron, A., Dershowitz, N., Rabinovich, A. (eds.) Pillars of Computer Science. LNCS, vol. 4800, pp. 230–255. Springer, Heidelberg (2008)

4. Dekhtyar, M., Dikovsky, A.: Categorial dependency grammars. In: Moortgat, M., Prince, V. (eds.) Proc. of Intern. Conf. on Categorial Grammars, Montpellier, pp. 76–91 (2004)
5. Dikovsky, A.: Multimodal categorial dependency grammars. In: Proc. of the 12th Conference on Formal Grammar, Dublin, Ireland, pp. 1–12 (2007)
6. Dikovsky, A.: Dependencies as categories. In: Duchier, D., Kruijff, G.-J.M. (eds.) Recent Advances in Dependency Grammars (COLING 2004) Workshop, pp. 90–97 (2004)
7. Duchier, D., Debusmann, R.: Topological dependency trees: A constraint-based account of linear precedence. In: Proc. of the 39th Intern. Conf. (ACL 2001), pp. 180–187. ACL & Morgan Kaufman (2001)
8. Hudson, R.A.: Word Grammar. Basil Blackwell, Oxford-New York (1984)
9. Joshi, A.K., Shanker, V.K., Weir, D.J.: The convergence of mildly context-sensitive grammar formalisms. In: Sells, P., Shieber, S., Wasow, T. (eds.) Foundational Issues in Natural Language Processing, pp. 31–81. MIT Press, Cambridge (1991)
10. Karlov, B.N.: Normal forms and automata for categorial dependency grammars. In: Vestnik Tverskogo Gosudarstvennogo Universiteta (Annals of Tver State University). Applied Mathematics, vol. 35 (95), pp. 23–43 (2008) (in Russ.)
11. Lambek, J.: On the calculus of syntactic types. In: Jakobson, R. (ed.) Structure of Languages and its Mathematical Aspects, pp. 166–178. American Mathematical Society, Providence (1961)
12. Mel'čuk, I.: Dependency Syntax. SUNY Press, Albany (1988)
13. Moortgat, M., Morrill, G.V.: Heads and phrases. Type calculus for dependency and constituent structure. Ms OTS, Utrecht (1991)
14. Morrill, G.V.: Type Logical Grammar. Categorial Logic of Signs. Kluwer, Dordrecht (1994)
15. Sleator, D., Temperly, D.: Parsing English with a Link Grammar. In: Proc. IWPT 1993, pp. 277–291 (1993)
16. Steedman, M., Baldridge, J.: Combinatory categorial grammar. In: Brown, K. (ed.) Encyclopedia of Language and Linguistics, vol. 2, pp. 610–622. Elsevier, Oxford (2006)
17. Tesnière, L.: Éléments de syntaxe structurale. Librairie C. Klincksieck, Paris (1959)
18. Vijay-Shanker, K., Weir, D.J.: The equivalence of four extensions of context-free grammars. Mathematical Systems Theory 27, 511–545 (1994)

Property Grammar Parsing Seen as a Constraint Optimization Problem

Denys Duchier, Thi-Bich-Hanh Dao, Yannick Parmentier, and Willy Lesaint

Centre, Val de Loire Université,
LIFO, Université d'Orléans, Bât. 3IA, Rue Léonard de Vinci,
F-45 067 Orléans Cedex 2, France
{denys.duchier,thi-bich-hanh.dao,
yannick.parmentier,willy.lesaint}@univ-orleans.fr
http://www.univ-orleans.fr/lifo

Abstract. Blache [1] introduced Property Grammar as a formalism where linguistic information is represented in terms of non hierarchical constraints. This feature gives it an adequate expressive power to handle complex linguistic phenomena, such as long distance dependencies, and also agrammatical sentences [2].

Recently, Duchier *et al.* [3] proposed a model-theoretic semantics for property grammar. The present paper follows up on that work and explains how to turn such a formalization into a constraint optimization problem, solvable using constraint programming techniques. This naturally leads to an implementation of a fully constraint-based parser for property grammars.

Keywords: Parsing, Property Grammar, Constraint Satisfaction.

1 Introduction

Formal grammars typically limit their scope to well-formed utterances. As noted by [4], formal grammars in the style of *generative-enumerative syntax*, as they focus on generating well-formed models, are intrinsically ill-suited for providing accounts of ill-formed utterances. Formal grammars in the style of *model-theoretic syntax*, on the contrary, as they focus on judging models according to the constraints that they satisfy, are naturally well-suited to accommodate *quasi-expressions*.

Blache [2] proposed *Property Grammars* (PG) as a constrained-based formalism for analyzing both grammatical and agrammatical utterances. Prost [5] developed an approach based on PG and capable not only of providing analyses for any utterances, but also of making accurate judgements of grammaticality about them. Duchier *et al.* [3] provided model-theoretical semantics for PG and a formal logical account of Prost's work. In this paper, we show how such a formalization can be converted into a Constraint Optimization Problem, thus yielding a constraint-based parser that finds optimal parses using classical constraint programming techniques, such as branch-and-bound.

P. de Groote and M.-J. Nederhof (Eds.): Formal Grammar 2010/2011, LNCS 7395, pp. 82–96, 2012.

The use of constraint-based techniques for parsing is not new in itself, one may cite the seminal work of Duchier [6] on Dependency Grammar parsing, or that of Debusmann *et al.* [7] on Tree Adjoining Grammar parsing, or more recently that of Parmentier and Maier [8], who proposed constraint-based extensions to Range Concatenation Grammar parsing. Nonetheless, PG parsing was lacking such a constraint-based axiomatization.[1]

The paper is organized as follows. We first introduce property grammars (section 2). In section 3, we then summarize the model-theoretic semantics of PG, as defined by Duchier *et al.* [3]. This semantics is used to define PG parsing as a Constraint Optimization Problem. This definition will be based on two types of constraints: tree-shapedness constraints, introduced in section 4, and property-related constraints, introduced in section 5. We then report on the implementation of a constraint-based parser using the Gecode library in section 6. In section 7, we compare our work with existing parsing environments for PG. Finally, in section 8, we conclude with some experimental results and hints about future work.

2 Property Grammars

Property grammar [2] is a grammatical formalism where the relations between constituents are expressed in terms of local constraints[2], called properties, which can be independently violated. This makes it possible to describe agrammatical utterances (that is, whose description would not respect the whole set of constraints), and also to associate a given description with a grammaticality score (ratio between satisfied and unsatisfied constraints).

These constraints rely on linguistic observations, such as linear precedence between constituents, coocurrency between constituents, exclusion between constituents, *etc.* As suggested by Duchier *et al.* [3], a property grammar can be usefully understood as exploding classical phrase structure rules into collections of fine-grained properties. Each property has the form $A : \psi$ meaning that, in a syntactic tree, for a node of category A, the constraint ψ applies to its children.

For each node of category A, we consider the following properties:

Obligation	$A : \triangle B$	at least one B child
Uniqueness	$A : B!$	at most one B child

[1] Note that a first experiment of constraint-based axiomatization of PG was done by Dahl and Blache [16], we give more information on this later in section 7.

[2] Several attemps at characterizing syntactic trees through a system of constraints were developed in the late nineties. Among these, one may cite D-Tree Substitution Grammar (DSG) [9], and Tree Description Grammar (TDG) [10]. The main difference between these formalisms and PG is that the latter has been designed to provide a way to handle agrammatical sentences. Furthermore, in DSG and TDG, constraints are expressed using dominance-based tree descriptions, while PG's constraints are applied to syntactic categories.

Linearity	$A : B \prec C$	B children precede C children
Requirement	$A : B \Rightarrow C$	if a B child, then also a C child
Exclusion	$A : B \not\Leftrightarrow C$	B and C children are mutually exclusive
Constituency	$A : S$	children must have categories in S

As an example, let us consider the context free rules $\mathtt{NP} \rightarrow \mathtt{D\ N}$ and $\mathtt{NP} \rightarrow \mathtt{N}$ describing the relation between a noun and a determiner. They are translated into the following 7 properties:

$$_{(1)}\mathtt{NP} : \{\mathtt{D},\mathtt{N}\},\ _{(2)}\mathtt{NP} : \mathtt{D}!,\ _{(3)}\mathtt{NP} : \triangle\mathtt{N},\ _{(4)}\mathtt{NP} : \mathtt{N}!,\ _{(5)}\mathtt{NP} : \mathtt{D} \prec \mathtt{N},\ _{(6)}\mathtt{D} : \{\},\ _{(7)}\mathtt{N} : \{\}.$$

(1) indicates that noun phrases only contain nouns or determiners, (2) states that in a noun phrase, there is at most one determiner. (3) and (4) say that in a noun phrase, there is exactly one noun. (5) indicates that, in a noun phrase, a determiner precedes a noun. Finally (6) and (7) state that determiners and nouns are leaf nodes in a valid syntactic tree.

In this context, if we only consider syntactic trees whose root has category NP, there are only two trees satisfying all properties:

We note that these syntactic trees are not lexicalized. In case we want to describe lexicalized trees, we can add some more lexical properties, such as

$$\mathsf{cat}(\mathit{apple}) = \mathtt{N}$$

which defines the word *apple* as being a noun.

About the complexity of PG parsing. Deep parsing with PG has been shown to be theoretically exponential in the number of categories of the grammar and the size of the sentence to parse [11]. As we shall see in section 7, existing approaches to deep parsing with PG usually rely on heuristics to reduce complexity in practice. In our approach, we want to avoid such heuristics. We are interested in studying the logical consequences of representation choices made in PG, while developing a parsing architecture for PG fully relying on constraint-satisfaction.

3 Model-Theoretic Semantics of Property Grammar

In this section, we give a summary of the model-theoretic semantics of PG developed by Duchier *et al.* [3].

First, recall that PGs are interpreted over syntactic tree structures. Two types of models are considered, according to whether we want to enforce the satisfaction of all grammatical properties or not: *strong* models and *loose* models (the latter corresponding to the modelization of agrammatical utterances).

Strong models. A syntax tree τ is a *strong* model of a grammar $\mathcal{G}$ iff for every node of τ and every property of $\mathcal{G}$, if that property is *pertinent* at that node, then it is also *satisfied.* The evaluation of the pertinence of a property depends on the type of the property. We consider 3 types of properties:

Type 1 : those which apply to a given node, such as *obligation.* For these properties, the pertinence only depends on the category of that node,

Type 2 : those which apply to a given couple of nodes (mother-daughter), such as *requirement* and *constituency.* For these properties, the pertinence depends on the category of the mother node, and that of its daughter nodes,

Type 3 : those which apply to a given triple of nodes (mother, daughter1, daughter2), such as *linearity, exclusion* and *uniqueness.* For these properties, the pertinence depends on the category of the mother node, and those of its daughter nodes.

Hence, when a node n has more than 2 children, a given property of type 3 has to be considered for every triple of nodes $(n, \_, \_)$. We call the pair consisting of a property ψ and such a tuple (*i.e.*, singleton, couple or triple of nodes), an *instance of property.* For example, the property `NP` : `D` $\prec$ `N` yields for every node as many instances as there are pairs of children. Such an instance is pertinent iff the mother node is of category `NP` and its children of categories `D` and `N` respectively. In addition, it is satisfied if the first child precedes the second one.

Later in this paper, we will represent the instance of a property $A : \psi$ at a node n using the following notation (n_x refers to a daughter node of n):

$$\begin{array}{ll} A : \psi@\langle n \rangle & \text{if } \psi \text{ is of } \textbf{type 1} \\ A : \psi@\langle n, n_1 \rangle & \text{if } \psi \text{ is of } \textbf{type 2} \\ A : \psi@\langle n, n_1, n_2 \rangle & \text{if } \psi \text{ is of } \textbf{type 3} \end{array}$$

To sum up, every property is instantiated in every possible way at every node. Furthermore, as mentioned above, in a strong model every property instantiation that is pertinent has to be satisfied.

Loose models. Unlike strong models, in a *loose* model of a grammar, for a given utterance, every instance of property which is pertinent does not have to be satisfied. More precisely, a loose model is a syntax tree of maximal *fitness*, where fitness is the ratio of satisfied instances among those which are pertinent.

For a more detailed definition of this model-theoretic semantics of PG, we refer the reader to [3] . In the next sections, we show how this formalization of PG can be converted into a Constraint Optimization Problem, thus yielding a constraint-based parser that finds optimal parses.

4 Representing Tree Models Using a Grid

Our approach needs to enumerate candidate tree models, and to retain only those of maximal *fitness.* Since we do not know *a priori* the number of nodes of our models, we propose to use a grid as a substrate, and to enumerate the trees which can be laid out on this grid.

For an utterance of m words, we know that each tree model has m leaves (PG do not use ϵ nodes). Unfortunately, we do not know the maximum depth of each tree model. We may use some heuristics to automatically assign a value n to the tree depth[3]. We chose to parametrize the associated parsing problem with a maximum tree depth n. Fixing this parameter allows us to layout a model over a subset of the nodes of an $n \times m$ grid. To represent our tree model, we will use a matrix $\mathcal{W}$ such that w_{ij} (with $1 \leq i \leq n$, and $1 \leq j \leq m$) refers to the node located at position (i, j) on the grid (rows and columns are numbered starting from 1, coordinate (1,1) being in the bottom-left corner). As an illustration of such a layout, see Fig. 1. We present in this section the constraints used to build a tree model on an $n \times m$ grid.

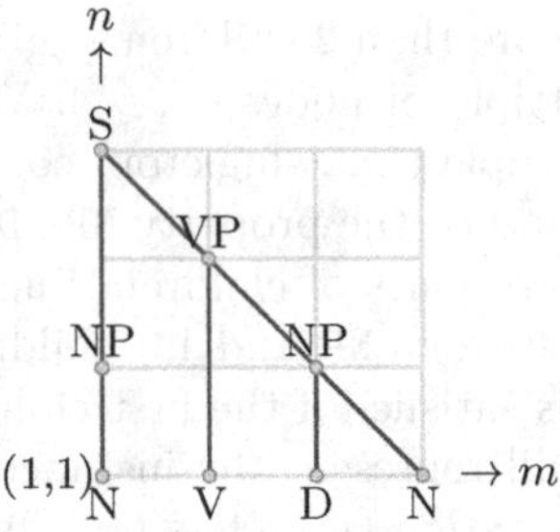

Fig. 1. Parse tree laid on a grid

Active nodes. Let V be the set of all nodes. A node is active if it is used by the model and inactive otherwise. We write V^+ for the set of active nodes and V^- for the rest. We have:

$$V = V^+ \uplus V^-$$

where $\uplus$ represents "disjoint union". Following the modeling technique of [12], for each node w, we write $\downarrow w$ for it's children, $\downarrow^+ w$ for its descendants, $\downarrow^* w$ for w and its descendants. Dually, we write $\uparrow w$ for w's parents, $\uparrow^+ w$ for its ancestors, $\uparrow^* w$ for w and its ancestors. Constraints relating these sets are:

$$\downarrow^+ w = \uplus\{\downarrow^* w' \mid w' \in \downarrow w\} \qquad \downarrow^* w = \{w\} \uplus \downarrow^+ w$$
$$\uparrow^+ w = \uplus\{\uparrow^* w' \mid w' \in \uparrow w\} \qquad \uparrow^* w = \{w\} \uplus \uparrow^+ w$$

[3] Due to the intrinsic recursive nature of language, the possibility to find an adequate depth value, *i.e.* not too big to prevent useless computations, and not too small to avoid missing solutions, is an open question.

Disjoint unions are justified by the fact that we are interested in tree models (*i.e.*, we do not allow for cycles). We additionally enforce the duality between ancestors and descendants:

$$w \in \uparrow w' \Leftrightarrow w' \in \downarrow w$$

and that each node has at most one parent:

$$|\uparrow w'| \leq 1$$

Inactive nodes have neither parents nor children:

$$w \in V^- \Rightarrow \downarrow w = \uparrow w = \emptyset$$

Since the root of the tree is still unknown, we write R for the set of root nodes. A tree model must have a single root:

$$|R| = 1$$

and the root node cannot be a child of any node:

$$V^+ = R \uplus (\uplus\{\downarrow w \mid w \in V\})$$

Projection. We write $\Downarrow w$ for the set of columns occupied by the tree anchored in w. As leaf nodes are located on the first row, their projection corresponds to their column, and only it:

$$\Downarrow w_{1j} = \{j\}$$

There are no interleaving projections (hence the disjoint union):

$$\Downarrow w_{ij} = \uplus\{\Downarrow w \mid w \in \downarrow w_{ij}\} \qquad 1 < j \leq m$$

There are no holes in the projection of any node (trees are projective):

$$\mathsf{convex}(\Downarrow w) \qquad \forall w \in V$$

Dealing with symmetries. There are many ways of laying out a given tree on a grid. For instance, a four-node and three-leaf tree has among others the following layouts :

 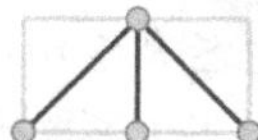 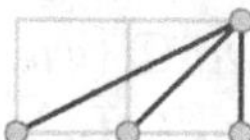 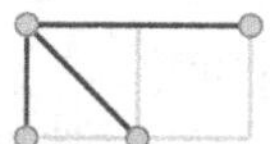

In order to have a unique way of laying out a tree, we add specific anti-symmetric constraints (the models satisfying these constraints are called *rectangular trees*) :

1. all leaves are located on the first row of the grid (*i.e.*, the bottom row),

2. the left-most daughter of any node is located on the same column as its mother node (this implies the subtree of a given node n occupies columns on the right of n),
3. every node is above any of its descendant nodes (this implies the subtree of a given node n occupies rows that are below that of n),
4. every internal node must have a daughter node in the row directly below (this implies there are no empty rows below the root node).

As an illustration, among the following trees, only the first one is a rectangular tree (the second tree violates condition 2, the third one condition 3 and the fourth one condition 4) :

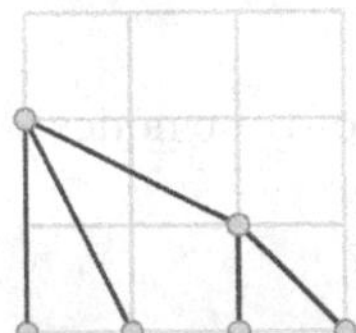 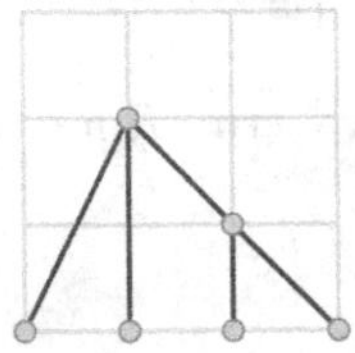 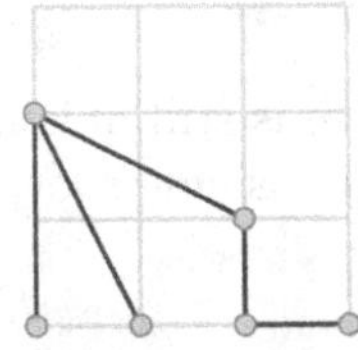 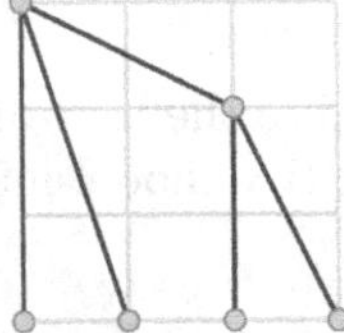

How these 4 conditions are represented in our axiomatization ? First, let us write $c(w)$ for the column of node w and $\ell(w)$ for its line:

$$c(w_{ij}) = j \qquad\qquad \ell(w_{ij}) = i$$

(1) The words are linked to the bottom row of the grid, which contains the leaves of the tree. These must all be active:

$$\{w_{1j} \mid 1 \leq j \leq m\} \subseteq V^+$$

(2) Any active node must be placed in the column of the left-most leaf of its subtree:

$$w_{ij} \in V^+ \quad \Leftrightarrow \quad j = \min \Downarrow w_{ij}$$

This stipulation and the fact (3) every node is above of its descendants are translated by constraints on the domains of variables. As mentioned above, the descendants of a node n are on the down-right part of the grid with respect to n. The dual holds, that is the ancestors of a node n are on the upper-left part of the grid with respect to n:

$$\downarrow^+ w_{ij} \subseteq \{w_{lk} \mid 1 \leq l < i,\ j \leq k \leq m\}$$
$$\uparrow^+ w_{ij} \subseteq \{w_{lk} \mid i < l \leq n,\ 1 \leq k \leq j\}$$

(4) Any active non-bottom node has at least one child at the level just below:

$$w_{ij} \in V^+ \quad \Leftrightarrow \quad i - 1 \in \{\ell(w) \mid w \in \downarrow w_{ij}\} \qquad 1 < i \leq n$$

Categories. In order to model syntax trees, we also need to assign to each active node a syntactic category. For simplicity, we will assign the category **none** to all and only the inactive nodes:

$$\mathsf{cat}(w) = \mathbf{none} \quad \Leftrightarrow \quad w \in V^-$$

For active nodes, the category will be assigned via property-related constraints, which are introduced in the next section. Finally, words are related to leaves via their category:

$$\mathsf{cat}(w_{1j}) = \mathsf{cat}(\text{word}_j)$$

where word_j refers to the jth word of the sentence to parse.

5 Handling Instances of Properties

Recall that each property has the form $A : \psi$, which means that for a node of category A, the constraint ψ applies to its children. For example the property $A : B \prec C$ is intended to mean that, for a non-leaf node of category A and any two daughters of this node labeled respectively with categories B and C, then the one labelled with B must precede the one labeled with C. Clearly, for each node of category A, this property must be checked for every pair of its daughters. This corresponds to the notion of instances of a property introduced earlier in section 3.

An instance of a property is a pair of the property and a tuple of nodes to which it is applied. An instance is pertinent if the node where it is instantiated is active (*i.e.*, belongs to V^+) and the parameter nodes of its tuple have the categories stipulated in the property. An instance is satisfied if the property is satisfied. For each instance I we define two boolean variables $P(I)$ and $S(I)$ denoting respectively its *pertinence* and its *pertinence and satisfaction.*

In the following paragraphs,we translate each property of PG into a set of constraints for our constraint optimization problem.

Properties of type 1. Let us start with properties of type 1, that is to say, whose instance depends on a single node. The only such property is obligation.

Obligation. The property $A : \triangle B$ yields instances I of the form:

$$(A : \triangle B)@\langle w_{i_0 j_0} \rangle$$

It is pertinent if $w_{i_0 j_0}$ is an active node labelled with A:

$$P(I) \quad \Leftrightarrow \quad (\, w_{i_0 j_0} \in V^+ \wedge cat(w_{i_0 j_0}) = A\,)$$

It is satisfied if at least one of its children is labelled with B:

$$S(I) \quad \Leftrightarrow \quad (\, P(I) \wedge \bigvee_{w_{ij} \in \downarrow w_{i_0 j_0}} cat(w_{ij}) = B\,)$$

Properties of type 2. Let us continue with properties of type 2, whose instance depends on a couple of nodes. These corresponds to requirement and constituency.

Requirement. The property $A : B \Rightarrow C$ yields instances I of the form:

$$(A : B \Rightarrow C)@\langle w_{i_0 j_0}, w_{i_1 j_1}\rangle$$

It is pertinent only if $w_{i_0 j_0}$ is active and $w_{i_1 j_1}$ is one of its children and their categories correspond:

$$P(I) \Leftrightarrow \begin{pmatrix} w_{i_0 j_0} \in V^+ \wedge w_{i_1 j_1} \in \downarrow w_{i_0 j_0} \wedge \\ \mathsf{cat}(w_{i_0 j_0}) = A \wedge \mathsf{cat}(w_{i_1 j_1}) = B \end{pmatrix}$$

It is satisfied if one of $w_{i_0 j_0}$'s children is labelled with C:

$$S(I) \quad \Leftrightarrow \quad (\, P(I) \wedge \bigvee_{w_{ij} \in \downarrow w_{i_0 j_0}} \mathsf{cat}(w_{ij}) = C\,)$$

Constituency. The property $A : S$ yields instances I of the form:

$$(A : B \prec C)@\langle w_{i_0 j_0}, w_{i_1 j_1}\rangle$$

It is pertinent only if $w_{i_0 j_0}$ is active and labelled with A and $w_{i_1 j_1}$ is one of its children:

$$P(I) \Leftrightarrow \begin{pmatrix} w_{i_0 j_0} \in V^+ \wedge w_{i_1 j_1} \in \downarrow w_{i_0 j_0} \wedge \\ \mathsf{cat}(w_{i_0 j_0}) = A \end{pmatrix}$$

It is satisfied if the category of $w_{i_1 j_1}$ is in S:

$$S(I) \quad \Leftrightarrow \quad (\, P(I) \wedge \mathsf{cat}(w_{i_1 j_1}) \in S\,)$$

Properties of type 3. Let us finish with properties of type 3, whose instance depends on a triple of nodes. These properties are linearity, uniqueness and exclusion.

Linearity. The property $A : B \prec C$ yields instances I of the form:

$$(A : B \prec C)@\langle w_{i_0 j_0}, w_{i_1 j_1}, w_{i_2 j_2}\rangle$$

I is pertinent if $w_{i_0 j_0}$ is active, $w_{i_1 j_1}$ and $w_{i_2 j_2}$ are its children, and each node is labelled with the corresponding category:

$$P(I) \Leftrightarrow \begin{pmatrix} w_{i_0 j_0} \in V^+ \wedge w_{i_1 j_1} \in \downarrow w_{i_0 j_0} \wedge \\ w_{i_2 j_2} \in \downarrow w_{i_0 j_0} \wedge \mathsf{cat}(w_{i_0 j_0}) = A \wedge \\ \mathsf{cat}(w_{i_1 j_1}) = B \wedge \mathsf{cat}(w_{i_2 j_2}) = C \end{pmatrix}$$

Its satisfaction depends on whether the node $w_{i_1 j_1}$ precedes $w_{i_2 j_2}$ or not. It is thus defined as:

$$S(I) \quad \Leftrightarrow \quad (\, P(I) \wedge j_1 < j_2\,)$$

Uniqueness. The property $A : B!$ yields instances I of the form:

$$(A : B!)@\langle w_{i_0 j_0}, w_{i_1 j_1}, w_{i_2 j_2}\rangle$$

It is active only if $w_{i_0 j_0}$ is active and labelled with A and $w_{i_1 j_1}$ and $w_{i_2 j_2}$ are its children and are labelled with B:

$$P(I) \Leftrightarrow \begin{pmatrix} w_{i_0 j_0} \in V^+ \wedge w_{i_1 j_1} \in {\downarrow} w_{i_0 j_0} \wedge \\ w_{i_2 j_2} \in {\downarrow} w_{i_0 j_0} \wedge \mathsf{cat}(w_{i_0 j_0}) = A \wedge \\ \mathsf{cat}(w_{i_1 j_1}) = B \wedge \mathsf{cat}(w_{i_2 j_2}) = B \end{pmatrix}$$

It is satisfied if $w_{i_1 j_1}$ and $w_{i_2 j_2}$ are the same node:

$$S(I) \quad \Leftrightarrow \quad (\, P(I) \wedge w_{i_1 j_1} = w_{i_2 j_2}\,)$$

Exclusion. The property $A : B \not\Leftrightarrow C$ yields instances I of the form:

$$(A : B \not\Leftrightarrow C)@\langle w_{i_0 j_0}, w_{i_1 j_1}, w_{i_2 j_2}\rangle$$

It is pertinent only if $w_{i_0 j_0}$ is active and labelled with A and $w_{i_1 j_1}$ and $w_{i_2 j_2}$ are its children where either $w_{i_1 j_1}$ is labelled with B or $w_{i_2 j_2}$ is labelled with C:

$$P(I) \Leftrightarrow \begin{pmatrix} w_{i_0 j_0} \in V^+ \wedge w_{i_1 j_1} \in {\downarrow} w_{i_0 j_0} \wedge \\ w_{i_2 j_2} \in {\downarrow} w_{i_0 j_0} \wedge \mathsf{cat}(w_{i_0 j_0}) = A \wedge \\ (\mathsf{cat}(w_{i_1 j_1}) = B \vee \mathsf{cat}(w_{i_2 j_2}) = C) \end{pmatrix}$$

Its satisfaction relies on the fact that the two children does not both have the incompatible categories:

$$S(I) \Leftrightarrow P(I) \wedge (\, \mathsf{cat}(w_{i_1 j_1}) \neq B \vee \mathsf{cat}(w_{i_2 j_2}) \neq C\,)$$

In other terms, if $w_{i_1 j_1}$ is labelled with B then $w_{i_2 j_2}$ cannot be labelled with C, and if $w_{i_2 j_2}$ is labelled with C then $w_{i_1 j_1}$ cannot be labelled with B.

As was mentioned in section 3, in the loose semantics of PG, we want to compute models with the best fitness. To do this, we add an optimization constraint.

Optimization Constraint. To account for the loose semantics of PG, a property instance counts if it is pertinent, it counts positively if satisfied, negatively otherwise. Let $\mathcal{I}$ be the set of all property instances, $\mathcal{I}^0$ the subset of pertinent instances and $\mathcal{I}^+$ the subset of positive instances. We want to find models which maximize the ratio $|\mathcal{I}^+|/|\mathcal{I}^0|$.

Since for each instance I, the variables $P(I)$ and $S(I)$ are boolean, their reified value is either 0 or 1. We can calculate the cardinality of these sets the following way:

$$|\mathcal{I}^0| = \sum_{I \in \mathcal{I}} P(I) \qquad\qquad |\mathcal{I}^+| = \sum_{I \in \mathcal{I}} S(I)$$

6 Implementation

The approach described so far has been implemented using the Gecode constraint programming library [13].

Each node w_{ij} of the grid is identified with an integer $k = (i-1) \times m + j$ (where m is the width of the grid). The set of nodes V is defined as $V = \{1, \ldots, n \times m\}$. V^+ and V^- are two set variables such that $V^+, V^- \subseteq V$. All the constraints related to these sets are implemented using Gecode's API. The relations $\downarrow, \downarrow^+, \downarrow^*$ and $\uparrow, \uparrow^+, \uparrow^*$ are encoded using arrays of set variables, whose indexes are nodes of the grid. We also use arrays of set variables to encode property-related constraints. As there are many types of constraints and many instances to consider, the computation of the indexes is slightly more complex than the ones used for tree-shapedness constraints. Definitions of $P(I)$ and $S(I)$ are realized using *reified constraints*. The search for an optimal parse is achieved using the *branch-and-bound* search strategy to maximize the ratio $|\mathcal{I}^+|/|\mathcal{I}^0|$.

An example search tree for the grammatical utterance "`Peter eats the apple`" is given in Fig. 2. The graphical representation of the search tree has been built by Gist, the Gecode Interactive Search Tool [14]. On this figure, round nodes represent choice points, square and triangle-shaped nodes failures, and diamond-shaped nodes solutions of the constraint optimization problem (in this example, the last diamond on the right refers to the optimal solution).

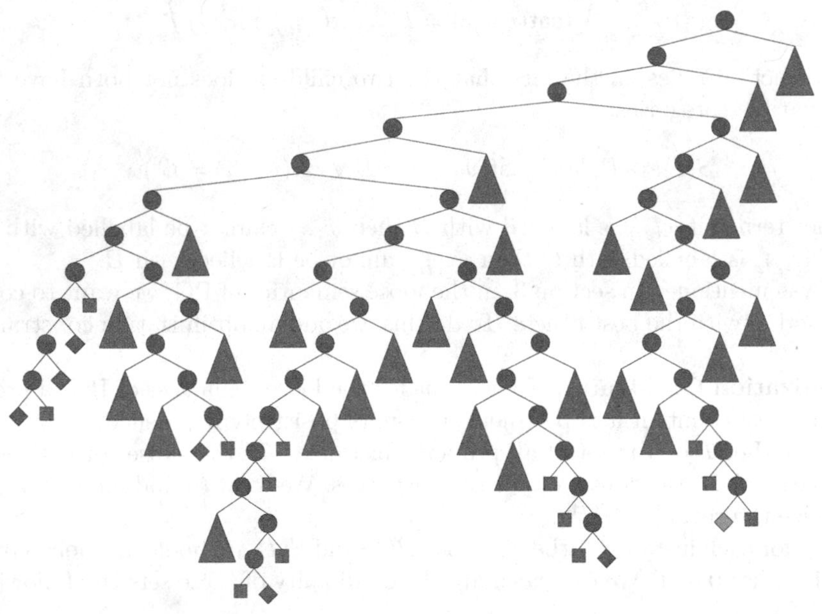

Fig. 2. Example search tree

To give an illustration of the complexity of the constraint optimization problem[4] for the parse example of Fig. 2 (sentence "`Peter eats the apple`", and grammar having 19 properties handling 6 categories), the search tree has about 450,000 nodes, among which 7 are solutions. The optimal syntactic tree is represented in Fig. 3.

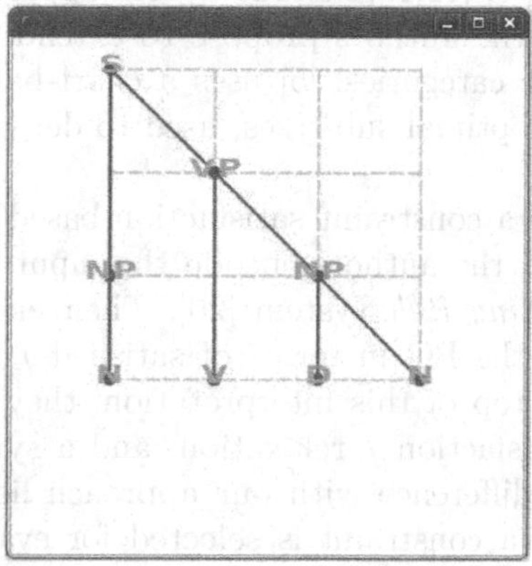

Fig. 3. Optimal syntactic tree for "Peter eats the apple"

As the parser is still in an early development stage, we do not have any benchmark. As mentioned above, there are many instances of property to handle, that is to say, many constraints to evaluate. In practice, our parser can relatively quickly find a syntactic tree (in less than a second for the example above), but the proof of optimality can take about a minute.[5] While we are mainly interested in exploring the logical consequences of representation choices made in PG (that is, without using any heuristic to reduce complexity), we are also interested in improving the computation of optimal parses, either by:

- parallelizing the exploration of the search tree,
- or enriching the information associated with lexical entries, for example by using a Part-Of-Speech tagger.

The parser is freely available on demand, and released under the terms of the GNU General Public License.

7 Comparison with Existing Work

Among the different approaches to PG parsing, one may cite the seminal work of Blache and Balfourier [15]. This work was later followed by a series of papers

[4] Our PG parsing algorithm using constraint-satisfaction is clearly exponential as all candidate trees are enumerated.

[5] These results are obtained on a 2.6 Ghz processor with 4 Gb of RAM.

by Dahl and Blache [16], Estratat and Henocque [17], Van Rullen [11,18], Blache and Rauzy [19], and more recently Prost [5].

The main difference between these approaches and our work, is that, apart from [16] and [17], they do not rely on a model-theoretic formal semantics of PG. They rather apply well-known efficient parsing techniques and heuristics to PG. Thus, [15] uses a constraint selection process to incrementally build the syntactic tree of a sentence. [11,18] include hybrid approaches mixing deep and shallow parsing. In [19], the authors propose to extend symbolic parsing with probabilities on syntactic categories. [5] uses a chart-based parsing algorithm, where the items contain optimal sub-trees, used to derive a complete syntactic tree.

A first attempt to use a constraint satisfaction-based approach to PG parsing is [16]. In their work, the authors encode the input PG into a set of rules for the *Constraint Handling Rule* system [20]. Their encoding makes it possible to directly interpret the PG in terms of satisfied / relaxed constraints on syntactic categories. On top of this interpretation, they use rewriting rules to propagate constraint satisfaction / relaxation, and a syntactic tree is built as a side effect. The main difference with our approach lies in the fact that the authors control the way a constraint is selected for evaluation, while we rely on classical constraint-based techniques such as branch-and-bound to select and propagate constraint evaluations. That is, we clearly distinguish the definition of the constraint satisfaction problem from its resolution.

Another constraint-based approach to PG parsing is [17]. In their work, the authors translate a PG into a model in the Object Constraint Language (OCL). This model is interpreted as a configuration problem, which is fed to a configurator. The latter solves the constraints lying in the input model. The result is a valid syntactic structure. Contrary to our approach, or that of [16], this OCL-encoding does not allow for relaxed constraints. Hence, it only computes syntactic structures that satisfy the whole set of constraints. In other terms, it cannot make full advantage of the PG formalism, which describes natural language in terms of local constraints that can be violated. This feature is particularly useful when dealing with agrammatical sentences such as those spoken language often contain.

8 Conclusion

Duchier *et al.* [3] provided precise model-theoretical semantics for property grammars. In this paper, we extend that work and show how such a formalization can be converted into a Constraint Optimization Problem, thus yielding a constraint-based parser capable of finding optimal parses using classical constraint-based techniques such as branch-and-bound. Furthermore, we have implemented this convertion and are able to experiment with analyzing both grammatical and agrammatical utterances.

The work described here is still at an early stage of development. It is not intended to compete with high-performance parsers, but rather to serve as an experimental platform for grammar development and linguistic modeling, where logical consequences are not accidentally hidden by the effect of performance-oriented heuristics.

In a near future, we plan to work on the definition and implementation of an extension to branch-and-bound, in order to keep not only one but all syntactic trees having the maximum fitness.

Acknowledgments. We are grateful to Sylvie Billot, Matthieu Lopez, Jean-Philippe Prost, Isabelle Tellier and three anonymous reviewers for useful comments on this work.

References

1. Blache, P.: Constraints, Linguistic Theories, and Natural Language Processing. In: Christodoulakis, D.N. (ed.) NLP 2000. LNCS (LNAI), vol. 1835, pp. 221–232. Springer, Heidelberg (2000)
2. Blache, P.: Property Grammars: A Fully Constraint-Based Theory. In: Christiansen, H., Skadhauge, P.R., Villadsen, J. (eds.) CSLP 2005. LNCS (LNAI), vol. 3438, pp. 1–16. Springer, Heidelberg (2005)
3. Duchier, D., Prost, J.-P., Dao, T.-B.-H.: A Model-Theoretic Framework for Grammaticality Judgements. In: De Groote, P., Egg, M., Kallmeyer, L., Penn, G. (eds.) Formal Grammar. LNCS, vol. 5591, pp. 17–30. Springer, Heidelberg (2011)
4. Pullum, G., Scholz, B.: On the Distinction between Model-Theoretic and Generative-Enumerative Syntactic Frameworks. In: de Groote, P., Morrill, G., Retoré, C. (eds.) LACL 2001. LNCS (LNAI), vol. 2099, pp. 17–43. Springer, Heidelberg (2001)
5. Prost, J.-P.: Modelling Syntactic Gradience with Loose Constraint-based Parsing. PhD Thesis, Macquarie University, Sydney, Australia, and Université de Provence. Aix-en-Provence, France (2008)
6. Duchier, D.: Axiomatizing Dependency Parsing Using Set Constraints. In: 6th Meeting on Mathematics of Language, Orlando, pp. 115–126 (1999)
7. Debusmann, R., Duchier, D., Kuhlmann, M., Thater, S.: TAG Parsing as Model Enumeration. In: 7th International Workshop on Tree-Adjoining Grammar and Related Formalisms - TAG+7, Vancouver, pp. 148–154 (2004)
8. Parmentier, Y., Maier, W.: Using Constraints over Finite Sets of Integers for Range Concatenation Grammar Parsing. In: Nordström, B., Ranta, A. (eds.) GoTAL 2008. LNCS (LNAI), vol. 5221, pp. 360–365. Springer, Heidelberg (2008)
9. Rambow, O.: D-tree Substitution Grammars. J. Comp. Ling. 27(1), 89–121 (2001)
10. Kallmeyer, L.: Local Tree Description Grammars: a Local Extension of TAG Allowing Underspecified Dominance Relations. J. Grammars 4, 85–137 (2001)
11. Van Rullen, T.: Vers une Analyse Syntaxique à Granularité Variable. PhD Thesis. Université de Provence, Aix-Marseille 1, France (2005)
12. Duchier, D.: Configuration of Labeled Trees Under Lexicalized Constraints And Principles. J. of Research on Lang. and Comp. 1(3/4), 307–336 (2003)
13. Gecode Team: Gecode: Generic Constraint Development Environment (2010), `http://www.gecode.org`
14. Schulte, C., Tack, G., Lagerkvist, M.Z.: Modeling and Programming with Gecode. Gecode documentation (2010), `http://www.gecode.org/doc-latest/MPG.pdf`

15. Blache, P., Balfourier, J.-M.: Property Grammars: a Flexible Constraint-Based Approach to Parsing. In: 7th International Workshop on Parsing Technologies - IWPT 2001, Beijing (2001)
16. Dahl, V., Blache, P.: Directly Executable Constraint Based Grammars. Programmation en Logique Avec Contraintes. Journées Francophones de Programmation Logique et par Contraintes 2004 - JFPLC 2004, Hermès (2004)
17. Estratat, M., Henocque, L.: Parsing Languages with a Configurator. In: European Conference for Artificial Intelligence - ECAI 2004, Valencia (2004)
18. Van Rullen, T., Blache, P., Balfourier, J.-M.: Constraint-Based Parsing as an Efficient Solution: Results from the Parsing Evaluation Campaign EASy. In: 5th International Conference on Language Resources and Evaluation - LREC 2006, Genoa (2006)
19. Blache, P., Rauzy, S.: Mécanismes de Contrôle pour l'Analyse en Grammaires de Propriétés. In: 13e Conférence sur le Traitement Automatique des Langues Naturelles - TALN 2006, Leuven (2006)
20. Frühwirth, T.: Constraint Handling Rules. Cambridge University Press (2009)

Reference-Set Constraints as Linear Tree Transductions via Controlled Optimality Systems

Thomas Graf

Department of Linguistics,
University of California, Los Angeles
tgraf@ucla.edu
http://tgraf.bol.ucla.edu

Abstract. Reference-set constraints are a special class of constraints used in Minimalist syntax. They extend the notion of well-formedness beyond the level of single trees: When presented with some phrase structure tree, they compute its set of competing output candidates and determine the optimal output(s) according to some economy metric. Doubts have frequently been raised in the literature whether such constraints are computationally tractable [4]. I define a subclass of Optimality Systems (OSs) that is sufficiently powerful to accommodate a wide range of reference-set constraints and show that these OSs are globally optimal [5], a prerequisite for them being computable by linear tree transducers. As regular and linear context-free tree languages are closed under linear tree transductions, this marks an important step towards showing that the expressivity of various syntactic formalisms is not increased by adding reference-set constraints. In the second half of the paper, I demonstrate the feasibility of the OS-approach by exhibiting an efficiently computable OS for a prominent reference-set constraint, Focus Economy [10].

Keywords: Optimality Systems, Tree Transducers, Reference-Set Constraints, Transderivationality, Modeling.

1 Introduction

Out of all the items in a syntactician's toolbox, reference-set constraints are probably the most peculiar one. When handed some syntactic tree, a reference-set constraint does not determine its well-formedness from inspection of the tree itself. Instead, it constructs a *reference set* — a set containing a number of trees competing against each other — and chooses the optimal candidate from said set.

Consider *Fewest Steps* [1]. The reference set that this constraint constructs for any given tree t consists of t itself and all the trees that were assembled from the same lexical items as t. All the trees in the reference set are then ranked by the number of movement steps that occurred during their assembly (this is usually identical to the number of traces they contain), and the tree(s) with the fewest instances of movement is (are) chosen as the winner. All other trees are flagged as ungrammatical, including t if it did not emerge as a winner.

P. de Groote and M.-J. Nederhof (Eds.): Formal Grammar 2010/2011, LNCS 7395, pp. 97–113, 2012.

Another reference-set constraint is *Focus Economy* [10], which accounts for the empirical fact that neutral stress is compatible with more discourse situations than shifted stress. Take a look at the utterances in (1), where main stress is indicated by **bold face**. Example (1a) can serve as an answer to various questions, among others "What's going on?" and "What did your neighbor buy?". Yet the virtually identical (1b), in which the main stress falls on the subject rather than the object, is compatible only with the question "Who bought a book?". These contrasts indicate a difference as to which constituents may be *focused*, i.e. can be interpreted as providing new information.

(1) a. My neighbor bought a **book**.
  b. My **neighbor** bought a book.

Focus Economy derives the relevant contrast by stipulating that first, any constituent containing the node carrying the sentential main stress can be focused, and second, in a tree in which stress was shifted from the neutral position, a constituent may be focused only if it cannot be focused in the original tree with unshifted stress. In (1a), the object, the VP and the entire sentence can be focused, since these are the constituents containing the main stress carrier. In (1b), the main stress is contained by the subject and the entire sentence, however, only the former may be focused because focusing of the the latter is already a licit option in the neutral stress counterpart (1a).

This esoteric behavior of reference-set constraints coupled with a distinct lack of formal work on their properties has led to various conjectures that they are computationally intractable [4]. In this paper, I refute these claims by showing how reference-set constraints can be emulated by a new variant of Optimality Systems (OSs), and I contend that this route paves the way for reference-set constraints to be implemented as finite-state devices; linear bottom-up tree transducers (lbutts), to be precise. Lbutts are of interest for theoretical as well as practical purposes because both regular and linear context-free tree languages are known to be closed under linear transductions, so applying a linear transducer to a regular/linear context-free tree language yields a regular/linear context-free tree language again. On a theoretical level, this provides us with new insights into the nature of reference-set constraints, while on a practical level, it ensures that adding reference-set constraints to a grammar does not jeopardize its computability. I support my claim by exhibiting a formal model of Focus Economy as an lbutt. My results shed new light on reference-set computation as well as on Optimality Systems and should be of interest to readers from various formal backgrounds, foremost computational phonology and Minimalist grammars.

The paper is laid out as follows: After the preliminaries section, which due to space restrictions has to be shorter than is befitting, I give a brief introduction to OSs before introducing my own variant, controlled OSs, in Sec. 4. The mathematical core results of this section are a new characterization of the important property of global optimality and a simplification of Jäger's theorem [5] regarding the properties of an OS that jointly ensure that it does not exceed the power of linear tree transducers. In the last section, I show how to model Focus Economy as such a restricted OS.

2 Preliminaries and Notation

Let me introduce some notational conventions first. Given a relation R, its *domain* is denoted by $\text{dom}(R)$, its *range* by $\text{ran}(R)$. For any $a \in \text{dom}(R)$, we let $aR := \{b \mid \langle a,b\rangle \in R\}$, unless R is a function, in which case $aR = R(a)$. The *composition* of two relations R and S is $R \circ S := \{\langle a,c\rangle \mid \langle a,b\rangle \in R, \langle b,c\rangle \in S\}$. The *diagonal* of some set A is $id(A) := \{\langle a,a\rangle \mid a \in A\}$.

Tree languages and tree transductions form an integral part of this paper, however, the technical machinery is mostly hidden behind the optimality-theoretic front-end so that only a cursory familiarity with the subject matter is required. Nevertheless the reader is advised to consult [3] and [7] for further details. I also assume that the reader is knowledgeable about string languages and generalized sequential machines.

Definition 1. *A* context-free tree grammar *(CFTG) is defined to be a 4-tuple $\mathcal{G} := \langle \Sigma, F, S, \Delta\rangle$, where Σ and F are disjoint, finite, ranked alphabets of terminals and non-terminals, respectively, and $S \in F$ is the start symbol. Furthermore, Δ is a finite set of productions of the form $F(x_1, \dots, x_n) \to t$, where F is of rank n, and t is a tree with the node labels drawn from $\Sigma \cup F \cup \{x_1, \dots, x_n\}$.*

A production is linear if each variable in its left-hand side occurs at most once in its right-hand side. A CFTG is *linear* if each production is linear. A CFTG is a *regular* tree grammar (RTG) if all non-terminals are of rank 0. A tree language is *regular* iff it is generated by an RTG, and every regular tree language has a context-free language as its string yield.

Definition 2. *A* bottom-up tree transducer *is a 5-tuple $\mathcal{A} := \langle \Sigma, \Omega, Q, Q', \Delta\rangle$, where Σ and Ω are finite ranked alphabets, Q is a finite set of states, and $Q' \subseteq Q$ the set of final states. By Δ we denote a set of productions of the form $f(q_1(x_1), \dots, q_n(x_n)) \to q(t(x_1, \dots, x_n))$, where $f \in \Sigma$ is of rank n, $q_1, \dots, q_n, q \in Q$, and $t(x_1, \dots, x_n)$ is a tree with the node labels drawn from $\Omega \cup \{x_1, \dots, x_n\}$.*

Definition 3. *A* top-down tree transducer *is 5-tuple $\mathcal{A} := \langle \Sigma, \Omega, Q, Q', \Delta\rangle$, where Σ, Ω and Q are as before, $Q' \subseteq Q$ is the set of initial states, and all productions in Δ are of the form $q(f(x_1, \dots, x_n)) \to t$, where $f \in \Sigma$ is of rank n, $q \in Q$ and t is a tree with the node labels drawn from $\Omega \cup \{q(x) \mid q \in Q, x \in \{x_1, \dots, x_n\}\}$.*

As with CFTGs, a production is linear if each variable in its left-hand side occurs at most once in its right-hand side. A transducer is *linear* if each production is linear. I denote a linear bottom-up/top-down tree transducer by lbutt/ltdtt. The class of ltdtts is properly contained in the class of lbutts, which in turn is closed under union and composition. The domain and the range of an lbutt are both recognizable, i.e. regular tree languages. The relation τ induced by a (linear) tree transducer is called a (linear) *tree transduction*. For a bottom-up tree transducer, the graph of τ consists of pairs $\langle s,t\rangle$ such that s and t are Σ- and Ω-labeled trees, respectively, and for some $q \in Q'$, $q(t)$ can be obtained from s by finitely many applications of productions $\delta \in \Delta$. The definition is almost unchanged for top-down tree transducers, except that we require that t can be obtained from $q(s)$. In a slight abuse of terminology, I call a relation *rational* iff it is a finite-state string transduction or a linear tree transduction. For any recognizable tree language L, $id(A)$ is a rational relation. Furthermore, both regular string/tree languages and linear context-free tree languages are closed under rational relations.

In Sec. 5.2, I make good use of $\mathcal{L}^2_{K,P}$ [11], an incarnation of monadic second-order logic (MSO) specifically designed for linguistic purposes. MSO is the extension of first-order logic with monadic second-order variables and predicates as well as quantification over them such that the first-order variables represent nodes in the tree and the second-order variables and predicates sets of nodes. A set of finite strings/trees is definable in MSO iff it is regular. Specifics of $\mathcal{L}^2_{K,P}$ will be briefly introduced in the relevant section. See [11] for further background.

3 Traditional Perspective on Optimality Systems

OSs were introduced independently by [2] and [6] as a formalization of Optimality Theory (OT). In OT, well-formed expressions are no longer derived from underlying representations through iterated applications of string rewrite rules, as was the case with SPE. Instead, underlying representations — which are usually referred to as *inputs* — are assigned a set of *output candidates* by a relation called *generator*, abbreviated GEN. This set is subsequently narrowed down by a sequence of constraints $c_1, \ldots, c_n$ until only the *optimal* output candidates remain. This narrowing-down process proceeds in a fashion such that only the candidates that incurred the least number of violations of constraint c_i are taken into account for the evaluation of c_{i+1}. Thus every constraint acts as a (violable) filter on the set of output candidates, with the important addendum that the order in which the filters are applied is crucial in determining optimality.

Consider the example in Fig. 1, which depicts an OT evaluation of output candidates using the tableau notation. Here some input i is assigned three output candidates o_1, o_2 and o_3. The OT grammar uses only three constraints c_1, c_2 and c_3. Constraint c_1 is applied first. Candidates o_2 and o_3 each violate it once, however, o_1 violates it twice. Thus o_2 and o_3 are the output candidates incurring the least number of violations of the constraint and are allowed to proceed to the next round of the evaluation. Candidate o_1, on the other hand, is ruled out and does not participate in further evaluations. Neither o_2 nor o_3 violate c_2 (nor does o_1, but this is immaterial since it has previously been discarded), so neither is filtered out. In the third round, o_2 and o_3 are evaluated with respect to c_3. Each of them violates the constraint once, but since there is no candidate that fares better than them (again, o_1 is not taken into consideration anymore), they also survive this round of the evaluation. Thus, o_2 and o_3 are the optimal output candidates for i. If c_3 had been applied before c_1, on the other hand, o_2 and o_3 would lose out against o_1.

With this intuitive understanding of OT grammars under our belt, the formal definitions of OSs and their *output language* (not to be confused with the *candidate language* ran(GEN)) are straight-forward.

i	c_1	c_2	c_3
o_1	2	0	0
o_2	1	0	1
o_3	1	0	1

Fig. 1. Example of an OT evaluation in tableau notation

Definition 4. *An* optimality system *over languages L and L' is a pair $\mathcal{O} := \langle \text{GEN}, C \rangle$ with $\text{GEN} \subseteq L \times L'$ and $C := \langle c_1, \ldots, c_n \rangle$ a linearly ordered sequence of functions $c_i \colon \text{GEN} \to \mathbb{N}$. For $a, b \in \text{GEN}$, $a <_{\mathcal{O}} b$ iff there is a $1 \leq k \leq n$ such that $c_k(a) < c_k(b)$ and for all $j < k$, $c_j(a) = c_j(b)$.*

Definition 5. *Given an optimality system $\mathcal{O} := \langle \text{GEN}, C \rangle$, $\langle i, o \rangle$ is* optimal *with respect to $\mathcal{O}$ iff both $\langle i, o \rangle \in \text{GEN}$ and there is no o' such that $\langle i, o' \rangle \in \text{GEN}$ and $\langle i, o' \rangle <_{\mathcal{O}} \langle i, o \rangle$. The output language of $\mathcal{O}$ is $\text{ran}(\{\langle i, o \rangle \mid \langle i, o \rangle$ is optimal with respect to $\mathcal{O}\})$.*

The important insight of [2] as well as [6], which was later improved upon by [5, 7, 13], is that an OS as defined above can be understood to define a transduction from a set of inputs to its set of optimal output candidates. Moreover, if the OS is suitably restricted, it is guaranteed to define a rational relation, which implies its efficient computability.

Theorem 6. *Let $\mathcal{O} := \langle \text{GEN}, C \rangle$ be an OS such that*

- *$\text{dom}(\text{GEN})$ is a regular string language, or a regular/linear context-free tree language, and*
- *GEN is a rational relation, and*
- *all $c \in C$ are output-markedness constraints, and*
- *each $c \in C$ defines a rational relation on $\text{ran}(\text{GEN})$, and*
- *$\mathcal{O}$ is globally optimal.*

Then the transduction τ induced by the OS is a rational relation and $\text{ran}(\tau)$ belongs to the same formal language class as $\text{dom}(\tau)$.

As the reader might have guessed, the use of τ here is carried over straightforwardly from tree transducers, meaning that $\tau := \{\langle i, o \rangle \mid \langle i, o \rangle$ is optimal with respect to $\mathcal{O}\}$. The theorem also makes reference to two notions we have not encountered yet at all, output-markedness and global optimality. The former is easily defined.

Definition 7. *Given an OS $\mathcal{O} := \langle \text{GEN}, C \rangle$, $c \in C$ is an* output-markedness constraint *iff $c(\langle i, o \rangle) = c(\langle i', o \rangle)$ for all $\langle i, o \rangle, \langle i', o \rangle \in \text{GEN}$.*

Global optimality, on the other hand, requires a lot of finite-state machinery to be in place before it can be made formally precise, which would lead us off track here. For our purposes it is sufficient to know that an OS is globally optimal iff for every optimal output candidate o it holds that there is no input i such that o is an output candidate for i but not an optimal one. The curious reader is referred to [5] for a more rigorous definition.

4 Controlled Optimality Systems

OSs are perfectly capable of modeling reference-set constraints. The reference set for any input i is defined by iGEN, and the evaluation metric can straightforwardly be implemented as a sequence of constraints. Fewest Steps, for instance, can be viewed as an OS in which a tree t is related by GEN to all the trees that were constructed from the same lexical items as t, including t itself. Besides that, the OS has only one constraint *TRACE, which punishes traces. As a consequence, only the trees with the least number of traces will be preserved, and these are the optimal output candidates for t. While this short example shows that OSs can get the job done, the way output candidates are specified for reference-set constraints actually relies on additional structure — the reference sets — that is only indirectly represented by GEN. In the following I introduce controlled OSs as a variant of standard OSs that is closer to reference-set computation by making reference sets prime citizens of OSs and demoting GEN to an ancillary relation that is directly computed from them.

We observe first that many reference-set constraints allow for distinct inputs to be assigned the same reference set. Hence it makes sense to map entire sets of inputs to reference sets, rather than the individual inputs themselves. Let us call such a set of inputs a *reference type*. An OS can then be defined by reference types and a function mapping them to reference sets:

Definition 8 (Controlled Optimality Systems). *An* $\mathcal{F}$-controlled optimality system *over languages* L, L' *is a 4-tuple* $\mathcal{O}[\mathcal{F}] := \langle \text{GEN}, C, \mathcal{F}, \gamma \rangle$, *where*

- GEN *and* C *are defined as usual,*
- $\mathcal{F}$ *is a family of non-empty subsets of* L, *each of which we call a* reference type,
- *the* control map $\gamma : \mathcal{F} \to \wp(L') \setminus \{\emptyset\}$ *associates every reference type with a* reference set, *i.e. a set of output candidates,*
- *the following conditions are satisfied*
  - exhaustivity*:* $\bigcup_{X \in \mathcal{F}} X = L$
  - bootstrapping*:* $x\text{GEN} = \bigcup_{x \in X \in \mathcal{F}} X\gamma$

Every controlled OS can be translated into a canonical OS by discarding its third and fourth component (i.e. $\mathcal{F}$ and the control map γ). In the other direction, a controlled OS can be obtained from every canonical OS by setting $\mathcal{F} := \{\{i\} \mid i \in L\}$ and $\gamma : \{i\} \mapsto i\text{GEN}$. So the only difference between the two is that controlled OSs modularize GEN by specifying it through reference types and the control map.

Now that OSs operate (at least partially) at the level of sets, it will often be interesting to talk about the set of optimal output candidates assigned to some reference type, rather than one particular input. But whereas the set of optimal output candidates is always well-defined for inputs — for any input i this set is given by $i\tau$ — we have to be more careful when lifting it to reference types, because distinct inputs that belong to the same reference type may not necessarily be assigned the same optimal output candidates. Such a situation might arise in OSs with faithfulness or input-markedness constraints, which are sensitive to properties of the input, or when two inputs i and j are of reference type X, but in addition j is also of reference type Y. Given this ambiguity, one has to distinguish between the set of output candidates that are optimal for at least one member of reference type X, and the set of output candidates that are optimal for all members of reference type X. The former is the *up-transduction* $X\tau^{\uparrow} := \bigcup_{x \in X} x\tau$, the latter the *down-transduction* $X\tau^{\downarrow} := \bigcap_{x \in X} x\tau$.

At this point it might be worthwhile to work through a simple example. Fig. 2 on the following page depicts a controlled OS and the distinct steps of its computation. We are given a collection of reference types consisting of $\textsc{Red} := \{i_1, i_2, i_3, i_4, i_5, i_6\}$, $\textsc{Sienna} := \{i_4\}$, $\textsc{Teal} := \{i_5, i_6, i_7\}$, $\textsc{Purple} := \{i_8\}$, and $\textsc{Lime} := \{i_7, i_9, i_{10}\}$. The reference sets are $\textsc{Blue} := \{o_1, o_2, o_3\}$, $\textsc{Orange} := \{o_3, o_4, o_5, o_6, o_7\}$, $\textsc{Green} := \{o_6, o_7\}$, and $\textsc{Brown} := \{o_8, o_9\}$. Finally, the graph of γ consists of the pairs $\langle \textsc{Red}, \textsc{Blue} \rangle$, $\langle \textsc{Sienna}, \textsc{Brown} \rangle$, $\langle \textsc{Teal}, \textsc{Green} \rangle$, $\langle \textsc{Purple}, \textsc{Brown} \rangle$, and $\langle \textsc{Lime}, \textsc{Orange} \rangle$. Note that a reference type may overlap with another reference type or even be a proper subset of it, and the same holds for reference sets. This means that an input can belong to several reference types at once. Consequently, $x\textsc{Gen}$ may be a superset of $X\gamma$ for every reference type X that contains x, as is the case for i_4, say, but not for i_7, even though both are assigned exactly two reference types. Input i_4 is related by $\textsc{Gen}$ to all outputs contained in $\textsc{Red}\gamma \cup \textsc{Sienna}\gamma = \textsc{Blue} \cup \textsc{Brown} = \{o_1, o_2, o_3, o_8, o_9\}$, whereas i_7 is related to $\textsc{Lime}\gamma \cup \textsc{Teal}\gamma = \textsc{Orange} \cup \textsc{Green} = \textsc{Orange} = \{o_3, o_4, o_5, o_6, o_7\}$. As soon as $\textsc{Gen}$ has been determined from the reference types and the control map, the computation proceeds as usual with the constraints of the OS filtering out all suboptimal candidates.

Interestingly, almost all reference-set constraints fall into two classes with respect to how reference types and reference sets are distributed. In the case of Fewest Steps, where the input language is also the candidate language, each reference type is mapped to itself, that is to say, there is no distinction between reference types and reference sets. A constraint like Focus Economy, on the other hand, requires not only the input language and the candidate language to be disjoint, but also all reference sets and reference types. A natural unification of these two subclasses is available in the form of *output joint preservation.*

Definition 9. *An $\mathcal{F}$-controlled optimality system is* output joint preserving *iff for all distinct $X, Y \in \mathcal{F}$, $X\gamma \cap Y\gamma \neq \emptyset \rightarrow X \cap Y \neq \emptyset$.*

The OS depicted in Fig. 2 on the next page fails output joint preservation. It is clearly violated by $\textsc{Sienna}$ and $\textsc{Purple}$, which are disjoint yet mapped to the same reference set, $\textsc{Brown}$. It isn't respected by $\textsc{Red}$ and $\textsc{Lime}$, either, which are mapped to $\textsc{Blue}$ and $\textsc{Orange}$, respectively, the intersection of which is non-empty even though $\textsc{Red}$ and $\textsc{Lime}$ are disjoint.

Output joint preservation is certainly general enough a property to encompass the kinds of controlled OSs we are interested in. In the following, I show that it is also sufficiently restrictive to establish a link to the crucial notion of global optimality.

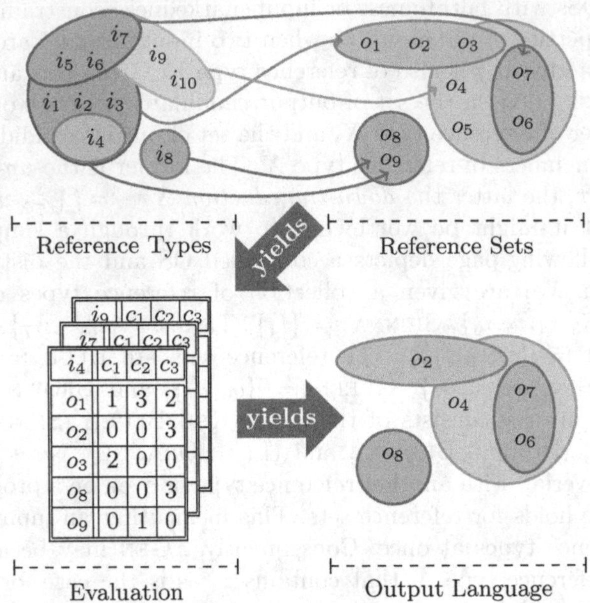

i_4	c_1	c_2	c_3
o_1	1	3	2
o_2	0	0	3
o_3	2	0	0
o_8	0	0	3
o_9	0	1	0

Fig. 2. Example of a controlled OS. Gen is defined in a modular fashion using reference types, reference sets, and the control map γ from reference types to reference sets.

Definition 10. *An $\mathcal{F}$-controlled OS is* type-level optimal *iff* $X\tau^{\uparrow} \upharpoonright X\gamma = X\tau^{\downarrow} \upharpoonright X\gamma$ *for all* $X \in \mathcal{F}$.

Lemma 11. *Let $\mathcal{O}[\mathcal{F}]$ an $\mathcal{F}$-controlled OS. Then $\mathcal{O}[\mathcal{F}]$ is type-level optimal only if it is globally optimal.*

Proof. We prove the contrapositive. If $\mathcal{O}[\mathcal{F}]$ is not type-level optimal, then it holds for some $X \in \mathcal{F}$ that $X\tau^{\uparrow} \upharpoonright X\gamma \neq X\tau^{\downarrow} \upharpoonright X\gamma$. But this implies that there are $x, y \in X$ and $z \in X\gamma$ such that $x\tau \ni z \notin y\tau$, which is an unequivocal violation of global optimality. □

Theorem 12. *Every output joint preserving OS is type-level optimal iff it is globally optimal.*

Proof. The right-to-left direction follows from Lemma 11. We prove the contrapositive of the other direction. If $\mathcal{O}[\mathcal{F}]$ fails global optimality, then there are $x, y \in L$ and $z \in L'$ such that $\langle x, z\rangle, \langle y, z\rangle \in$ Gen yet $x\tau \ni z \notin y\tau$. W.l.o.g. let $x \in X$ and $y \in Y$, $X, Y \in \mathcal{F}$, whence $z \in X\gamma \cap Y\gamma$. As $\mathcal{O}[\mathcal{F}]$ is output joint preserving, $X\gamma \cap Y\gamma \neq \emptyset$ entails $X \cap Y \neq \emptyset$. Pick some $p \in X \cap Y$. Now if $\mathcal{O}[\mathcal{F}]$ is type-level optimal, then it holds that $X\tau^{\uparrow} \upharpoonright X\gamma = X\tau^{\downarrow} \upharpoonright X\gamma$ and $Y\tau^{\uparrow} \upharpoonright Y\gamma = Y\tau^{\downarrow} \upharpoonright Y\gamma$, so $z \in x\tau$ implies $z \in p\tau$, whereas $z \notin y\tau$ implies $z \notin p\tau$. Contradiction. It follows that $\mathcal{O}[\mathcal{F}]$ is not type-level optimal. □

Intuitively, type-level optimality ensures that optimality is fixed for entire reference types, so the individual inputs can be ignored for determining optimality. However, it is too weak a restriction to rule out disagreement between reference types that are mapped to overlapping reference sets, so output joint preservation has to step in; it guarantees that if two reference types X and Y share at least one output candidates, there exists some input p belonging to both X and Y that will be faced by conflicting requirements if X and Y disagree with respect to which candidates in $X\gamma \cap Y\gamma$ they deem optimal (since the OS is type-level optimal, optimality can be specified for entire reference types rather than their members). It should be easy to see that the conditions jointly imply global optimality.

Given our interest in using controlled OS to investigate the computability of reference-set constraints, it would be advantageous if we could read off the constraints right away whether they yield type-level optimality. This is indeed very easy to do thanks to the following entailment.

Lemma 13. *Let* $\mathcal{O}[\mathcal{F}] := \langle \text{Gen}, C, \mathcal{F}, \gamma \rangle$ *an* $\mathcal{F}$*-controlled OS such that every* $c \in C$ *is an output-markedness constraint. Then* $\mathcal{O}[\mathcal{F}]$ *is type-level optimal.*

Proof. Assume the opposite. Then for some $X \in \mathcal{F}$, $X\tau^{\uparrow} \restriction X\gamma \neq X\tau^{\downarrow} \restriction X\gamma$, whence there are $x, y \in X$ and $z \in X\gamma$ with $x\tau \ni z \notin y\tau$. But this is the case only if there is some $c \in C$ such that $c(\langle x, z\rangle) \neq c(\langle y, z\rangle)$, i.e. c isn't an output-markedness constraint. □

Corollary 14. *Let* $\mathcal{O}[\mathcal{F}] := \langle \text{Gen}, C, \mathcal{F}, \gamma \rangle$ *an output joint preserving OS such that every* $c \in C$ *is an output-markedness constraint. Then* $\mathcal{O}[\mathcal{F}]$ *is globally optimal.*

Combining these results, we arrive at the equivalent of Thm. 6 for $\mathcal{F}$-controlled OSs.

Corollary 15. *Let* $\mathcal{O}[\mathcal{F}] := \langle \text{Gen}, C, \mathcal{F}, \gamma \rangle$ *an* $\mathcal{F}$*-controlled OS such that*

- dom(Gen) *is a regular string language, or a regular/linear context-free tree language, and*
- Gen *is a rational relation, and*
- *all* $c \in C$ *are output-markedness constraints, and*
- *each* $c \in C$ *defines a rational relation on* ran(Gen), *and*
- $\mathcal{O}[\mathcal{F}]$ *is output joint preserving.*

Then the transduction τ *induced by the OS is a rational relation and* ran(τ) *belongs to the same formal language class as* dom(τ).

In sum, then, not only do output joint preserving OSs look like a solid base for modeling reference-set constraints, they also have the neat property that the global optimality check is redundant, thanks to Lem. 13. As it is pretty easy to determine for any given reference-set constraint whether it can be modeled by output-markedness constraints alone, the decisive factor in their implementation are the transducers for the constraints and Gen. If those transducers each define a rational relation, so does the entire optimality system.

5 Application to Reference-Set Computation

5.1 Focus Economy Explained

Focus Economy [10] was briefly discussed in the introduction. It is invoked in order to account for the fact that sentences such as (2a), (2b) and (2c) below differ with respect to what is given and what is new information. Once again main stress is marked by **boldface**.

(2) a. My friend bought a red **car**.
b. My friend **bought** a red car.
c. My friend bought a **red** car.

That these utterances are associated to different information structures is witnessed by the fact that for instance only (2a) is compatible with the question "What happened?".

The full-blown Focus Economy system (rather than the simplified sketch given in the introduction) accounts for the data as follows. First, the *Main Stress Rule* demands that in every pair of sister nodes, the "syntactically more embedded" node [10, p.133] is assigned strong stress, its sister weak stress (marked in the phrase structure tree by subscripted S and W, respectively). If a node has no sister, it is always assigned strong stress (in Minimalist syntax, this will be the case only for the root node, as all Minimalist trees are strictly binary branching). Main stress then falls on the unique leaf node that is connected to the root node by a path of nodes that have an S-subscript. See Fig. 3 for an example.

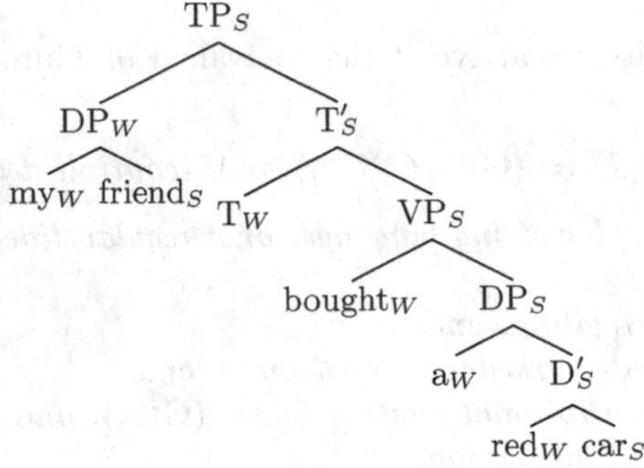

Fig. 3. The stress-annotated phrase structure tree for (2a)

The notion of being syntactically more embedded isn't explicitly defined in the literature. It is stated in passing, though, that "...main stress falls on the most embedded constituent on the recursive side of the tree" [10, p.133]. While this is rather vague, it is presumably meant to convey that, at least for English, in which complements follow the heads they are introduced by, the right sister node is assigned strong stress as long as it isn't an adjunct. This interpretation seems to be in line with the empirical facts.

The second integral part of the proposal is the operation *Stress Shift*, which shifts the main stress to some leaf node n by assigning all nodes on the path from n to the root strong stress and demoting the sisters of these nodes to weakly stressed nodes. For instance, the tree for "**My** friend bought a red car" is obtained from the tree in Fig. 3 by changing my_W and friend_S to my_S and friend_W, respectively, and DP_W and T'_S to DP_S and T'_W, respectively.

While Stress Shift could be invoked to move stress from anaphoric elements to their left sister as in (3), this burden is put on a separate rule, for independent reasons. The rule in question is called *Anaphoric Destressing* and obligatorily assigns weak stress to anaphoric nodes, where a node is anaphoric iff it is "...D[iscourse]-linked to an accessible discourse entity" [10, p.147]. Thus Anaphoric Destressing not only accounts for the unstressed anaphor in (3), but also for the special behavior of stress in cases of parallelism.

(3) John **killed** her.

(4) First Paul bought a red **car**.
  a. Then **John** bought one.
  b. * Then John bought **one**.

The overall system now works as follows. Given a phrase structure tree that has not been annotated for stress yet, one first applies Anaphoric Destressing to make sure that all d-linked constituents are assigned weak stress and thus cannot carry main stress. Next the Main Stress Rule is invoked to assign every node in the tree either W or S. Note that the Main Stress Rule cannot overwrite previously assigned labels, so if some node n has been labeled W by Anaphoric Destressing, the Main Stress Rule has to assign S to the sister of n. Now that the tree is fully annotated, we compute its *focus set*, the set of constituents that may be focused.

(5) *Focus Projection*
Given some stress-annotated tree t, its focus set is the set of nodes reflexively dominating the node carrying main stress.

The focus set of "My friend bought a red **car**", for instance, contains [car], [.D′ red car], [.DP a red car], [.VP bought a red car] and [.TP My friend bought a red car]. For "Then **John** bought one", on the other hand, it consists only of [John] and [.TP Then John bought one].

At this point, Stress Shift may optionally take place. After the main stress has been shifted, however, the focus set has to be computed all over again, and this time the procedure involves reference-set computation.

(6) *Focus Projection Redux*
Given some stress-annotated tree t' that was obtained from tree t by Stress Shift, the focus set of t' contains all the nodes reflexively dominating the node carrying main stress which aren't already contained in the focus set of t.

So if "Then **John** bought one" had been obtained by Stress Shift from [.TP Then John bought one] rather than Anaphoric Destressing, its focus set would have contained only [John], because [.TP Then John bought one] already belongs to the focus set of "Then John bought **one**". As an easy exercise, the reader may want to draw annotated trees for the examples in (2) and compute their focus sets.

5.2 A Model of Focus Economy

After this general overview, the time has come to formalize Focus Economy. In order to precisely model Focus Economy, though, I have to make some simplifying assumptions, for reasons that are entirely independent from the restrictions of OSs. First, I stipulate that adjuncts are explicitly marked as such by a subscript A on their label. This is simply a matter of convenience, as it reduces the complexity of the transducers and makes my model independent from the theoretical status of adjuncts in syntax.

Second, I decided to take movement out of the picture, because the interaction of focus and movement is not touched upon in [10], so there is no original material to formalize. Incidentally, movement seems to introduce several interesting complications (e.g. mandatory main stress for topicalized constituents). The end of this section contains a brief discussion as to whether my model can be extended to capture theories involving movement.

The last simplification concerns Anaphoric Destressing. While the destressing of pronominal (and possibly covert) elements is easy to accommodate, the invoked notion of d-linking is impossible to capture in any model that operates on isolated syntactic trees. Devising a working model of discourse structure vastly exceeds the scope of this contribution. Also, the role of d-linking in anaphoric destressing is of little importance to this paper, which focuses on the reference-set computational aspects of Focus Economy. Thus my implementation will allow almost any constituent to be anaphorically distressed and leave the task of matching trees to appropriate discourse contexts to an external theory of d-linking that remains to be specified.

With these provisions made explicit, the formalization of Focus Economy as a controlled OS can commence. The input language is supposedly derived by some movement-free Minimalist grammar $\mathcal{E}$ for English [12] in which interior nodes are given explicit category labels (once more for the sake of convenience). As Minimalist grammars without remnant movement generate regular tree languages [8], it is safe to assume that Minimalist grammars without any kind of movement do so, too.

Next I define GEN as the composition of four linear transducers corresponding to Anaphoric Distressing, the Main Stress Rule, Stress Shift, and Focus Projection, respectively. Given a tree t derived by $\mathcal{E}$, the transducer cascade computes all logically possible variants of t with respect to stress assignment and then computes the focus in a local way. This means that GEN actually overgenerates with respect to focus, a problem that we have to take care of at a letter step.

Anaphoric Distressing is modeled by a non-deterministic ltdtt that may randomly add a subscript D to a node's label in order to mark it as anaphoric. The only condition is that if a node is labeled as anaphoric, all the nodes it properly dominates must be marked as such, too.

Definition 16. *Let* $\Sigma := \Sigma_L \cup \Sigma_A$ *be the vocabulary of the Minimalist grammar* $\mathcal{E}$ *that generated the input language, where* Σ_L *contains all lexical items and category labels and* Σ_A *their counterparts explicitly labeled as adjuncts.* Anaphoric Destressing *is the ltdtt* $\mathcal{D}$ *where* $\Sigma_{\mathcal{D}} := \Sigma$, $\Omega_{\mathcal{D}}$ *is the union of* Σ *and* $\Sigma_D := \{\sigma_D \mid \sigma \in \Sigma\}$, $Q := \{q_i, q_d\}$, $Q' := \{q_i\}$, *and* $\Delta_{\mathcal{D}}$ *contains the rules below, with* $\sigma \in \Sigma$ *and* $\sigma_D \in \Sigma_D$ *and* $\alpha_{\{x,y\}}$ *to be read as "*α_x *or* α_y*":*

$$q_i(\sigma(x,y)) \to \sigma(q_i(x), q_i(y)) \qquad q_i(\sigma) \to \sigma$$
$$q_{\{i,d\}}(\sigma(x,y)) \to \sigma_D(q_d(x), q_d(y)) \qquad q_{\{i,d\}}(\sigma) \to \sigma_D$$

The transducer for the Main Stress Rule is non-deterministic, too, but it proceeds in a bottom-up manner. It does not alter nodes subscripted by A or D, but if it encounters a leaf node without a subscript, it randomly adds the subscript S or W to its label. However, W is allowed to occur only on left sisters, whereas S is mostly restricted to right sisters and may surface on a left sister just in case the right sister is already marked by A or D. Note that we could easily define a different stress pattern, maybe even parametrized with respect to category labels, to incorporate stress assignment rules from other languages.

Definition 17. Main Stress *is the lbutt* $\mathcal{M}$ *where* $\Sigma_{\mathcal{M}} := \Omega_{\mathcal{D}}$, $\Omega_{\mathcal{M}}$ *is the union of* Σ, Σ_D *and* $\Sigma_* := \{\sigma_S, \sigma_W \mid \sigma \in \Sigma\}$, $Q := \{q_s, q_u, q_w\}$, $Q' := \{q_s\}$ *and* $\Delta_{\mathcal{M}}$ *contains the following rules, with* $\sigma \in \Sigma$, $\sigma_A \in \Sigma_A$, $\sigma_x \in \{\sigma_x \mid \sigma \in \Sigma\}$ *for* $x \in \{D, S, W\}$, *and* $\alpha_{a \ldots z}(\beta_{a' \ldots z'}, \ldots, \zeta_{a'' ,\ldots, z''})$ *to be read as "*$\alpha_a(\beta_{a'}, \ldots, \zeta_{a''})$ *or* $\ldots$ *or* $\alpha_z(\beta_{z'}, \ldots, \zeta_{z''})$*":*

$$\sigma_A \to q_u(\sigma_A) \qquad \sigma_A(q_u(x), q_u(y)) \to q_u(\sigma_A(x,y))$$
$$\sigma_D \to q_u(\sigma_D) \qquad \sigma_D(q_u(x), q_u(y)) \to q_u(\sigma_D(x,y))$$
$$\sigma \to q_{sw}(\sigma_{SW}) \qquad \sigma(q_{\{u,w\}}(x), q_s(y)) \to q_{sw}(\sigma_{SW}(x,y))$$
$$\sigma(q_s(x), q_u(y)) \to q_{sw}(\sigma_{SW}(x,y))$$

Stress Shift is best implemented as a non-deterministic ltdtt that may randomly switch the subscripts of two S/W-annotated sisters.

Definition 18. Stress Shift *is the ltdtt* $\mathcal{S}$ *where* $\Sigma_{\mathcal{S}} = \Omega_{\mathcal{S}} = \Omega_{\mathcal{M}}$, $Q := \{q_i, q_s, q_w\}$, $Q' := \{q_s\}$, *and* $\Delta_{\mathcal{S}}$ *contains the rules below, with* $\sigma \in \Sigma_{\mathcal{S}}$ *and* $\sigma_* \in \Sigma_*$*:*

$$q_s(\sigma_*(x,y)) \to \sigma_{SSS}(q_{isw}(x), q_{iws}(y)) \qquad q_s(\sigma_*) \to \sigma_S$$
$$q_w(\sigma_*(x,y)) \to \sigma_W(q_i(x), q_i(y)) \qquad q_w(\sigma_*) \to \sigma_W$$
$$q_i(\sigma(x,y)) \to \sigma(q_i(x), q_i(y)) \qquad q_i(\sigma) \to \sigma$$

The last component is Focus Projection, a non-deterministic ltdtt with two states, q_f and q_g. The transducer starts at the root in q_f. Whenever a node

n is subscripted by W, the transducer switches into q_g at this node and stays in the state for all nodes dominated by n. As long as the transducer is in q_f, it may randomly add a superscript F to a label to indicate that it is focused. Right afterward, it changes into q_g and never leaves this state again. Rather than associating a stress-annotated tree with a set of constituents that can be focused, Focus Projection now generates multiple trees that differ only with respect to which constituent along the path of S-labeled nodes is focus-marked.

Definition 19. Focus Projection *is the ltdtt* $\mathcal{F}$, *where* $\Sigma_{\mathcal{F}} = \Omega_S$, $\Omega_{\mathcal{F}}$ *is the union of* Ω_S *and* $\Omega_S^F := \{\omega^F \mid \omega \in \Omega_S\}$, $Q := \{q_f, q_g\}$, $Q' := \{q_f\}$, *and* $\Delta_{\mathcal{F}}$ *contains the rules below, with* $\sigma \in \Sigma_{\mathcal{F}}$ *and* $\sigma_{\overline{S}} \in \Sigma_{\mathcal{F}} \setminus \{\sigma_S \mid \sigma \in \Sigma\}$*:*

$$
\begin{aligned}
q_f(\sigma_S(x,y)) &\to \sigma_S(q_f(x), q_f(y)) & & \\
q_f(\sigma_S(x,y)) &\to \sigma_S^F(q_g(x), q_g(x)) & q_f(\sigma_S) &\to \sigma_S^F \\
q_f(\sigma_{\overline{S}}(x,y)) &\to \sigma_{\overline{S}}(q_g(x), q_g(x)) & q_f(\sigma_{\overline{S}}) &\to \sigma_{\overline{S}} \\
q_g(\sigma(x,y)) &\to \sigma(q_g(x), q_g(y)) & q_g(\sigma) &\to \sigma
\end{aligned}
$$

All four transducers are linear, whence they can be composed into a single linear transducer modeling GEN. Expanding on what was said above about the inner workings of GEN, we now see that for any tree t in the input language, tGEN is the set of stress-annotated trees in which, first, some subtrees may be marked as adjuncts or anaphorical material (or both) and thus do not carry stress information, second, there is exactly one path from the root to some leaf such that every node in the path is labeled by S, and third, exactly one node belonging to this path is marked as focused. The reader should have no problem verifying that in terms of controlled OSs, all reference types are singleton and their reference-sets do not overlap, i.e. output joint preservation and type-level optimality are satisfied.

Now it only remains for us to implement *Focus Projection Redux*. In the original account, Focus Projection Redux applied directly to the output of Stress Shift, i.e. trees without focus information, and the task at hand was to assign the correct focus. In my system, on the other hand, every tree is fed into Focus Projection and marked accordingly for focus. This leads to overgeneration for trees in which Stress Shift has taken place — a node may carry focus even if it could also do so in the tree without shifted main stress. Consequently, the focus set of "**John** died", for instance, turns out to contain both [John] and [.TP John died] rather than just the former. Under my proposal, then, Focus Projection Redux is faced with the burden of filtering out focus information instead of assigning it. In other words, Focus Projection Redux is a constraint. This is accomplished by defining a regular tree language L_c such that when GEN is composed with the diagonal of L_c (which is guaranteed to be a linear transduction), only trees with licit focus marking are preserved; said regular language is easily specified in the monadic second-order logic $\mathcal{L}^2_{K,P}$ [11].

First one defines two predicates, *StressPath* and *FocusPath*. The former picks out the path from the root to the leaf carrying main stress, whereas the latter refers to the path from the root to the leaf that would carry main stress in the absence of stress shift. This implies that *FocusPath* replicates some of the information that is already encoded in the Main Stress transducer. Note that in the definitions below, $A(x)$, $D(x)$ and $S(x)$ are predicates picking out all nodes with subscript A, D, S, respectively, $x \triangleleft y$ denotes "x is the parent of y", $x \prec y$ "x is the left sibling of y", and $\triangleleft^*$ the reflexive transitive closure of $\triangleleft$.

$$\text{Path}(X) \leftrightarrow \exists x\Big[X(x) \wedge \neg\exists y[y \triangleleft x]\Big] \wedge \exists! x\Big[X(x) \wedge \neg\exists y[x \triangleleft y]\Big] \wedge$$

$$\forall x, y, z\Big[\big(X(x) \wedge X(y) \rightarrow x \triangleleft^* y \vee y \triangleleft^* x\big) \wedge \big(X(x) \wedge \neg X(z) \rightarrow \neg(z \triangleleft^* x)\big)\Big]$$

$$\text{StressPath}(X) \leftrightarrow \text{Path}(X) \wedge \forall x[X(x) \rightarrow S(x)]$$

$$\text{FocusPath}(X) \leftrightarrow \text{Path}(X) \wedge \forall x, y, z\Big[X(x) \wedge x \triangleleft y \wedge x \triangleleft z \rightarrow \big((A(y) \vee D(y)) \rightarrow X(z)\big) \wedge \big(\neg A(y) \wedge \neg D(y) \wedge y \prec z \rightarrow X(z)\big)\Big]$$

In a tree where no stress shift has taken place, StressPath and FocusPath are true of the same subsets and any node contained by them may be focused. After an application of the Stress Shift rule, however, the two paths are no longer identical, although their intersection is never empty (it has to contain at least the root node). In this case, then, the only valid targets for focus are those nodes of the StressPath that are not contained in the FocusPath. This is formally expressed by the $\mathcal{L}^2_{K,P}$ sentence ϕ below. Just like $A(x)$, $D(x)$ and $S(x)$ before, $F(x)$ is a predicate defining a particular set of nodes, this time the set of nodes labeled by some $\omega^F \in \Omega^F_{\mathcal{S}}$. I furthermore use $X \approx Y$ as a shorthand for $\forall x[X(x) \leftrightarrow Y(x)]$.

$$\phi := \forall x, X, Y[F(x) \wedge X(x) \wedge \text{StressPath}(X) \wedge \text{FocusPath}(Y) \rightarrow (Y(x) \rightarrow X \approx Y)]$$

Note that ϕ by itself does not properly restrict the distribution of focus. First of all, there is no requirement that exactly one node must be focused. Second, nodes outside StressPath may carry focus, in which case no restrictions apply to them at all. Finally, StressPath and FocusPath may be empty, because we have not made any assumptions about the distribution of labels. Crucially, though, ϕ behaves as expected over the trees in the candidate language. Thus taking the diagonal of the language licensed by ϕ and composing it with GEN filters out all illicit foci, and only those. Since the diagonal over a regular language is a linear transduction, the transduction obtained by the composition is too. This establishes the computational feasibility of Focus Economy when the input language is a regular tree language.

So far I have left open the question, though, how movement fits into the picture. First of all, it cannot be ruled out *a priori* that the interaction of movement and focus are so intricate on a linguistic level that significant modifications have to be made to the original version of Focus Economy. On a formal level, this would mean that the transduction itself would have to be changed. In this case, it makes little sense to speculate how my model could be extended to accommodate movement, so let us instead assume that Focus Economy can remain virtually unaltered and it is only the input language that has to be modified. In my model, the input language is a regular tree language by virtue of being generated by a Minimalist grammar without movement. But note that Minimalist grammars with movement generate regular tree languages, too, in the presence of a ban against more exotic kinds of movement such as remnant movement or head movement [8]. Thus the restriction to regular tree languages itself does not preclude us from accommodating most instances of movement.[1]

6 Conclusion

I have shown that despite claims to the contrary, reference-set constraints aren't computationally intractable. Controlled OSs were introduced as a formal model for reference-set constraints. I gave a new characterization of global optimality for the subclass of output joint preserving OSs, which is general enough to accommodate most reference-set constraints. The shift in perspective induced by controlled OSs made it apparent that out of the five conditions that jointly guarantee for an OS to stay within the limits of linear tree transductions, three are almost trivially satisfied by reference-set constraints, with the only problematic areas being the power of GEN and the rankings induced by the constraints on the range of GEN. A model of a prominent reference-set constraint, Focus Economy, showed that at least for some constraints those point aren't problematic, either. These new results suggest that reference-set constraints are significantly better behaved than is usually believed.

Acknowledgements. I am greatly indebted to Ed Stabler, Ed Keenan and Uwe Mönnich as well as the two anonymous reviewers for their motivational comments and helpful criticism. The research reported herein was supported by a DOC-fellowship of the Austrian Academy of Sciences.

References

[1] Chomsky, N.: The Minimalist Program. MIT Press, Cambridge (1995)
[2] Frank, R., Satta, G.: Optimality theory and the generative complexity of constraint violability. Computational Linguistics 24, 307–315 (1998)
[3] Gécseg, F., Steinby, M.: Tree Automata. Academei Kaido, Budapest (1984)
[4] Johnson, D., Lappin, S.: Local Constraints vs. Economy. CSLI, Stanford (1999)

[1] If we want the full expressive power of Minimalist grammars, then the best strategy is to express Focus Economy as a constraint over derivation trees, since for every Minimalist grammar the set of derivation trees it licenses forms a regular language that fully determines the tree yield of the grammar [9]. The only difference between Minimalist derivation trees and movement-free phrase structure trees as derived above is that the latter are unordered. Hence, if we require that linear order (which can be easily determined from the labels of the leaves) is directly reflected in the derivation trees, the formalization above carries over unaltered to derivation trees and may be extended as desired to deal with instances of movement.

[5] Jäger, G.: Gradient constraints in finite state OT: The unidirectional and the bidirectional case. In: Kaufmann, I., Stiebels, B. (eds.) More than Words. A Festschrift for Dieter Wunderlich, pp. 299–325. Akademie Verlag, Berlin (2002)
[6] Karttunen, L.: The proper treatment of optimality in computational phonology (1998); manuscript, Xerox Research Center Europe
[7] Kepser, S., Mönnich, U.: Closure properties of linear context-free tree languages with an application to optimality theory. Theoretical Computer Science 354, 82–97 (2006)
[8] Kobele, G.M.: Without Remnant Movement, mGs are Context-Free. In: Ebert, C., Jäger, G., Michaelis, J. (eds.) MOL 10. LNCS, vol. 6149, pp. 160–173. Springer, Heidelberg (2010)
[9] Kobele, G.M., Retoré, C., Salvati, S.: An automata-theoretic approach to minimalism. In: Rogers, J., Kepser, S. (eds.) Model Theoretic Syntax at 10, pp. 71–80 (2007); Workshop Organized as Part of the Europen Summer School on Logic, Language and Information (ESSLLI 2007), Dublin, Ireland, August 6-17 (2007)
[10] Reinhart, T.: Interface Strategies: Optimal and Costly Computations. MIT Press, Cambridge (2006)
[11] Rogers, J.: A Descriptive Approach to Language-Theoretic Complexity. CSLI, Stanford (1998)
[12] Stabler, E.P., Keenan, E.: Structural similarity. Theoretical Computer Science 293, 345–363 (2003)
[13] Wartena, C.: A note on the complexity of optimality systems. In: Blutner, R., Jäger, G. (eds.) Studies in Optimality Theory, pp. 64–72. University of Potsdam, Potsdam (2000)

Hyperintensional Dynamic Semantics*
Analyzing Definiteness with Enriched Contexts

Scott Martin and Carl Pollard

Department of Linguistics
Ohio State University Columbus, OH 43210 USA
{scott,pollard}@ling.ohio-state.edu

Abstract. We present a dynamic semantic theory formalized in higher order logic that synthesizes aspects of de Groote's continuation-based dynamics and Pollard's hyperintensional semantics. In this theory, we rely on an enriched notion of discourse context inspired by the work of Heim and Roberts. We show how to use this enriched context to improve on de Groote's treatment of English definite anaphora by modeling it as presupposition fulfillment.

Keywords: discourse, context, presupposition, definite anaphora, higher order logic.

1 Introduction

As Muskens [14,15] showed, many of the insights of dynamic semantic theories such as Kamp's discourse representation theory (DRT, [9,10]) and Heim's file change semantics (FCS, [6,7]) can be formalized within the well-understood framework of classical higher order logic (HOL) of Church [1], Henkin [8], and Gallin [2] that is familiar to Montague semanticists. Also working within HOL, de Groote [4] showed that the description of context update could be streamlined by modeling right contexts by analogy with the *continuations* employed in programming language semantics [27].

Though both Muskens' and de Groote's work are positive developments in the sense of helping to integrate dynamic notions into mainstream semantic theory, both fall short in modeling how definite anaphora works in discourse, which was one of the two central problems that Kamp and Heim originally set out to solve. (The other was to characterize the novelty of indefinite descriptions.) For Muskens, definite pronouns are simply ambiguous with respect to which accessible and sortally appropriate discourse referent they 'pick up'. The trouble with this theory is that, empirically, definite anaphora is generally *not* ambiguous; if it were, it would fail to serve its communicative purpose.

* For advice and clarifying discussion, we are especially grateful to Craige Roberts, E. Allyn Smith, Michael White, to our comrades in the Ohio State Logic, Language, Information and Computation discussion group, and to the participants of the CAuLD Workshop on Logical Methods for Discourse.

P. de Groote and M.-J. Nederhof (Eds.): Formal Grammar 2010/2011, LNCS 7395, pp. 114–129, 2012.

De Groote fares no better. On his account, the antecedent of a pronoun is picked out by an oracular choice function `sel` from among the sortally correct candidate entities (de Groote does not have discourse referents as distinct from entities in his theory). At first one might think the weakness of this account is that it doesn't tell us anything about *what* choice function this oracle actually is. But in fact it is worse than that, because it is easy to show that *no* choice function is the right one. That is because, in general, the antecedent of a definite pronoun fails to be uniquely determined by the set of sortally appropriate candidate entities. Consider, e.g., the following two narratives:

(A) 1. A donkey and a mule walked in.
2. The donkey was sad.
3. It brayed.

(B) 1. A donkey and a mule walked in.
2. The mule was sad.
3. It brayed.

In both (A) and (B), the set of candidate antecedents consists of two nonhuman entities, a donkey and a mule. In each narrative, one of the candidates has been rendered more salient by virtue of having been re-invoked by a definite description. And in each narrative, the anaphora resolves to the more salient, 'definitized' entity. But the choice function only 'knows about' the members of the set of candidates and their sortal properties, not about their relative salience in the discourse at hand. So it cannot pick the right antecedent both times.

In our view, the weaknesses of Muskens' and de Groote's theories arise because they fail to build in a notion of context that is sufficiently rich to support a satisfactory account of *presuppositions*, the conditions on contexts that must be satisfied in order for utterances to be felicitous. In this paper, we suggest a revision and extension of de Groote's theory that copes with definite anaphora by building in a (slightly) more articulated discourse model inspired by proposals due to Craige Roberts [20]. However, the work reported in this paper is part of a larger research program, joint with Roberts and E. Allyn Smith, aimed at constructing formally explicit, categorially-based, natural language grammars that deal effectively with projective aspects of meaning [22]. Besides Roberts' work on modeling discourse contexts, this research program also builds on Pollard's [19,18] hyperintensional semantics (relevant aspects of which are sketched below), and on 'pheno-tecto grammar' (PTG; [17,3,16,24]) the line of development in categorial grammar that distinguishes concrete syntax from combinatorics (not touched upon in this paper).

The rest of this paper is organized as follows. We present some facts about English definiteness presuppositions in Sect. 2. In Sect. 3, we formally lay out our hyperintensional dynamic semantic theory of discourse. Specific machinery for dealing with definiteness is introduced in Sect. 4 that handles some of the cases introduced in Sect. 2. Section 5 concludes and promises future work.

2 Facts about Definiteness Presuppositions in English

We take discourse contexts to be comprised of information that is mutual to the interlocutors participating in a discourse, made available either linguistically (i.e. what has been previously uttered by the interlocutors) or non-linguistically (i.e. sense data and world knowledge). Our view of presuppositions is influenced by the work of Stalnaker [25] and later Heim [6,7] in which the presuppositions of a sentence are taken to be those conditions that must be met by any discourse context in which it is to be felicitously interpreted. For Stalnaker, these conditions are based on what the interlocutors are able to infer from the context. Similarly, Heim formally expresses this notion by modeling discourse contexts as the conjunction of the propositions asserted by the interlocutors, and dynamic meanings as partial functions from contexts to contexts. The domains of these partial functions are determined by the presuppositions of the sentences they interpret.

A wide range of presuppositional phenomena has been discussed in the philosophical and linguistic literature, including (to name just a few) factivity (i.e. verbs like *suck*, *know* and *realize* that take sentences as their complements), presupposition 'projection' and 'cancellation', and the presuppositions associated with definite descriptions (i.e. expressions such as proper names, the English definite article *the*, and pronouns such as *it* that presuppose that a suitable anaphoric antecedent is available in the discourse context in which they occur [21]). Here, we consider only presuppositions of **definiteness**, which encompasses—in senses to be made precise—familiarity and unique greatest salience. To take the most simple example, if

(C) # It brayed

is uttered out of the blue and in the absence of some perceptible nonhuman entity in the immediate surroundings that might plausibly have brayed, the presupposition of the pronoun is not fulfilled and so the utterance is infelicitous. Such a familiarity presupposition need not be globally satisfied, but instead can be satisfied locally (roughly: within the scope of one or more operators) as in

(D) No donkey denies it brays.

For this reason the familiarity presuppositions associated with definite anaphora are usually taken to have to do with the availability not of entities *per se* but rather of "discourse referents", a notion which the ambient theory must make precise.

Definites also presuppose more than just the familiarity of the intended discourse referent:

(E) 1. I saw the donkey.
2. What donkey?
3. # Oh, just some donkey out in a field on the way to Upper Sandusky.

Even if the utterances in (E) are situated within a discourse where donkeys have been mentioned or in the presence of one or more donkeys that could serve as resolution targets for *the donkey*, the use of the definite article is infelicitous unless there is one that is uniquely most salient among all the others in the discourse context.

Another dimension to uniqueness is related to the fact that both *it* and *the* also presuppose as their respective anaphoric antecedents a discourse referent that meets certain sortal restrictions. The discourses in (F) and (G) demonstrate these presuppositions:

(F) 1. A donkey had a red blanket.
2. A mule had a blue blanket.
3. $\left\{\begin{array}{l}\text{The donkey}\\ \#\ \text{It}\end{array}\right\}$ snorted.

(G) 1. A donkey had a red blanket.
2. Another donkey had a blue blanket.
3. $\left\{\begin{array}{l}\text{The donkey with the blue blanket}\\ \#\ \text{The donkey}\\ \#\ \text{It}\end{array}\right\}$ snorted.

In (F), salience alone is not enough to decide whether *a donkey* or *a mule* antecedes the pronoun *it*. The identity of the antecedent must also be uniquely determinable from the discourse context, but since *it* can be anteceded by any nonhuman entity in English, the noun phrase *the donkey* is used instead to unambiguously single out the pronoun's unique antecedent. The discourse in (G) is a variant of (F) where the property of being a donkey is not enough to disambiguate the antecedent because of the uniqueness presuppositions associated with *the* and *it*. These examples show that sortal restrictions, even potentially complex ones, play a role in determining the unique greatest salience of DRs when they are considered as targets for anaphora resolution.

Finally, the absurd discourse in (H) shows that the definite article identifies not just a familiar and uniquely most salient individual, but one with a certain specified property:

(H) 1. I saw the donkey.
2. What donkey?
3. # That llama we always see on the way to Findlay.

Here, it is infelicitous to use the noun phrase *the donkey* to identify the llama even if it is the most salient individual in the utterance context because it does not have the property of being a donkey.

3 Hyperintensional Dynamic Semantics

Our formalization of discourse dynamics builds on the hyperintensional theory of (static) meaning given by Pollard [19,18]. Like Montague semantics [13], this semantic theory is couched in HOL and has a basic type e for entities as well as the truth-value type t provided by the underlying logic.[1]

But unlike Montague semantics, we follow Thomason [28] in assuming a basic type p for (static) propositions (but no basic type for worlds).[2] Following Lambek and Scott [11], we also assume (1) a natural number type ω; (2) the type constructors U (unit type) and $\times$ (cartesian product) in addition to the usual $\rightarrow$ (exponential); and (3) separation-style subtyping.[3] Subtypes are usually written in the form $\{x \in T \mid \varphi[x]\}$ where $\varphi[x]$ is a formula (boolean term) possibly with x free. Additionally, we make use of dependent coproduct types parameterized by the natural number type, written $\coprod_n T_n$.

From the typed lambda calculus that underlies the HOL, we have the usual pairing and projection functions; applications are written $(f\,a)$ rather than $f(a)$. Successive applications associate to the left; e.g $(f\ a\ b)$ abbreviates $((f\ a)\ b)$.

The type of propositions is axiomatized as a preboolean algebra (like a boolean algebra, but without antisymmetry), preordered by the entailment relation $\mathsf{entails} : \mathrm{p} \rightarrow \mathrm{p} \rightarrow \mathrm{t}$. The propositional connectives and quantifiers are written as boldface versions of the usual (boolean) connectives of the underlying logic: i.e. $\boldsymbol{\neg}$, $\boldsymbol{\wedge}$, $\boldsymbol{\vee}$, $\boldsymbol{\rightarrow}$, $\boldsymbol{\exists}$, and $\boldsymbol{\forall}$; true denotes a greatest element relative to the entailment preorder (a necessary truth).

Following Heim, we use natural numbers (type ω) as discourse referents (hereafter, DRs). The type ω is equipped with the usual linear order $<$ and the successor function $\mathbf{suc} : \omega \rightarrow \omega$. Additionally, for each natural number n, we define the type of the first n natural numbers as a subtype of ω:

$$\omega_n =_{\text{def}} \{i \in \omega \mid i < n\}$$

These types will be used for the domains of anchors (functions from DRs to entities; we eschew the term 'assignment function' commonly used in the linguistic literature because arguments of these functions are not object-language variables).

We adopt the convention that applications and pairings associate to the left and abstractions associate to the right. Parentheses are sometimes abbreviated using . in the usual way (e.g., $\lambda_x.M$) or omitted altogether when no confusion can arise. When a term contains multiple embedded λ-abstractions of the form $\lambda_a\lambda_b\lambda_c M$, we collapse them together as $\lambda_{abc}M$.

---

[1] For expository simplicity, we depart from Pollard in not distinguishing between the extensional type e and the corresponding hyperintensional type i (individual concepts).

[2] Using separation subtyping, we can define the type of worlds as a certain subtype of the type $\mathrm{p} \rightarrow \mathrm{t}$ (sets of propositions), but this will not be needed here.

[3] Thus if A is a type and a an A-predicate (closed term of type $A \rightarrow \mathrm{t}$), then there is a type A_a interpreted as the subset of the interpretation of A that has the interpretation of a as its characteristic function; and there is a constant μ_a that denotes the subset embedding.

3.1 Information Structures

To advance from static to dynamic semantics, we need to extend our ontology to model contexts. Our notion of context is a simplified version of Roberts' *discourse information structures* [20], which are in turn inspired by the work of Lewis [12], here called simply *structures*.[4] A structure is a tuple consisting of (1) an *anchor* of entities to a set of DR's, (2) a salience preorder on those DR's called the *resolution* preorder (so-called because it will be used to resolve definite anaphora), and (3) a proposition, the *common ground*, which (following Stalnaker [26]) is the conjunction of all the propositions that are taken by the interlocutors to be mutually agreed upon. The common ground includes not only propositions explicitly asserted and accepted in the discourse, but also encyclopedic knowledge about the world that is assumed as shared background.

To make this notion of structure more precise, we begin by defining the type of n-ary anchors α_n to be the type of functions from the first n discourse referents to entities:

$$\alpha_n =_{\text{def}} \omega_n \rightarrow \mathrm{e}$$
$$\alpha =_{\text{def}} \coprod_n \alpha_n$$

An n-ary anchor can be extended to include a new DR mapped to a specified entity using the function $\bullet_n : \alpha_n \rightarrow \mathrm{e} \rightarrow \alpha_{(\mathbf{suc}\ n)}$ (written infix), that is subject to the axiom schema

$$\vdash \forall_{n:\omega} \forall_{a:\alpha_n} \forall_{x:\mathrm{e}} \forall_{m:\omega_{(\mathbf{suc}\ n)}} .(a \bullet_n x)\, m = \begin{cases} x & \text{if } m = n \\ (a\, m) & \text{otherwise} \end{cases}$$

To track the relative salience of the DRs in the domain ω_n of an anchor, we use a preorder on ω_n. For arbitrary $n : \omega$, an n-ary *resolution* is just a preorder (reflexive, transitive relation) on the set of DRs:

$$\rho_n =_{\text{def}} \{r \in \omega_n \rightarrow \omega_n \rightarrow \mathrm{t} \mid (\mathsf{preorder}_n\ r)\}$$

Note that this is a subtype of the type of binary relations on ω_n; here $(\mathsf{preorder}_n r)$ is a formula which says of the binary relation r on ω_n that it *is* a preorder.

$$\vdash \forall_{n:\omega} \forall_{r:\omega_n \rightarrow \omega_n \rightarrow \mathrm{t}} .(\mathsf{preorder}_n\ r) = \forall_{i,j,k:\omega_n} .(i\, r\, i \wedge ((i\, r\, j \wedge j\, r\, k) \rightarrow i\, r\, k))$$

Below, we will see that DRs which are "higher" in the resolution preorder are "better" candidates for subsequent definite anaphora.

The function $\star_n : \rho_n \rightarrow \rho_{(\mathbf{suc}\ n)}$ is used to extend a resolution to the next larger domain, subject to the following schemata:

$$\vdash \forall_{n:\omega} \forall_{r:\rho_n} .n\, (\star_n\, r)\, n$$
$$\vdash \forall_{n:\omega} \forall_{r:\rho_n} \forall_{l,m:\omega_n} .(\neg(m\, (\star_n\, r)\, n)) \wedge (l\, (\star_n\, r)\, m = l\, r\, m)$$

[4] At this stage, we omit Roberts' moves, domain goals, QUD stack, etc.

Resolution extension thus occurs in a noncommittal way: for a resolution $r : \rho_n$, the extended resolution $(\star_n\ r)$ has n reflexively as high as itself, but leaves n incomparable to every $m < n$.

Information structures combine an n-ary anchor and resolution with a proposition, the common ground of the discourse. The type σ_n, mnemonic for *structure*, is a triple defined as:

$$\sigma_n =_{\text{def}} \alpha_n \times \rho_n \times \mathrm{p}$$
$$\sigma =_{\text{def}} \coprod_n \sigma_n$$

The type σ combines all the types σ_n together as a single type. The type σ plays a role analogous to that of γ (left contexts) in de Groote's type-theoretic dynamics [4], but enriches his notion of left context to include salience and a common ground in addition to a set of DRs.[5]

The function $\mathbf{next}_n : \sigma_n \to \omega$ gives the length (size of the domain) of the anchor of an n-ary structure:

$$\vdash \forall_{n:\omega} \forall_{s:\sigma_n} . (\mathbf{next}_n\ s) = n$$

In dynamic interpretations, the size of an anchor's domain is used as the "next" DR.

The functions $\mathbf{a} : \sigma \to \alpha$ (for *anchor*), $\mathbf{r} : \sigma \to \rho$ (for *resolution*) and $\mathbf{c} : \sigma \to \mathrm{p}$ (for *common ground*) are just the projections from σ to its three components. As an abbreviation, we write $[n]_s$ to denote the entity $(\mathbf{a}\ s\ n)$ that is the image of the DR n under the anchor of the structure s. When no confusion can arise, we usually drop the subscript s and write simply $[n]$.

To extend a structure with a new entity (i.e., introduce a new discourse referent and anchor it to a certain entity), we use the function $::_n : \sigma_n \to \mathrm{e} \to \sigma_{(\mathbf{suc}\ n)}$:

$$\vdash \forall_{n:\omega} .\ ::_n = \lambda_{sx} \langle (\mathbf{a}\ s) \bullet_n x, \star_n\ (\mathbf{r}\ s), (\mathbf{c}\ s) \rangle$$

This enriched replacement for de Groote's :: extends both a structure's anchor and its resolution, whereas de Groote's version just adds an entity to an existing set of entities. The function $+ : \sigma \to \mathrm{p} \to \sigma$ adds the ability to update the common ground of a structure with a new proposition:

$$\vdash + = \lambda_{sp} \langle (\mathbf{a}\ s), (\mathbf{r}\ s), (\mathbf{c}\ s) \wedge p \rangle$$

where $\wedge$ is conjunction or propositions, not truth values. Together, these two functions share the work of adding a DR to a context (::) and mutually-accepted information about a context's DRs (+). They play a central role in the definitions of the dynamic existential quantifier EXISTS and the dynamic counterparts of static propositions, discussed below. We often omit parentheses around applications involving :: and + since they both associate to the left.

[5] Actually, de Groote's theory has no DRs as distinct from entities, but we have been unable to see how to manage without such a distinction.

3.2 Continuations and Dynamic Semantics

The type κ of continuations is the type of functions from structures to propositions:

$$\kappa =_{\mathrm{def}} \sigma \to \mathrm{p}$$

Modulo replacement of de Groote's γ (left contexts) and o (truth values) by σ and p respectively, our continuations are direct analogs of his right contexts $(\gamma \to o)$. The **null continuation** is $\lambda_s \mathsf{true}$.

We use the following to recursively notate n-ary static properties, for each $n : \omega$:

$$\mathrm{r}_0 =_{\mathrm{def}} \mathrm{p}$$
$$\mathrm{r}_{(\mathbf{suc}\ n)} =_{\mathrm{def}} \mathrm{e} \to \mathrm{r}_n$$

Note that nullary static properties are simply (static) propositions. A **dynamic proposition** (type π), also known as an **update**, maps a structure and a continuation to a (static) proposition:

$$\pi =_{\mathrm{def}} \sigma \to \kappa \to \mathrm{p}$$

This is a direct analog of de Groote's type Ω. Extending Muskens [14,15], we define the type of n-ary **dynamic relations** in an way analogous to static properties:

$$\delta_0 =_{\mathrm{def}} \pi$$
$$\delta_{(\mathbf{suc}\ n)} =_{\mathrm{def}} \omega \to \delta_n$$

In terms of their types, the difference between static and dynamic properties is that the base type for dynamic properties is π rather than p, and the type of arguments to dynamic properties is ω (of DRs) rather than e (of entities). Note that nullary dynamic properties are dynamic propositions. We abbreviate δ_1, the type of unary dynamic properties, as simply δ.

The **dynamicizer** functions **dyn** take an n-ary static property to its dynamic counterpart:

$$\vdash \mathbf{dyn}_0 = \lambda_{psk}.p \wedge (k\,(s + p)) : \mathrm{r}_0 \to \delta_0$$
$$\vdash \forall_{n:\omega}.\mathbf{dyn}_{(\mathbf{suc}\ n)} = \lambda_{Rn}.\mathbf{dyn}_n\,(R\,[n]) : \mathrm{r}_{(\mathbf{suc}\ n)} \to \delta_{(\mathbf{suc}\ n)}$$

This definitions of the **dyn** functions reflect the fact that utterances in natural language (modeled in our theory by dynamic propositions) update the discourse context with their content. As a mnemonic, we abbreviate the dynamicization of a static property by the same name as its static counterpart except that the name is written in smallcaps. For instance:

$$\vdash \textsc{rain} = (\mathbf{dyn}_0\ \mathsf{rain}) = \lambda_{sk}.\mathsf{rain} \wedge (k\,(s + \mathsf{rain}))$$
$$\vdash \textsc{snow} = (\mathbf{dyn}_0\ \mathsf{snow}) = \lambda_{sk}.\mathsf{snow} \wedge (k\,(s + \mathsf{snow}))$$
$$\vdash \textsc{donkey} = (\mathbf{dyn}_1\ \mathsf{donkey}) = \lambda_{nsk}.(\mathsf{donkey}\,[n]) \wedge (k\,(s + (\mathsf{donkey}\,[n])))$$
$$\vdash \textsc{bray} = (\mathbf{dyn}_1\ \mathsf{bray}) = \lambda_{nsk}.(\mathsf{bray}\,[n]) \wedge (k\,(s + (\mathsf{bray}\,[n])))$$
$$\vdash \textsc{own} = (\mathbf{dyn}_2\ \mathsf{own}) = \lambda_{nmsk}.(\mathsf{own}\,[n]\,[m]) \wedge (k\,(s + (\mathsf{own}\,[n]\,[m])))$$

These examples show how dynamic propositions and properties interact with the structure of the utterance context they are situated inside. The common ground is always updated by a dynamic meaning via + with the proffered content (cf. [20]) and passed to the rest of the discourse (in the form of the continuation k). In the case of the dynamic properties DONKEY, BRAY and OWN, these expect an argument that is not an entity but instead a natural number (i.e., DR) which is mapped to an entity by the anchor ($\mathbf{a}\ s$).

The static propositional content of a dynamic proposition in context is retrieved using the **staticizer** function $\mathbf{stat} : \sigma \to \pi \to \mathrm{p}$, a direct analog of de Groote's READ [5]:

$$\vdash \mathbf{stat} = \lambda_{sA}.A\ s\ \lambda_s \mathsf{true}$$

This function gives the dynamic proposition A access to the context s, but then "throws away" the rest of the discourse by specifying the null continuation as its κ-type argument. For example, assuming that the context $s : \sigma$ is such that $[n]_s = x$ for some $n : \omega$, we calculate the static content of ($\text{DONKEY}\, n$) as follows:

$$\begin{aligned}
\mathbf{stat}\ s\ (\text{DONKEY}\ n) &= \mathbf{stat}\ s\ \lambda_{sk}.(\mathsf{donkey}\ [n]) \wedge (k\ (s + (\mathsf{donkey}\ [n]))) \\
&= \mathbf{stat}\ s\ \lambda_{sk}.(\mathsf{donkey}\ x) \wedge (k\ (s + (\mathsf{donkey}\ x))) \\
&= \lambda_{sk}((\mathsf{donkey}\ x) \wedge (k\ (s + (\mathsf{donkey}\ x))))\ s\ \lambda_s \mathsf{true} \\
&= \lambda_{k}((\mathsf{donkey}\ x) \wedge (k\ (s + (\mathsf{donkey}\ x))))\ \lambda_s \mathsf{true} \\
&= (\mathsf{donkey}\ x) \wedge (\lambda_s \mathsf{true}\ (s + (\mathsf{donkey}\ x))) \\
&= (\mathsf{donkey}\ x) \wedge \mathsf{true} \\
&\equiv \mathsf{donkey}\ x
\end{aligned}$$

where $\equiv$ denotes propositional equivalence (mutual entailment). This example shows why **stat** is not defined for every structure s and dynamic proposition A, only those where A can be resolved to a static proposition based on the contents of s. Here, DONKEY must accesses the structure passed to **stat** to determine the entity that its anchor maps n to.

Dynamic Conjunction. For conjoining dynamic propositions, we likewise follow de Groote in defining dynamic $\text{AND} : \pi \to \pi \to \pi$ to compose the meanings of two dynamic propositions over a structure and a discourse continuation:

$$\vdash \text{AND} = \lambda_{ABsk}.A\ s\ (\lambda_s.B\ s\ k) \tag{1}$$

The continuation passed to the first conjunct A is the second conjunct B with its structure (type σ) argument abstracted over. In addition to conjoining utterances to form discourses, AND plays a central role in our dynamic indefinite, given in (3), below.

To demonstrate dynamic conjunction in action, we take RAIN $= (\mathbf{dyn}_0\, \mathsf{rain})$ and SNOW $= (\mathbf{dyn}_0\, \mathsf{snow})$. The conjunction of RAIN and SNOW is then as follows:

$$\begin{aligned}
&\vdash \text{RAIN AND SNOW} : \pi \\
&= \lambda_{sk}.\text{RAIN}\ s\ (\lambda_s.\text{SNOW}\ s\ k) \\
&= \lambda_{sk}(\lambda_{sk}(\mathsf{rain} \wedge (k\ (s + \mathsf{rain})))\ s\ (\lambda_s.\text{SNOW}\ s\ k)) \\
&= \lambda_{sk}(\lambda_k(\mathsf{rain} \wedge (k\ (s + \mathsf{rain})))\ (\lambda_s.\text{SNOW}\ s\ k)) \\
&= \lambda_{sk}.\mathsf{rain} \wedge (\lambda_s(\text{SNOW}\ s\ k)\ (s + \mathsf{rain})) \\
&= \lambda_{sk}.\mathsf{rain} \wedge (\lambda_{sk}(\mathsf{snow} \wedge (k\ (s + \mathsf{snow})))\ (s + \mathsf{rain})\ k) \\
&= \lambda_{sk}.\mathsf{rain} \wedge (\lambda_k(\mathsf{snow} \wedge (k\ (s + \mathsf{rain} + \mathsf{snow})))\ k) \\
&= \lambda_{sk}.\mathsf{rain} \wedge \mathsf{snow} \wedge (k\ (s + \mathsf{rain} + \mathsf{snow})) \\
&= \lambda_{sk}.\mathsf{rain} \wedge \mathsf{snow} \wedge (k\ \langle(\mathbf{a}\ s), (\mathbf{r}\ s), (\mathbf{c}\ s) \wedge \mathsf{rain} \wedge \mathsf{snow}\rangle)
\end{aligned}$$

It is important to note that, at this example shows, AND ensures that the content proffered by RAIN is available in the common ground of the structure $(s + \mathsf{rain})$ that is passed to SNOW.

The Dynamic Existential Quantifier. Our replacement for de Groote's Σ is EXISTS $: \delta \to \pi$:

$$\vdash \text{EXISTS} = \lambda_{Dsk}.\exists\, \lambda_x.D\ (\mathbf{next}\ s)\ (s :: x)\ k \tag{2}$$

This version of the dynamic existential quantifier introduces a DR using :: to extend both the anchor and resolution of the current structure. We examine EXISTS DONKEY for an example of the behavior of EXISTS:

$$\begin{aligned}
&\vdash \text{EXISTS DONKEY} : \delta \\
&= \lambda_{sk}.\exists\, \lambda_x.\text{DONKEY}\ (\mathbf{next}\ s)\ (s :: x)\ k \\
&= \lambda_{sk}.\exists\, \lambda_x.(\mathsf{donkey}\ [(\mathbf{next}\ s)]_{s::x}) \wedge (k\ (s :: x + (\mathsf{donkey}\ [(\mathbf{next}\ s)]_{s::x}))) \\
&= \lambda_{sk}.\exists\, \lambda_x.(\mathsf{donkey}\ x) \wedge (k\ (s :: x + (\mathsf{donkey}\ x))) \\
&= \lambda_{sk}.\exists\, \lambda_x.(\mathsf{donkey}\ x) \wedge (k\ \langle(\mathbf{a}\ s) \bullet x, \star\, (\mathbf{r}\ s), (\mathbf{c}\ s) \wedge (\mathsf{donkey}\ x)\rangle)
\end{aligned}$$

Note that $[(\mathbf{next}\ s)]_{s::x}$ necessarily reduces to the entity variable x because the anchor of $(s :: x)$ always maps $(\mathbf{next}\ s)$ to x, the complexity of s itself notwithstanding.

We use EXISTS and AND to model the English indefinite article *a* as the dynamic generalized determiner A $: \delta \to \delta \to \pi$:

$$\vdash \text{A} = \lambda_{DE}.\text{EXISTS}\ \lambda_n.(D\ n)\ \text{AND}\ (E\ n) \tag{3}$$

This definition ensures, via AND, that the scope E inherits whatever extensions are made to the structure by the restriction D. We illustrate the effects of the indefinite article A by applying it to DONKEY = $(\mathbf{dyn}_1\ \mathsf{donkey})$ to yield:

$$\begin{aligned}
&\vdash \text{A DONKEY} : \delta \to \pi\\
&= \lambda_E.\text{EXISTS}\ \lambda_n.(\text{DONKEY}\ n)\ \text{AND}\ (E\ n)\\
&= \lambda_{Esk}.\exists\ \lambda_x.(\lambda_n((\text{DONKEY}\ n)\ \text{AND}\ (E\ n))\ (\mathbf{next}\ s))\ (s :: x)\ k\\
&= \lambda_{Esk}.\exists\ \lambda_x.((\text{DONKEY}\ (\mathbf{next}\ s))\ \text{AND}\ (E\ (\mathbf{next}\ s)))\ (s :: x)\ k\\
&= \lambda_{Esk}.\exists\ \lambda_x.(\mathsf{donkey}\ x) \wedge (E\ (\mathbf{next}\ s)\ (s :: x + (\mathsf{donkey}\ x))\ k)\\
&= \lambda_{Esk}.\exists\ \lambda_x.(\mathsf{donkey}\ x) \wedge (E\ (\mathbf{next}\ s)\ \langle(\mathbf{a}\ s) \bullet x, \star\ (\mathbf{r}\ s), \mathbf{c}\ s \wedge (\mathsf{donkey}\ x)\rangle\ k)
\end{aligned}$$

Note that the sortal restriction imposed by the noun on the new DR is part of the common ground passed to the scope. We next apply A DONKEY to the dynamic property WALK = $(\mathbf{dyn}_1\ \mathsf{walk})$:

$$\begin{aligned}
&\vdash (\text{A DONKEY WALK}) : \pi\\
&= \text{EXISTS}\ \lambda_n.(\text{DONKEY}\ n)\ \text{AND}\ (\text{WALK}\ n)\\
&= \lambda_{sk}.\exists\ \lambda_x.((\text{DONKEY}\ (\mathbf{next}\ s))\ \text{AND}\ (\text{WALK}\ (\mathbf{next}\ s)))\ (s :: x)\ k\\
&= \lambda_{sk}.\exists\ \lambda_x.(\mathsf{donkey}\ x) \wedge (\mathsf{walk}\ x) \wedge (k\ (s :: x + (\mathsf{donkey}\ x) + (\mathsf{walk}\ x)))\\
&= \lambda_{sk}.\exists\lambda_x.(\mathsf{donkey}\ x) \wedge (\mathsf{walk}\ x) \wedge k\ \langle(\mathbf{a}\ s) \bullet x, \star\ (\mathbf{r}\ s), (\mathbf{c}\ s) \wedge \mathsf{donkey}\ x \wedge \mathsf{walk}\ x\rangle
\end{aligned}$$

In this example, the division of labor between EXISTS and AND in the definition of A is apparent, with EXISTS extending the anchor and resolution and AND accumulating the additions to the CG made by the dynamic properties DONKEY and WALK.

4 Modeling Definiteness

With our hyperintensional dynamic semantic theory in place, we are ready to extend it to handle definiteness presuppositions in English. We examine both definite pronominal anaphora with *it* and the definite determiner *the*.

4.1 Definite Anaphora with *It*

Rather than adopting an analog of de Groote's `sel` to model English *it*, which cannot possibly select the "right" DR from a left context (see Sect. 1, above), we define dynamic IT : $\delta \to \pi$ as follows:

$$\vdash \text{IT} = \lambda_{Ds}.D\ (\mathbf{def}\ s\ \text{NONHUMAN})\ s \tag{4}$$

where NONHUMAN = $(\mathbf{dyn}_1\ \mathsf{nonhuman})$ and $\mathbf{def}_n : \sigma_n \to \delta \to \omega_n$ is the definiteness operator, defined as follows:

$$\vdash \mathbf{def}_n = \lambda_{sD}.\bigsqcup_{(\mathbf{r}\ s)} \lambda_{i:\omega_n}.(\mathbf{c}\ s)\ \mathsf{entails}\ (\mathbf{stat}\ s\ (D\ i)) \tag{5}$$

(Recall that $\mathsf{entails} : \mathrm{p} \to \mathrm{p} \to \mathrm{t}$ is the entailment relation between (static) propositions.) For each $r : \rho_n$, the operator $\bigsqcup_r : (\omega_n \to \mathrm{t}) \to \omega_n$ takes a subset of the first n DRs and returns the unique greatest element (if any) with respect to the (restriction of the) preorder r on ω_n. Thus for a structure s and a dynamic property D, the **def** operator returns the highest DR (if any) in the resolution $(\mathbf{r}\ s)$ whose image under the current anchor $(\mathbf{a}\ s)$ can be inferred from the common ground $(\mathbf{c}\ s)$ to have the staticized counterpart of the property D.

As defined in (4), IT selects the most salient inferably NONHUMAN discourse referent from a given structure. This is because, as (5) shows, IT is equivalent to

$$\lambda_{Ds}.D\,(\bigsqcup_{(\mathbf{r}\ s)} \lambda_{i:\omega_n}((\mathbf{c}\ s)\ \mathsf{entails}\ (\mathsf{nonhuman}\ (\mathbf{a}\ s\ i))))\ s$$

We can assume that the static proposition (every donkey nonhuman) reflecting the common world knowledge that every donkey is nonhuman is in every common ground we would consider, where every is as given in [19]. This ensures that **def** will allow IT to select a donkey as its antecedent, as desired.

We demonstrate by applying this definition of IT in the interpretation of the following example:

(I) 1. A donkey enters.
 2. It brays.

With DONKEY $= (\mathbf{dyn}_1 \mathsf{donkey})$, ENTER $= (\mathbf{dyn}_1 \mathsf{enter})$, and BRAY $= (\mathbf{dyn}_1 \mathsf{bray})$, we analyze (I) as:

$$\vdash (\text{A DONKEY ENTER})\ \text{AND}\ (\text{IT BRAY}) : \pi$$

To see how IT retrieves its antecedent from context, we examine the rightmost application:

$$\begin{aligned} &\vdash \text{IT BRAY} : \pi \\ &= \lambda_s.\text{BRAY}\,(\mathbf{def}\ s\ \text{NONHUMAN})\ s \\ &= \lambda_s.\lambda_{nsk}((\mathsf{bray}\ [n]) \wedge (k\ (s + (\mathsf{bray}\ [n])))))\ (\mathbf{def}\ s\ \text{NONHUMAN})\ s \\ &= \lambda_{sk}.(\mathsf{bray}\ [(\mathbf{def}\ s\ \text{NONHUMAN})]) \wedge (k\ (s + (\mathsf{bray}\ [(\mathbf{def}\ s\ \text{NONHUMAN})]))) \end{aligned} \tag{6}$$

Here, IT ensures that the argument to BRAY is the most salient nonhuman entity in the discourse context. Recall from the example analysis given above of (A DONKEY WALK) that A DONKEY updates the common ground passed to IT BRAY with the proposition that $[n]$ is a donkey. The full reduction of the dynamic interpretation of (I) in is then:

$$\begin{aligned} &\vdash (\text{A DONKEY ENTER})\ \text{AND}\ (\text{IT BRAY}) : \pi \\ &= (\text{EXISTS}\ \lambda_n((\text{DONKEY}\ n)\ \text{AND}\ (\text{ENTER}\ n)))\ \text{AND}\ (\text{IT BRAY}) \\ &= \lambda_{sk}.\exists\,\lambda_x.(\mathsf{donkey}\ x) \wedge (\mathsf{enter}\ x) \\ &\qquad \wedge (\mathsf{bray}\ [(\mathbf{def}\ \varsigma\ \text{NONHUMAN})]) \wedge (k\ (\varsigma + (\mathsf{bray}\ [(\mathbf{def}\ \varsigma\ \text{NONHUMAN})]))) \end{aligned} \tag{7}$$

where $\varsigma = s :: x + (\mathsf{donkey}\, x) + (\mathsf{enter}\, x)$ is the structure passed to IT BRAY (shown in (6)). Since the CG of ς contains the proposition $(\mathsf{donkey}\, x)$, the definiteness operator **def** is able to select the DR that is the preimage of x as the most salient nonhuman in the context it is passed. With $[(\mathbf{def}\ \varsigma\ \text{NONHUMAN})]_{\varsigma} = x$, we can reduce the term in (7) interpreting the discourse in (I) to:

$$\lambda_{sk}.\exists\, \lambda_x.(\mathsf{donkey}\, x) \wedge (\mathsf{enter}\, x) \wedge (\mathsf{bray}\, x) \wedge (k\, (\varsigma + (\mathsf{bray}\, x))) : \pi$$

Thus IT selects its antecedent based on its definiteness presuppositions, yielding the desired truth conditions for (I). The definition of IT in (4) also captures the infelicity of (C), where *it* is used without a salient antecedent, because there is no DR that can be inferred from context to be nonhuman.

4.2 The Definite Determiner

We also use **def** to model the English definite determiner *the* as THE : $\delta \to \delta \to \pi$:

$$\vdash \text{THE} = \lambda_{DEs}.(\lambda_n((D\, n)\ \text{AND}\ (E\, n))\, (\mathbf{def}\ s\ D))\ s \qquad (8)$$

This translation of *the* resembles the indefinite determiner A in (3) in that the meanings of the dynamic properties D and E are composed via AND. The main difference is that while A uses EXISTS to introduce a new DR, THE uses **def** from (5) to select the most salient DR from the discourse context with the property D. Using AND to pass this DR to both properties ensures that any modifications to the structure that result from D are inherited by E, as would be necessary in the interpretation of an utterance like *The donkey that has a red blanket chews it.*

In the discourse in (J), which is a simplification of (A), (B), and (F), the noun phrase *the donkey* can only refer to one of the discourse referents introduced prior to its use:

(J) 1. A donkey enters.
2. A mule enters.
3. The donkey brays.

To model this discourse, we first define the dynamic properties MULE = $(\mathbf{dyn}_1\ \mathsf{mule})$ and ENTER = $(\mathbf{dyn}_1\ \mathsf{enter})$. With A and THE as in (3) and (8), and AND as in (1) used to conjoin utterances, we have the following dynamic meaning for the discourse in (J):

$$\vdash ((\text{A DONKEY ENTER})\ \text{AND}\ (\text{A MULE ENTER}))\ \text{AND}\ (\text{THE DONKEY BRAY}) : \pi \qquad (9)$$

We start with the leftmost conjunct of the discourse:

$$\begin{aligned} &\vdash (\text{A DONKEY ENTER}) : \pi \qquad (10)\\ &= \lambda_{sk}.\exists\, \lambda_x.(\mathsf{donkey}\, x) \wedge (\mathsf{enter}\, x) \wedge (k\, (s :: x + (\mathsf{donkey}\, x) + (\mathsf{enter}\, x)))\\ &= \lambda_{sk}.\exists\, \lambda_x.(\mathsf{donkey}\, x) \wedge (\mathsf{enter}\, x)\\ &\qquad \wedge (k\ \langle (\mathbf{a}\, s) \bullet x, \star\, (\mathbf{r}\, s), (\mathbf{c}\, s) \wedge (\mathsf{donkey}\, x) \wedge (\mathsf{enter}\, x)\rangle) \end{aligned}$$

Combining the term in (10) with the entire left conjunct of the discourse, we have:

$$\vdash (\text{A DONKEY ENTER}) \text{ AND } (\text{A MULE ENTER}) : \pi \tag{11}$$
$$= \lambda_{sk}.\exists\, \lambda_x.(\mathsf{donkey}\, x) \wedge (\mathsf{enter}\, x) \wedge (\exists\, \lambda_y.(\mathsf{mule}\, y) \wedge (\mathsf{enter}\, y) \wedge (k\, \varsigma))$$

where ς represents the structure that results from the application of AND in (11):

$$\varsigma = s :: x + (\mathsf{donkey}\, x) + (\mathsf{enter}\, x) :: y + (\mathsf{mule}\, y) + (\mathsf{enter}\, y)$$
$$= \langle (\mathbf{a}\, s) \bullet x \bullet y, \star\, (\star\, (\mathbf{r}\, s)), (\mathbf{c}\, s) \wedge (\mathsf{donkey}\, x) \wedge (\mathsf{enter}\, x) \wedge (\mathsf{mule}\, y) \wedge (\mathsf{enter}\, y) \rangle$$

The right conjunct then uses THE to select the most salient DONKEY from the preceding discourse, applying BRAY to it:

$$\vdash (\text{THE DONKEY BRAY}) : \pi \tag{12}$$
$$= \lambda_{sk}.(\lambda_n((\text{DONKEY}\, n) \text{ AND } (\text{BRAY}\, n))\, (\mathbf{def}\, s\, \text{DONKEY}))\, s\, k$$
$$= \lambda_{sk}.((\text{DONKEY}\, (\mathbf{def}\, s\, \text{DONKEY})) \text{ AND } (\text{BRAY}\, (\mathbf{def}\, s\, \text{DONKEY})))\, s\, k$$
$$= \lambda_{sk}.(\mathsf{donkey}\, [(\mathbf{def}\, s\, \text{DONKEY})]) \wedge (\mathsf{bray}\, [(\mathbf{def}\, s\, \text{DONKEY})])$$
$$\wedge\, (k\, (s + (\mathsf{donkey}\, [(\mathbf{def}\, s\, \text{DONKEY})]) + (\mathsf{bray}\, [(\mathbf{def}\, s\, \text{DONKEY})])))$$

The structure ς passed to (THE DONKEY BRAY) by the preceding discourse in (11) is such that $[(\mathbf{def}\, \varsigma\, \text{DONKEY})]_\varsigma = x$ because the CG of ς contains the proposition $(\mathsf{donkey}\, x)$ but does not contain any proposition in which the property donkey is applied to an entity other than x. Hence we arrive at the final reduction of the term in (9) that models the entirety of the discourse (J):

$$\lambda_{sk}.\exists\, \lambda_x.\mathsf{donkey}\, x \wedge \mathsf{enter}\, x \wedge (\exists\, \lambda_y.\mathsf{mule}\, y \wedge \mathsf{enter}\, y \wedge \mathsf{donkey}\, x \wedge \mathsf{bray}\, x \wedge k\, \varsigma')$$

where $\varsigma' = \varsigma + (\mathsf{donkey}\, x) + (\mathsf{bray}\, x)$ is the structure extending ς that results from applying AND to the left (11) and right (12) conjuncts of the discourse.

Note that, in this example, the dynamic definite determiner THE picks out the most salient DR from the structure it is given that has the property specified as its argument (i.e., DONKEY). However, as (F3) shows, substituting the pronoun *it* for *the donkey* makes the discourse infelicitous. Our theory captures this infelicity because IT only requires its antecedent to have the property $\mathsf{nonhuman}$, which is weaker than the property donkey with respect to entailment. With IT replacing THE DONKEY in the right conjunct of (9), **def** would be incapable of selecting a unique nonhuman DR from ς since two DRs would then have the property NONHUMAN (namely, the mule and the donkey). The infelicitous examples in (E) through (H) can be ruled out for similar reasons.

5 Conclusion and Future Work

We have presented a dynamic theory of discourse meaning formulated in higher order logic that incorporates aspects of de Groote's continuation-based theory and Pollard's hyperintensional semantics. Drawing on the work of Heim and Roberts, our theory provides an enriched notion of discourse context that includes discourse referents ordered by relative salience and a common ground of mutually accepted content. We have shown how this enriched context allows the definiteness presuppositions in English associated with the pronoun *it* and the determiner *the* to be captured in a way that is faithful to the facts. The resulting theory repairs the inadequate treatment of anaphora resolution in de Groote's work based on the oracular `sel` function.

In future work, we will continue our collaboration with Roberts and Smith on developing a general, categorially based theory of projective meaning. The next avenues for this research include spelling out how relative salience is adjusted by re-invoking a previously introduced DR (see the discussion of (A) and (B) in Sect. 1) and integrating the hyperintensional dynamic semantic theory introduced here with a fully compositional theory of English grammar that takes e.g. quantifier scope, unbounded dependencies, and prosodically encoded information structure into account. We will then apply this theory to a wider range of presuppositional phenomena, including the factivity of certain sentential complement verbs (e.g. *suck, know, realize*), the 'projection' of presuppositions occurring within e.g. the scope of a negation, and the phenomenon known as *farmer-donkey asymmetry* [23] that is associated with the celebrated 'donkey sentences' (e.g. *Most farmers who own a donkey beat it*).

References

1. Church, A.: A formulation of the simple theory of types. Journal of Symbolic Logic 5, 56–68 (1940)
2. Gallin, D.: Intensional and Higher Order Modal Logic. North-Holland, Amsterdam (1975)
3. de Groote, P.: Towards abstract categorial grammars. In: 39th Annual Meeting and 10th Conference of the European Chapter, Proceedings of the Conference on Association for Computational Linguistics (2001)
4. de Groote, P.: Towards a Montagovian account of dynamics. In: Proceedings of Semantics and Linguistic Theory, vol. 16 (2006)
5. de Groote, P.: Typing binding and anaphora: Dynamic contexts as $\lambda\mu$-terms. Presented at the ESSLLI Workshop on Symmetric Calculi and Ludics for Semantic Interpretation (2008)
6. Heim, I.: The Semantics of Definite and Indefinite Noun Phrases. Ph.D. thesis. University of Massachusetts, Amherst (1982)
7. Heim, I.: File change semantics and the familiarity theory of definiteness. In: Meaning, Use and the Interpretation of Language. Walter de Gruyter, Berlin (1983)
8. Henkin, L.: Completeness in the theory of types. Journal of Symbolic Logic 15, 81–91 (1950)
9. Kamp, H.: A theory of truth and semantic representation. In: Groenendijk, J., Janssen, T., Stokhof, M. (eds.) Formal Methods in the Study of Language. Mathematisch Centrum, Amsterdam (1981)
10. Kamp, H., Reyle, U.: From Discourse to Logic. Kluwer Academic Publishers, Dordrecht (1993)

11. Lambek, J., Scott, P.: Introduction to Higher-Order Categorical Logic. Cambridge University Press, Cambridge (1986)
12. Lewis, D.: Scorekeeping in a language game. In: Baüerle, R., Egli, U., von Stechow, A. (eds.) Semantics from a Different Point of View. Springer, Berlin (1979)
13. Montague, R.: The proper treatment of quantification in ordinary English. In: Hintikka, K., Moravcsik, J., Suppes, P. (eds.) Approaches to Natural Language, D. Reidel, Dordrecht (1973)
14. Muskens, R.: Categorial grammar and discourse representation theory. In: Proceedings of COLING (1994)
15. Muskens, R.: Combining Montague semantics and discourse representation theory. Linguistics and Philosophy 19, 143–186 (1996)
16. Muskens, R.: Separating syntax and combinatorics in categorial grammar. Research on Language and Computation 5, 267–285 (2007)
17. Oehrle, R.T.: Term-labeled categorial type systems. Linguistics and Philosophy 17(6), 633–678 (1994)
18. Pollard, C.: Hyperintensional Questions. In: Hodges, W., de Queiroz, R. (eds.) WoLLic 2008. LNCS (LNAI), vol. 5110, pp. 272–285. Springer, Heidelberg (2008)
19. Pollard, C.: Hyperintensions. Journal of Logic and Computation 18(2), 257–282 (2008)
20. Roberts, C.: Information structure in discourse: Towards an integrated formal theory of pragmatics. In: Papers in Semantics. Working Papers in Linguistics, vol. 49, Ohio State University, Department of Linguistics (1996)
21. Roberts, C.: Pronouns as definites. In: Reimer, M., Bezuidenhout, A. (eds.) Descriptions and Beyond, pp. 503–543. Oxford University Press (2004)
22. Roberts, C., Simons, M., Beaver, D., Tonhauser, J.: Presupposition, conventional implicature, and beyond: A unified account of projection. In: Klinedinst, N., Rothschild, D. (eds.) Proceedings of New Directions in the Theory of Presupposition. ESSLLI workshop (2009)
23. Rooth, M.: NP interpretation in Montague grammar, file change semantics, and situation semantics. In: Gärdenfors, P. (ed.) Generalized Quantifiers. Reidel, Dordrecht (1987)
24. Smith, E.A.: Correlational Comparison in English. Ph.D. thesis. Ohio State University (2010)
25. Stalnaker, R.: Presuppositions. Journal of Philosophical Logic 2(4) (1973)
26. Stalnaker, R.: Assertion. Syntax and Semantics 9: Pragmatics, 315–332 (1978)
27. Strachey, C., Wadsworth, C.P.: Continuations: A mathematical semantics for handling full jumps. Programming Research Group Technical Monograph PRG-11, Oxford University Computing Lab (1974)
28. Thomason, R.: A model theory for propositional attitudes. Linguistics and Philosophy 4, 47–70 (1980)

Distinguishing Phenogrammar from Tectogrammar Simplifies the Analysis of Interrogatives

Vedrana Mihaliček and Carl Pollard

The Ohio State University, Department of Linguistics, Columbus OH 43210, USA
{vedrana,pollard}@ling.ohio-state.edu

Abstract. Oehrle (1994) introduced a categorial grammar architecture in which word order is represented using the terms of a typed λ-calculus and the syntactic type system is based on linear logic. In this paper, we use a variant of this architecture to analyze interrogatives in English and Chinese. We show that separating word order (phenogrammar) and syntactic combinatorics (tectogrammar) in this way brings out the underlying similarities between different question-forming strategies. In particular, the difference between *wh* extraction (overt movement) and *wh in situ* (covert movement) turns out to be purely phenogrammatical.

1 Introduction

Oehrle (1994) introduced a categorial grammar (CG) architecture in which word order (not just meaning) is represented using the terms of a typed λ-calculus. Variants of this architecture have since been employed in a variety of CG frameworks, including abstract categorial grammar (ACG, de Groote 2001), λ-grammar (λG, Muskens 2003, 2007b), higher-order grammar (HOG, Pollard 2004), and pheno-tecto distinguished CG (PTDCG, Smith 2010). Some salient commonalities of these approaches include the following: (i) a clear separation of tectogrammar (roughly, abstract syntactic combinatorics) and phenogrammar (roughly, word order)[1]; (ii) an implementation of Montague's (1974) 'lowering' analysis of quantification in terms of β-reduction in the phenogrammatical calculus (exemplified by (12) below); (iii) uniform treatment of medial and peripheral extraction by phenogrammatical lowering of the null string into the 'trace' position; and (iv) a tectogrammatical type system based on linear logic, made possible by the 'outsourcing' to the phenogrammar of much of the work done by directionality and/or multimodality of the tectogrammar in other CG frameworks (e.g. Moortgat 1997, Morrill et al 2007, Baldridge 2002).

Works such as Oehrle 1994, de Groote 2001 and Muskens 2003, 2007b are largely programmatic in nature. However, Smith 2010 shows by example that

[1] The terms 'tectogrammar' and 'phenogrammar' are meant to suggest an affinity with the programmatic suggestions of Curry (1961), who employed the terms 'tectogrammatical structure' and 'phenogrammatical structure'. The analogous ACG (λG) notions are 'abstract syntax' ('combinatorics') and 'concrete syntax' ('syntax').

P. de Groote and M.-J. Nederhof (Eds.): Formal Grammar 2010/2011, LNCS 7395, pp. 130–145, 2012.

the pheno-tecto-distinguished style of CG (hereafter, PTG) can also be a practical framework for describing highly complex linguistic phenomena (specifically, remnant comparative and correlational comparative constructions) in a way that highlights the underlying simplicity of the combinatorics and compositional semantics, by representing the difference between lowering and extraction as purely phenogrammatical.

In this paper, we try to make the same point, with respect to a different set of phenomena, namely cross-linguistic variation in the form of interrogative sentences, with special attention to multiple constituent questions and so-called Baker ambiguities. Because of space limitations, we here omit inessential details and discuss only two languages, English and Chinese.

Unlike other CG approaches, both mainstream and PTG, we drop the traditional requirement that there be a function mapping tectogrammatical types to semantic types. Demanding that every semantic difference be reflected in tectogrammar, in our opinion, lacks empirical motivation and unnecessarily complicates the tectogrammar. Additionally, we foresake the standard (and usually, mostly extensional) Montague semantics in favor of a hyperintensional form of possible-worlds semantics with propositions (rather than worlds) as a basic type.[2]

What emerges is a surprisingly simple and uniform analysis of interrogatives in the two languages. On our analysis, English and Chinese constituent questions are essentially identical semantically and tectogrammatically, with phenogrammar identified as the sole locus of variation. That is, the difference between *wh* fronting and *wh in situ* is analyzed as a purely phenogrammatical difference.

The rest of the paper is organized as follows. In Section 2 we introduce the framework in more detail, including a brief review of some standard PTG features. In Section 3 we present the relevant data, which are analyzed in Section 4. Section 5 evaluates the framework and the analysis, and outlines some directions for future research.

2 An Overview of the Framework

The version of PTG we employ resembles λG (not ACG), in making no use of tectogrammatical terms. Thus, our 'signs' (the things the grammar proves) are triples consisting of a typed pheno term, a tecto type, and a typed semantic term (the pheno and semantic types are suppressed whenever no confusion results from so doing). However we depart from λG (as described in Muskens (2003 2007b) by (i) adopting a slightly different notation, (ii) dropping Muskens' Kripke models for the phenogrammatical terms in favor of standard (Henkin) models for the higher-order theory of a free monoid; and (iii) allowing for a relational interface between tectogrammar and semantics whereby a single tectogrammatical type may correspond to multiple semantic types.

[2] We think this choice greatly simplifies the compositional semantics of interrogatives, but space limitations prevent us from defending that belief here.

2.1 Phenogrammar

Phenogrammar is implemented as a classical higher-order theory (and therefore a typed λ calculus) with one basic type for strings, `Str` (besides the truth-value type `t` provided by higher-order logic (HOL)).

The pheno components of words (lexical signs) are treated in terms of non-logical constants of type `Str`, e.g. `chris`, `robin`, `liked`, `slept`, `who`, `whether`, etc. The constants $\circ$:`Str`→`Str`→`Str` and `e` : `Str` are axiomatized as the (binary, associative) operation of a free monoid and its two-sided identity respectively.

We use p, q, r as variables of type `Str`, and f, g as variables of type `Str`→`Str`. We call terms of this calculus *pheno terms* (cf. Oehrle's (1994) ϕ *terms*). A transitive English verb such as *liked* is associated with the following pheno term:

$$\vdash \lambda_p \lambda_q q \circ \texttt{liked} \circ p : \texttt{Str} \rightarrow \texttt{Str} \rightarrow \texttt{Str}$$

This pheno term requires that the first argument of *liked* – its object – concatenate to the right of `liked`, and its second argument – its subject – to the left of `liked`, so we get the expected subject–verb–object order.

2.2 Tectogrammar

The tectogrammatical signature is obtained by closing the set of basic types **N**, **NP**, **S** and $\bar{\mathbf{S}}$ under the linear implication $\multimap$. Type **N** is associated with common nouns (*book*, *dog*), and **NP** with noun phrases (*Chris*, *Robin*). Type **NP**$\multimap$**S** corresponds to intransitive finite verbs and verb phrases (*slept*, *liked Robin*), and type **NP**$\multimap$**NP**$\multimap$**S** to transitive finite verbs (*liked*). Type **S** is reserved for root clauses, declarative (*Chris liked Robin.*) or interrogative (*Who did Chris like?*), while $\bar{\mathbf{S}}$ corresponds to embedded clauses, again — declarative (*that Chris liked Robin*) or interrogative (*who Chris liked*).

Note that the tecto type of a transitive verb merely requires that it combine with two noun phrases, but does not determine the relative word order of the verb and its arguments since this is handled entirely within the phenogrammar.

We ignore the distinction between declaratives and interrogatives in the tectogrammar because we assume a nonfunctional relation between tecto types and semantic types. So it is possible for, say, the tectogrammatical type **S** to correspond to the semantic type of either declaratives or interrogatives. The two kinds of utterances are then distinguished in terms of semantic types and that is how overgeneration is prevented. At the same type, the tectogrammar is kept maximally simple.

2.3 Semantics

We assume a hyperintensional semantic theory along the lines of Pollard 2008a.[3] While we believe this choice to be well motivated (we direct the reader to Pollard 2008a for a detailed discussion of problems with traditional possible world

[3] See Thomason 1980 and Muskens 2005, 2007a for versions of hyperintensional semantics with somewhat different technical assumptions.

semantics), the analysis of interrogatives presented here is also compatible with a more mainstream Montague-style possible world semantics.

The most important departure from the standard possible world semantics is treating propositions as primitive and constructing possible worlds as certain sets of propositions, instead of the other way around. On our approach, propositions are modelled as members of a *pre*-boolean algebra *pre*-ordered by entailment. Entailment is axiomatized as a reflexive, transitive, but *not* antisymmetric relation on propositions. This way, it is possible for equivalent (mutully entailing) propositions to be distinct.

This hyperintensional semantic theory is expressed in a classical HOL with e (entities) and p (propositions)[4] as the basic types (other than the truth-value type t provided by the logic).[5] The kind of HOL we employ follows Lambek and Scott (1986) in also having a basic type n (natural numbers) and machinery for forming (separation-style) subtypes.[6] Additionally, we make use of dependent product and coproduct types parametrized by the natural number type.

We recursively define the function Ext mapping hyperintensional types (i.e. e, p and any implicative types constructed out of these) to the corresponding extensional types. Here, A and B are metavariables over hyperintensional types:

(1) a. $\mathsf{Ext}(\mathsf{e}) = \mathsf{e}$
 b. $\mathsf{Ext}(\mathsf{p}) = \mathsf{t}$
 c. $\mathsf{Ext}(A \to B) = A \to \mathsf{Ext}(B)$

The type of possible worlds w is constructed out of the basic types in such a way that the interpretation of the type w is the set of ultrafilters of the pre-boolean prealgebra that interprets the type p. Specifically, $\mathsf{w} =_{def} [\mathsf{p} \to \mathsf{t}]_u$, where $u : (\mathsf{p} \to \mathsf{t}) \to \mathsf{t}$ is a predicate on sets of propositions that picks out those sets of propositions that are ultrafilters (see Pollard (2008a) for details of this construction).

Concomitantly, we introduce a family of constants $\mathsf{ext}_A : A \to \mathsf{w} \to \mathsf{Ext}(A)$ (where the type variable A ranges over the hyperintensional types) interpreted as a polymorphic function that maps a hyperintension and a world to the extension of that hyperintension at that world, as follows:

(2) a. $\vdash \forall_{x:\mathsf{e}} \forall_{w:\mathsf{w}} [\mathsf{ext}_{\mathsf{e}}(x)(w) = x]$
 b. $\vdash \forall_{p:\mathsf{p}} \forall_{w:\mathsf{w}} [\mathsf{ext}_{\mathsf{p}}(p)(w) = p@w]$

[4] The type p is axiomatized so as to form a boolean preorder where the constants and, or and not denote the greatest lower bound, least upper bound and complement (involutive negation) operation, respectively.

[5] For expository simplicity, we depart from Pollard 2008a in not distinguishing between the extensional type e and the corresponding hyperintensional type i (individual concepts). In particular, the meaning of a name is the same as its reference.

[6] Thus if A is a type and a an A-predicate (closed term of type $A \to \mathsf{t}$), then there is a type A_a interpreted as the subset of the intepretation of A that has the interpretation of a as its characteristic function; and there is a constant μ_a that denotes the subset embedding.

c. $\vdash \forall_{f:A \to B} \forall_{w:\mathrm{w}} [\mathsf{ext}_{A \to B}(f)(w) = \lambda_{x:A} \mathsf{ext}_B(f(x))(w)]$

Here the notation '$p@w$' abbreviates $\mu_u(w)(p)$, where μ_u denotes the embedding of the set of worlds into the set of sets of propositions.

As for question meanings, we adopt an elaboration of the general approach of Pollard 2008b, which in turn modifies and refines the 'set of true answers' approach of Karttunen 1977. We present the details in Section 4.

2.4 A Small Example

Before moving on to the analysis of questions, we illustrate how the grammar works with a toy example. As in λG, a representation of a linguistic expression (or a *sign*) consists of a pheno term, a tecto type, and a semantic term. Lexical entries are written in the form:

(3) $\vdash$ **pheno term**; **TectoType**; semantic term

We make use of the following logical rules whose tecto components are the Gentzen sequent-style natural deduction rules for the implicative fragment of linear logic, while the pheno (semantic) labels are just recipes for compositionally constructing the word order (meaning) representations.[7]

(4) $$\frac{}{v; \mathbf{T}; v \vdash v; \mathbf{T}; v} \text{[Ax]}$$

(5) $$\frac{\Gamma \vdash f; \mathbf{T} \multimap \mathbf{T}'; f \qquad \Delta \vdash a; \mathbf{T}; a}{\Gamma, \Delta \vdash f(a); \mathbf{T}'; f(a)} [\multimap\text{E}]$$

(6) $$\frac{\Gamma, a; \mathbf{T}; a \vdash f; \mathbf{T}'; f}{\Gamma \vdash \lambda_a f; \mathbf{T} \multimap \mathbf{T}'; \lambda_a f} [\multimap\text{I}]$$

Here is a toy lexicon for English:

(7) $\vdash$ chris; **NP**; chris

$\vdash$ robin; **NP**; robin

$\vdash \lambda_p p \circ \mathtt{slept}; \mathbf{NP} \multimap \mathbf{S}; \mathsf{sleep}$

$\vdash \lambda_r \lambda_p p \circ \mathtt{liked} \circ r; \mathbf{NP} \multimap \mathbf{NP} \multimap \mathbf{S}; \mathsf{like}$

$\vdash$ dog; **N**; dog

$\vdash \lambda_p \lambda_f f(\mathtt{every} \circ p); \mathbf{N} \multimap (\mathbf{NP} \multimap \mathbf{S}) \multimap \mathbf{S}; \mathsf{every}$

$\vdash \lambda_p \lambda_f f(\mathtt{a} \circ p); \mathbf{N} \multimap (\mathbf{NP} \multimap \mathbf{S}) \multimap \mathbf{S}; \mathsf{exists}$

$\vdash \lambda_f f(\mathtt{everyone}); (\mathbf{NP} \multimap \mathbf{S}) \multimap \mathbf{S}; \mathsf{every(person)}$

$\vdash \lambda_f f(\mathtt{someone}); (\mathbf{NP} \multimap \mathbf{S}) \multimap \mathbf{S}; \mathsf{exists(person)}$

[7] Compare [⊸E] to *pointwise application* and [⊸I] to *pointwise abstraction* in Muskens 2007b. The three rules also roughly correspond to Trace, Merge and Move respectively in mainstream generative grammar.

Given the rules and these lexical entries we can derive the following by means of [$\multimap$ E] and β reduction in the pheno logic (to enhance readability, we freely β-reduce pheno and semantic terms in derivations):

(8) a. $\vdash \mathtt{chris} \circ \mathtt{slept}; \mathbf{S}; \mathsf{sleep}(\mathsf{chris})$

b. $\vdash \mathtt{chris} \circ \mathtt{liked} \circ \mathtt{robin}; \mathbf{S}; \mathsf{like}(\mathsf{robin})(\mathsf{chris})$

The hyperintensional generalized quantifiers $\vdash \mathsf{every} : (\mathsf{e} \to \mathsf{p}) \to (\mathsf{e} \to \mathsf{p}) \to \mathsf{p}$ and $\vdash \mathsf{exists} : (\mathsf{e} \to \mathsf{p}) \to (\mathsf{e} \to \mathsf{p}) \to \mathsf{p}$ given in the lexicon above are related to their extensional counterparts via the following meaning postulates:

(9) a. $\forall_P \forall_Q \forall_w [\mathsf{every}(P)(Q)@w = \forall_x (P(x)@w \to Q(x)@w)]$

b. $\forall_P \forall_Q \forall_w [\mathsf{exists}(P)(Q)@w = \exists_x (P(x)@w \wedge Q(x)@w)]$

Below we show the entire proof of *Robin liked a dog*, to illustrate the mechanism for scoping *in situ* semantic operators which will be relevant to our analysis of interrogatives. First we assemble the quantificational noun phrase *a dog*:

(10) $$\frac{\vdash \lambda_p \lambda_f f(\mathsf{a} \circ \mathsf{p}); \mathbf{N} \multimap (\mathbf{NP} \multimap \mathbf{S}) \multimap \mathbf{S}; \mathsf{exists} \qquad \vdash \mathsf{dog}; \mathbf{N}; \mathsf{dog}}{\vdash \lambda_f f(\mathsf{a} \circ \mathsf{dog}); (\mathbf{NP} \multimap \mathbf{S}) \multimap \mathbf{S}; \mathsf{exists}(\mathsf{dog})}[\multimap E]$$

We make use of [Ax] to introduce a hypothesis corresponding to the object argument of the verb. Intuitively, this is the slot that the quantificational object will eventually lower itself into.

(11) $$\frac{\vdash \lambda_r \lambda_p p \circ \mathtt{liked} \circ r; \mathbf{NP} \multimap \mathbf{NP} \multimap \mathbf{S}; \mathsf{like} \qquad \overline{q; \mathbf{NP}; x \vdash q; \mathbf{NP}; x}^{[Ax]}}{q; \mathbf{NP}; x \vdash \lambda_p p \circ \mathtt{liked} \circ q; \mathbf{NP} \multimap \mathbf{S}; \mathsf{like}(x)}[\multimap E]$$

Then we proceed to combine the verb phrase missing its object, with its subject *Robin*. In a step of [$\multimap$I] we discharge the object hypothesis, λ abstracting on the free variables in the pheno and the semantic term. The quantificational noun phrase *a dog* can now scope over $\lambda_x \mathsf{see}(x)(\mathsf{robin})$, but its pheno term ensures that it is lowered into the object gap in one step of β reduction in the pheno logic. By 'gap' we simply mean a λ bound variable in a pheno term.

(12) $$\frac{(10) \qquad \dfrac{\dfrac{(11) \qquad \vdash \mathtt{robin}; \mathbf{NP}; \mathsf{robin}}{q; \mathbf{NP}; x \vdash \mathtt{robin} \circ \mathtt{liked} \circ q; \mathbf{S}; \mathsf{like}(x)(\mathsf{robin})}[\multimap E]}{\vdash \lambda_q \mathtt{robin} \circ \mathtt{liked} \circ q; \mathbf{NP} \multimap \mathbf{S}; \lambda_x \mathsf{like}(x)(\mathsf{robin})}[\multimap I]}{\vdash \mathtt{robin} \circ \mathtt{liked} \circ \mathtt{a} \circ \mathtt{dog}; \mathbf{S}; \mathsf{exists}(\mathsf{dog})(\lambda_x \mathsf{like}(x)(\mathsf{robin}))}[\multimap E]$$

We easily predict the ambiguity of a sentence with two quantificational expressions such as *Everyone saw a dog*. Since the context is a multiset and not a list, the subject and the object hypothesis that would be introduced in the proof of this sentence can be discharged in either order. If the object hypothesis is discharged first, we get the reading in (13). If the subject hypothesis is discharged first, we get the reading in (14). In both cases, the word order is the same since the quantificational expressions just lower themselves into the appropriate gap of their argument.

(13) $\vdash$ everyone $\circ$ saw $\circ$ a $\circ$ dog; $\mathbf{S}$; every(person)($\lambda_x$exists(dog)($\lambda_y$saw(y)(x)))

(14) $\vdash$ everyone $\circ$ saw $\circ$ a $\circ$ dog; $\mathbf{S}$; exists(dog)($\lambda_y$every(person)($\lambda_x$saw(y)(x)))

3 The Data

In this section, we briefly describe the data we will account for in Section 4. Due to considerations of space, we mainly focus on embedded interrogatives.

3.1 Interrogatives in Chinese

Like English, Chinese is an SVO language:

(15) Zhangsan xihuan Lisi.
Zhangsan like Lisi
'Zhangsan likes Lisi.'

Unlike English, Chinese has distinct interrogative verb forms which reduplicate the first syllable of the verb, with the morpheme *bu* 'not' separating the two copies. These forms are employed in polar questions, both root and embedded. The only difference between declaratives and polar interrogatives is the form of the finite verb (e.g. *xihuan* vs. *xi-bu-xihuan*).

(16) a. Zhangsan xi-bu-xihuan Lisi?
Zhangsan like? Lisi
'Does Zhangsan like Lisi?'

b. Chunsheng xiang-zhidao Zhangsan xi-bu-xihuan Lisi?
Chunsheng wonder Zhangsan like? Lisi
'Chunsheng wonders whether Zhangsan likes Lisi.'

Constituent questions contain interrogative (*wh*) expressions such as *shenme* 'what' or *shei* 'who'. These interrogative expressions appear *in situ*, i.e. in the same place in the clause where their non-interrogative counterparts appear. This is true of both main and embedded clauses:

(17) a. Zhangsan xihuan shenme?
Zhangsan like what
'What does Zhangsan like?'

b. Shei xihuan shenme?
who like what
'Who likes what?'

c. Zhangsan xiang-zhidao Lisi xihuan shei.
Zhangsan wonder Lisi like who
'Zhansang wonders who Lisi likes.'

Chinese *wh*-expressions can have arbitrarily wide scope, constrained solely by the properties of the embedding verb(s).

(18) Zhangsan xiang-zhidao shei xihuan shenme./?
Zhangsan wonder who like what
'Zhangsan wonders who likes what.'
'Who does Zhangsan wonder what (that person) likes?'
'What does Zhangsan wonder who likes?'

The preceding example is three-ways ambiguous. Both embedded *wh* expressions can have embedded scope. On this interpretation, the main clause is declarative, and the embedded clause is a binary constituent question. Alternatively, either of the embedded *wh* expressions can have root scope, resulting in an interpretation where both the main and the embedded clause are unary constituent questions. It is impossible, however, for both embedded *wh*-expressions to have root scope, since the embedding verb *xiang-zhidao* 'wonder' can only take interrogative but not declarative complements (much like its English counterpart).

3.2 Interrogatives in English

In English, embedded polar interrogatives are formed by means of the interrogative 'complementizer' *whether*, which takes a sentential complement, e.g. *Chris wonders whether Robin likes Sandy.*

In constituent questions, in contrast to Chinese, *wh*-expressions are not all *in situ*. Rather, a *wh* expression must occur on the extreme left periphery of a clause, and take scope at that clause, in order for the clause to be interpreted as a constituent question. Adopting an HPSG usage, we call such a left-peripheral *wh* expression a *filler.*[8]

(19) a. Chris wonders who Robin likes.
b. * Chris wonders Robin likes who.

(20) a. Chris wonders who Robin gave what.
b. * Chris wonders Robin gave who what.
c. * Chris wonders who what Robin gave.

By definition, a filler *wh*-expression can only have surface scope. By contrast, an *in situ wh*-expression can scope at or wider than the minimal clause in which it occurs, but the latter option is available only if the clause at which it scopes also has a filler:

(21) Chris wonders who$_x$ likes what$_y$.

[8] In mainstream generative grammar, such *wh*-expressions are analyzed as having undergone overt *wh* movement (string-vacuous movement in case the *wh*-expression in question is the subject of the root clause). Note that not every extreme left-peripheral *wh* expression is a filler. For example, in *Who thought which dog barked?*, *which dog* is not a filler, but rather an *in situ wh* expression with root scope.

a. 'Chris wonders who likes what'

b. # 'For which person x does Chris wonder which thing y is such that x likes y?'

c. # 'For which thing y does Chris wonder which person x is such that x likes y?'

(22) Who_x wonders who_y likes what_z?

a. 'Which person x is such that x wonders which person y and which thing z are such that y likes z?'
possible answer: Chris.

b. 'Which person x and which thing z are such that x wonders which person y is such that y likes z?'
possible answer: Chris wonders who likes beer.

c. #'Which person x and which person y are such that x wonders which thing z is such that y likes z?'
impossible answer: Chris wonders what Robin likes.

4 The Analysis

4.1 Polar Questions

Semantic Assumptions. Like Karttunen 1977, we analyze polar questions (meanings of both root and embedded polar interrogative clauses) as having extensions which are singleton sets of true answers. On our hyperintensional approach, this means that polar questions have the type $\mathsf{p} \rightarrow \mathsf{p}$, so that the extension at some world w is then a set of propositions $(\mathsf{p} \rightarrow \mathsf{t})$ – intuitively, the set of true answers to it. Thus, e.g. *whether Chris slept* or *Did Chris sleep?* denotes at some w a set with exactly one member: either the proposition that Chris slept or that he didn't, whichever is true at w. We abbreviate the polar question type $\mathsf{p} \rightarrow \mathsf{p}$ as k_0.

(23) $\mathsf{k}_0 = \mathsf{p} \rightarrow \mathsf{p}$

Now we introduce the constant $\vdash \mathsf{whether} : \mathsf{p} \rightarrow \mathsf{k}_0$ together with the following meaning postulate (nonlogical axiom):

(24) $\vdash \mathsf{whether} = \lambda_q \lambda_p [p \ \mathsf{and} \ ((p \ \mathsf{eq_p} \ q) \ \mathsf{or} \ (p \ \mathsf{eq_p} \ (\mathsf{not} \ q)))]$

In the definition of whether we made use of the propositional connectives and, or and not that translate the English sentential connectives *and*, *or* and *it's not the case that*. The following theorems (which follow directly from the facts that (i) the propositions form a preboolean algebra, and (ii) worlds are ultrafilters) relate these propositional connectives to their extensional counterparts:

(25) a. $\vdash \forall_p \forall_q \forall_w [(p \ \mathsf{and} \ q)@w = (p@w \wedge q@w)]$

b. $\vdash \forall_p \forall_q \forall_w [(p \ \mathsf{or} \ q)@w = (p@w \vee q@w)]$

c. $\vdash \forall_p \forall_w [(\mathsf{not}\ p)@w = \neg(p@w)]$

We also made use of the constant eq_p (we omit the subscript when the type is clear from the context). This is one of a family of constants eq_A of type $A \to A \to \mathsf{p}$ which are used to express, for each hyperintensional meaning type A, propositions that two meanings of type A are one and the same meaning. The following meaning postulate states that at any world w, the extension of eq_A at w is the ordinary equality relation on things of type A:

(26) $\vdash \forall_w \forall_x \forall_y [(x\ \mathsf{eq}\ y)@w = (x = y)]$

English Polar Questions. The constant whether is used as the semantics of the English interrogative complementizer *whether*, which has the following lexical entry:

$\vdash \lambda_p(\texttt{whether} \circ p); \mathbf{S} \multimap \bar{\mathbf{S}}; \mathsf{whether}$

Now we can generate embedded polar questions in English, such as:

(27) $\vdash \texttt{whether} \circ \texttt{chris} \circ \texttt{slept}; \bar{\mathbf{S}}; \mathsf{whether(sleep(chris))}$

The semantic term denotes a singleton set of propositions, as desired:

(28) $\vdash \forall_w \mathsf{whether(sleep(chris))}@w =$
$\lambda_p[p@w \wedge ((p = \mathsf{sleep(chris)}) \vee (p = (\mathsf{not(sleep(chris)})))))]$

Embedded interrogatives in English are assigned a distinct tectogrammatical type from root questions, since they are not interchangeable - *whether Chris slept* cannot be a root question, and *Did Chris sleep?* cannot be an embedded question, hence we must distinguish between $\bar{\mathbf{S}}$ and $\mathbf{S}$. (Of course, the same tectogrammatical distinction is made for declarative clauses.)

Chinese Polar Questions. In Chinese, root and embedded polar interrogatives are interchangeable so they are both assigned to the same tectogrammatical type $\mathbf{S}$[9]. Since there is no interrogative complementizer in Chinese, whether is packaged into the semantic term of the interrogative verb forms, which as we saw are distinct form from their declarative forming counterparts. We give the following toy lexicon for Chinese:

$\vdash \texttt{zhangsan}; \mathbf{NP}; \mathsf{zhangsan}$

$\vdash \texttt{lisi}; \mathbf{NP}; \mathsf{lisi}$

$\vdash \lambda_p \lambda_q q \circ \texttt{xihuan} \circ p; \mathbf{NP} \multimap \mathbf{NP} \multimap \mathbf{S}; \mathsf{like}$

$\vdash \lambda_p \lambda_q q \circ \texttt{xi-bu-xihuan} \circ p; \mathbf{NP} \multimap \mathbf{NP} \multimap \mathbf{S}; \lambda_x \lambda_y \mathsf{whether(like}(x)(y))$

[9] This is true of reduplicative interrogatives in Chinese, discussed in this paper. Polar interrogatives formed using the particle *ma* can only be root interrogatives.

Now we can derive the following examples:

(29) a. $\vdash$ **zhangsan** $\circ$ **xihuan** $\circ$ **lisi**; **S**; like(lisi)(zhangsan)

b. $\vdash$ **zhangsan** $\circ$ **xi-bu-xihuan** $\circ$ **lisi**; **S**; whether(like(lisi)(zhangsan))

In sum, the difference between English and Chinese embedded polar interrogatives is that (i) English distinguishes (tectogrammatically) between embedded and root interrogatives, while Chinese does not, and (ii) the form that contributes the interrogative meaning is the interrogative complementizer *whether* in English, while in Chinese it is the distinct verbal form.

4.2 Constituent Questions

Semantic Assumptions. We analyze n-ary questions as having extensions which are (curried) functions from n individuals to a singleton set of propositions. Recall that k_0 is the type of polar questions. We define the type of n-ary constituent question as follows:

(30) $\mathsf{k}_{n+1} = \mathsf{e} \to \mathsf{k}_n$

For example, the hyperintensional meaning type of a unary constituent question such as *who slept* is a function from individuals to properties of propositions (type $\mathsf{e} \to \mathsf{k}_0$). Such a question denotes at any world w a function from individuals to a singleton set of propositions (type $\mathsf{e} \to \mathsf{p} \to \mathsf{t}$), mapping each individual x to the singleton set whose member is either the proposition that x slept or the proposition that x didn't sleep, whichever is true at w.

The type of questions, k, is then defined to be the dependent coproduct of all question types, indexed by the natural numbers:

(31) $\mathsf{k} =_{def} \coprod_{n:\mathsf{n}} \mathsf{k}_\mathsf{n}$

Like quantificational expressions, *wh* expressions take scope. But unlike quantificational expressions, which bind an entity variable in a proposition to yield a proposition, *wh* expressions do not have a fixed result type. For example, in a unary constituent question, the unique *wh* expression binds an entity variable in proposition to yield a term of type k_1; in a binary constituent question, one *wh* expression will yield a term of type k_1 while the other one will bind an entity variable in that term to yield a term of type k_2.

We first define the *wh* expression that scopes over propositions to yield a unary wh question. Its semantic argument type is the same as for quantificational noun phrases ($\mathsf{e} \to \mathsf{p}$), but its result type is k_1.

(32) $\vdash \mathsf{who}' = \lambda_P \lambda_x \lambda_p[(\mathsf{person}\ x)\ \mathsf{and}\ (\mathsf{whether}\ (Px)\ p)]$

Combining who′ with $\lambda_x(\mathsf{sleep}\ x) : \mathsf{e} \to \mathsf{p}$ we get the desired meaning:

(33) $\vdash \mathsf{who}'(\lambda_x(\mathsf{sleep}\ x)) = \lambda_x \lambda_p[(\mathsf{person}\ x)\ \mathsf{and}\ (\mathsf{whether}\ (\mathsf{sleep}\ x)\ p)]$

Unlike who′ which combines with individual properties to yield unary questions, *wh* expressions that form (n+2)-ary constituent questions must combine with (n+1)-ary constituent questions 'missing' an entity argument. So they combine with terms of type $\mathsf{e} \rightarrow \mathsf{k}_{\mathsf{n+1}}$ to yield terms of type $\mathsf{k}_{\mathsf{n+2}}$. Note that $\mathsf{e} \rightarrow \mathsf{k}_{\mathsf{n+1}}$ is exactly the type $\mathsf{k}_{\mathsf{n+2}}$.

More formally, we recursively define a family of constants $\mathsf{who}_n : \mathsf{k}_{n+2} \rightarrow \mathsf{k}_{n+2}$ for *wh* expressions that scope over constituent questions and yield constituent questions. In the recursion clause, we make use of the polymorphic function $\mathsf{per}_{A,B,C} : (A \rightarrow B \rightarrow C) \rightarrow (B \rightarrow A \rightarrow C)$ that permutes the first two arguments of a function:

(34) $\vdash \mathsf{per} = \lambda_f \lambda_x \lambda_y (f\ y\ x)$

(35) a. $\vdash \mathsf{who}_0 = \lambda_k \lambda_x \lambda_y \lambda_p [(\mathsf{person}\ x)\ \mathsf{and}\ (k\ x\ y\ p)]$
b. $\vdash \mathsf{who}_{n+1} = \lambda_k [\mathsf{per}\ \lambda_x.\mathsf{who}_n (\mathsf{per}\ k\ x)]$

Essentially, all that who_n does is require of its argument's first argument that it be a person. We package all the constants who_n together into a single dependent product type:

(36) $\vdash \mathsf{who} = \lambda_{n:\mathrm{n}}.\mathsf{who}_n : \prod_{n:\mathrm{n}} (\mathsf{k}_{n+2} \rightarrow \mathsf{k}_{n+2})$

For ease of exposition, we started with the meanings of the interrogative pronoun *who*. The generalization to the interrogative determiner *which* is straightforward:

(37) $\vdash \mathsf{which}' = \lambda_Q \lambda_P \lambda_x \lambda_p [(Qx)\ \mathsf{and}\ (\mathsf{whether}\ (Px)\ p)]$

(38) a. $\vdash \mathsf{which}_0 = \lambda_Q \lambda_k \lambda_x \lambda_y \lambda_p [(Q\ x)\ \mathsf{and}\ (k\ x\ y\ p)]$
b. $\vdash \mathsf{which}_{n+1} = \lambda_Q \lambda_k [\mathsf{per}\ \lambda_x \mathsf{which}_n (Q)(\mathsf{per}\ k\ x)]$

(39) $\vdash \mathsf{which} = \lambda_{n:\mathrm{n}}.\mathsf{which}_n : \prod_{n:\mathrm{n}} [(\mathrm{e} \rightarrow \mathrm{p}) \rightarrow (\mathrm{k}_{n+2} \rightarrow \mathrm{k}_{n+2})]$

We leave it to the reader to formulate the semantic constants needed for the interrogative pronoun *what*.

English Constituent Questions. Recall that in English there is exactly one filler *wh* expression per constituent question. Any other ones appear *in situ*. We straightforwardly account for this fact by associating the filler with the semantic term who′, and the *in situ* ones with the semantic term who, intuitively one of the who_n constants.

Since the semantic argument and result type of who′ are distinct ($(\mathsf{e} \rightarrow \mathsf{p})$ vs. k_1), it is impossible for two fillers to occur in the same clause. And since all who_n constants require that there already be a constituent question for them to scope over, and who′ is the unique constant that turns propositions into constituent questions, the presence of a filler *wh* expression is necessary for any *in situ* ones to occur. So, we guarantee that there is exactly one filler per constituent question.

Following Muskens 2007b, the fronting of the filler to the left periphery is accomplished entirely in the phenogrammar. We give the following lexical entries for English *wh* expressions:

(40) English filler *wh*-expressions:

$\vdash \lambda_f[\mathtt{who} \circ f(\mathsf{e})]; (\mathbf{NP} \multimap \mathbf{S}) \multimap \bar{\mathbf{S}}; \mathsf{who}'$

$\vdash \lambda_f[\mathtt{what} \circ f(\mathsf{e})]; (\mathbf{NP} \multimap \mathbf{S}) \multimap \bar{\mathbf{S}}; \mathsf{what}'$

$\vdash \lambda_p\lambda_f[\mathtt{which} \circ p \circ f(\mathsf{e})]; \mathbf{N} \multimap (\mathbf{NP} \multimap \mathbf{S}) \multimap \bar{\mathbf{S}}; \mathsf{which}'$

(41) English *in situ wh*-expressions:

$\vdash \lambda_f[f(\mathtt{who})]; (\mathbf{NP} \multimap \mathbf{S}) \multimap \bar{\mathbf{S}}; \mathsf{who}$

$\vdash \lambda_f[f(\mathtt{what})]; (\mathbf{NP} \multimap \mathbf{S}) \multimap \bar{\mathbf{S}}; \mathsf{what}$

$\vdash \lambda_p\lambda_f[f(\mathtt{which} \circ p)]; \mathbf{N} \multimap (\mathbf{NP} \multimap \mathbf{S}) \multimap \bar{\mathbf{S}}; \mathsf{which}$

Note that the pheno terms of *in situ wh*-expressions have the same structure as those of quantificational noun phrases. So, the *in situ wh*-expressions just lower themselves into the appropriate gap of their argument.

The pheno terms of the filler expressions are different. Instead of lowering themselves into the gap of their argument, they concatenate themselves to the left of their argument after feeding it the empty string e which effectively plugs the existing gap. This is how preposing of the fillers is accomplished. So, here are the kinds of embedded questions that our grammar can now generate:

(42) unary constituent questions

a. $\vdash \mathtt{who} \circ \mathtt{liked} \circ \mathtt{robin}; \bar{\mathbf{S}}; \mathsf{who}'(\lambda_x(\mathsf{like}(\mathsf{robin})(x))$

b. $\vdash \mathtt{which} \circ \mathtt{dog} \circ \mathtt{slept}; \bar{\mathbf{S}}; \mathsf{which}'(\mathsf{dog})(\lambda_x(\mathsf{sleep}(x))$

c. $\vdash \mathtt{which} \circ \mathtt{dog} \circ \mathtt{chris} \circ \mathtt{liked}; \bar{\mathbf{S}}; \mathsf{which}'(\mathsf{dog})(\lambda_x(\mathsf{like}(x)(\mathsf{chris}))$

(43) binary constituent questions

a. $\vdash \mathtt{who} \circ \mathtt{liked} \circ \mathtt{who}; \bar{\mathbf{S}}; \mathsf{who}_0(\lambda_y \mathsf{who}'(\lambda_x(\mathsf{like}(y)(x))))$

b. $\vdash \mathtt{which} \circ \mathtt{dog} \circ \mathtt{who} \circ \mathtt{liked}; \bar{\mathbf{S}}; \mathsf{who}_0(\lambda_y \mathsf{which}'(\mathsf{dog})(\lambda_x(\mathsf{like}(x)(y))))$

Given the pheno terms of filler *wh*-expressions, we automatically predict that they must scope over the clause on whose left periphery they occur. The *in situ* expressions, on the other hand, can scope higher than their surface position would suggest, since they can lower themselves into the right gap from virtually anywhere.

However, because of the semantic typing, the *in situ wh*-expressions are dependent on there already being some filler in the clause over which they are to scope. So we predict that they can only scope higher than their surface position would suggest in case the matrix clause already contains a filler *wh*-expression (in accordance with the data laid out in the preceding section).

Suppose we have the following lexical entry for *wonders*:

$$\vdash \lambda_p\lambda_q q \circ \mathtt{wonders} \circ p; \ \bar{\mathbf{S}} \multimap \mathbf{NP} \multimap \mathbf{S}; \mathsf{wonder}$$

where wonder has type $\mathsf{k} \rightarrow \mathsf{e} \rightarrow \mathsf{p}$. Then we correctly predict the ambiguity of embedded clauses such as *who wonders which dog liked what* (as in *Robin asked me who wonders which dog liked what*), depending on whether *what* has embedded or matrix scope. Below we show the two derivable semantic terms:

(44) *who wonders which dog liked what*

a. $\vdash$ $\mathsf{who'}(\lambda_z \mathsf{wonder}(\mathsf{what_0}(\lambda_y(\mathsf{which'}(\mathsf{dog})(\lambda_x(\mathsf{like}(y)(x))))(z)))$: $\mathsf{k_1}$

b. $\vdash$ $\mathsf{what_0}(\lambda_y \mathsf{who'}(\lambda_z \mathsf{wonder}((\mathsf{which'}(\mathsf{dog})(\lambda_x(\mathsf{like}(y)(x))))(z)))$: $\mathsf{k_2}$

Chinese Constituent Questions. Unlike English, Chinese has no filler *wh*-expressions, but only *in situ* ones. So, all *wh* expressions are assigned to the same kind of pheno term and are systematically ambiguous between who′ and who (what′ and what). We add the following lexical entries:

(45) Chinese *wh* pronouns

$\vdash \lambda_f[f(\texttt{shei})]; (\mathbf{NP} \multimap \mathbf{S}) \multimap \mathbf{S}; \mathsf{who'}$

$\vdash \lambda_f[f(\texttt{shei})]; (\mathbf{NP} \multimap \mathbf{S}) \multimap \mathbf{S}; \mathsf{who}$

$\vdash \lambda_f[f(\texttt{shenme})]; (\mathbf{NP} \multimap \mathbf{S}) \multimap \mathbf{S}; \mathsf{what'}$

$\vdash \lambda_f[f(\texttt{shenme})]; (\mathbf{NP} \multimap \mathbf{S}) \multimap \mathbf{S}; \mathsf{what}$

Now we can generate examples like the following:

(46) a. 'what Zhangsan likes'/'What does Zhangsan like?'

$\vdash \texttt{zhangsan} \circ \texttt{xihuan} \circ \texttt{shenme}; \mathbf{S}; \mathsf{what'}(\lambda_x \mathsf{like}(x)(\mathsf{zhangsan}))$

b. 'who likes Lisi'/'Who likes Lisi?'

$\vdash \texttt{shei} \circ \texttt{xihuan} \circ \texttt{lisi}; \mathbf{S}; \mathsf{who'}(\lambda_x \mathsf{like}(\mathsf{lisi})(x))$

c. 'who likes what'/'Who likes what?'

$\vdash \texttt{shei} \circ \texttt{xihuan} \circ \texttt{shenme}; \mathbf{S}; \mathsf{what_0}(\lambda_y \mathsf{who'}(\lambda_x \mathsf{like}(y)(x)))$

$\vdash \texttt{shei} \circ \texttt{xihuan} \circ \texttt{shenme}; \mathbf{S}; \mathsf{who_0}(\lambda_x \mathsf{what'}(\lambda_y \mathsf{like}(y)(x)))$

Note the insignificant ambiguity of binary questions such as the one in (46c) depending on which *wh* expression is scoped first.[10]

Since all *wh* expressions in Chinese lower themselves into their argument's gap, they can scope arbitrarily high. We give the following lexical entry for *xiang-zhidao* 'wonder':

$$\vdash \lambda_p \lambda_q q \circ \texttt{xiang-zhidao} \circ p; \mathbf{S} \multimap \mathbf{NP} \multimap \mathbf{S}; \mathsf{wonder}$$

Our grammar predicts that embedded *wh* expressions can in fact have root scope. All of the following semantic terms are derivable for the sentence in (47):

(47) Zhangsan xiang-zhidao shei xihuan shenme./?

a. 'Zhangsan wonders who likes what.'

$\vdash \mathsf{wonder}(\mathsf{what_0}(\lambda_y \mathsf{who'}(\lambda_x \mathsf{like}(y)(x))))(\mathsf{zhangsan})$

$\vdash \mathsf{wonder}(\mathsf{who_0}(\lambda_x \mathsf{what'}(\lambda_y \mathsf{like}(y)(x))))(\mathsf{zhangsan})$

b. 'Who does Zhangsan wonder what (that person) likes?'

[10] Insignificant in the sense that, at any world w, the two readings have the same extension.

$\vdash \mathsf{who_0}(\lambda_x \mathsf{wonder}(\mathsf{what}'(\lambda_y \mathsf{like}(y)(x)))(\mathsf{zhangsan}))$

c. 'What does Zhangsan wonder who likes?'

$\vdash \mathsf{what_0}(\lambda_y \mathsf{wonder}(\mathsf{who}'(\lambda_x \mathsf{like}(y)(x)))(\mathsf{zhangsan}))$

It is, however, impossible for both embedded *wh* expressions to have root scope, not because of the tectogrammatical type of *xiang-zhidao*, but because of its *semantic* type: it needs an argument of type k.

5 Discussion and Conclusion

We hope to have shown that PTG is a suitable framework for analyses of complex linguistic phenomena such as interrogatives. The explicit separation of phenogrammar and tectogrammar allows for a surprisingly simple analysis of vastly different strategies for forming *wh* questions, bringing out uniformities in combinatorics and interpretation of questions in English and Chinese, while identifying phenogrammar as the locus of cross-linguistic variation. The difference between overt and covert movement is analyzed as a purely phenogrammatical lexical difference.

In comparison to Vermaat's (2005) multi-modal approach, our tectogrammar is considerably simpler, with a single order-insensitive type constructor and just a handful of linguistically motivated types. While in this paper we cannot even approach the empirical breadth of Vermaat (2005), we would like to extend the analysis of interrogatives presented here to languages with different question-forming strategies, e.g. multiple-fronting languages such as Serbo-Croatian.

Japanese also presents an interesting case because, while it is a *wh in situ* language, it makes use of question markers (e.g. *ka*) which occur both in polar and constituent questions. It has been claimed that *wh*-expressions must scope at the minimal clause that contains them and is marked as a question (e.g. Nishigauchi 1990), but this is not uncontroversial (see Takahashi 1993, Kitagawa 2005).

If the scope of Japanese *wh*-expressions really is thus constrained, the architecture we employ in this paper will have to be elaborated in some way to allow for restrictions on the scope of *in situ* operators. We would like to suggest that this may be accomplished by recoding our analysis into a 'direct-style' framework where *in situ* operators are not type-raised in phenogrammar, but rather are scoped via (nonconfluent) reduction in the semantic calculus using polymorphic shift operators. Some such mechanism may then also be used to account for any other scope island effects.

References

Baldridge, Jason, Lexically Specied Derivational Control in Combinatory Categorial Grammar. Ph.D. thesis. University of Edinburgh (2002)

Curry, H.: Some logical aspects of grammatical structure. In: Jakobson, R. (ed.) Structure of Language and Its Mathematical Aspects (1961)

de Groote, P.: Towards Abstract Categorial Grammar. In: Proceedings of ACL (2001)
Karttunen, L.: Syntax and semantics of questions. Linguistics and Philosophy 1, 1 (1977)
Kitagawa, Y.: Prosody, syntax and pragmatics of WH-questions. English Linguistics 22, 2 (2005) (in Japanese)
Joachim, L., Scott, P.J.: Introduction to Higher Order Categorical Logic. Cambridge University Press (1986)
Montague, R.: The proper treatment of quantification in English. In: Thomason, R. (ed.) Formal Philosophy: Selected Papers of Richard Montague, pp. 247–270. Yale University Press, New Haven (1974)
Moortgat, M.: Generalized quantification and discontinuous type constructors. In: Bunt, H., van Horck, A. (eds.) Discontinuous Constituency. Mouton de Gruyter (1996)
Moortgat, M.: Categorial type logics. In: van Benthem, J., ter Meulen, A. (eds.) Handbook of Logic and Language. Elsevier, Amsterdam (1997)
Glyn, M., Fada, M., Valentin, O.: Nondeterministic discontinuous Lambek calculus. In: Proceedings of the Seventh International Workshop on Computational Semantics (IWCS 2007), Tilburg (2007)
Muskens, R.: Language, lambdas, and logic. In: Kruijff, G.-J., Oehrle, R. (eds.) Resource Sensitivity in Binding and Anaphora. Studies in Linguistics and Philosophy. Kluwer (2003)
Muskens, R.: Sense and the computation of reference. Linguistics and Philosophy 28, 4 (2005)
Muskens, R.: Intensional models for the theory of types. The Journal of Symbolic Logic 72, 1 (2007a)
Muskens, R.: Separating syntax and combinatorics in categorial grammar. Research on Language and Computation 5, 3 (2007b)
Nishigauchi, T.: Quantification in the Theory of Grammar. Kluwer, Dordrecht (1990)
Oehrle, R.: Term-labeled categorial type systems. Linguistics and Philosophy 17, 6 (1994)
Pollard, C.: Type-Logical HPSG. In: Jäger, G., Monachesi, P., Penn, G., Wintner, S. (eds.) Proceedings of Formal Grammar 2004, pp. 107–124. European Summer School in Language, Logic, and Information, Nancy (2004)
Pollard, C.: Hyperintensions. Journal of Logic and Computation 18, 2 (2008a)
Pollard, C.: Hyperintensional Questions. In: Hodges, W., de Queiroz, R. (eds.) WoLLic 2008. LNCS (LNAI), vol. 5110, pp. 272–285. Springer, Heidelberg (2008)
Smith, E.A.: Correlational Comparison in English. Ph.D. dissertation, Department of Linguistics. The Ohio State University (2010)
Takahashi, D.: Movement of WH-phrases in Japanese. Natural Language and Linguistic Theory 11, 4 (1993)
Thomason, R.: A model theory for propositional attitudes. Linguistics and Philosophy 4, 1 (1980)
Vermaat, W.: The Logic of Variation. A Cross-Linguistic Account of Wh-Question Formation. Ph.D. thesis. Utrecht Institute of Linguistics OTS, Utrecht University (2005)

Generalized Discontinuity

Glyn Morrill[1] and Oriol Valentín[2]

[1] Departament de Llenguatges i Sistemes Informàtics
Universitat Politècnica de Catalunya
morrill@lsi.upc.edu
http: //www-lsi.upc.edu/~morrill/
[2] Barcelona Media, Centre d'Innovació
Universitat Pompeu Fabra
oriol.valentin@upf.edu

Abstract. We define and study a calculus of discontinuity, a version of displacement calculus, which is a logic of segmented strings in exactly the same sense that the Lambek calculus is a logic of strings. Like the Lambek calculus, the displacement calculus is a sequence logic free of structural rules, and enjoys Cut-elimination and its corollaries: the subformula property, decidability, and the finite reading property. The foci of this paper are a formulation with a finite number of connectives, and consideration of how to extend the calculus with defined connectives while preserving its good properties.

1 Introduction: Architecture of Logical Grammar

An *argument* in logic comprises some premises and a conclusion; for example:[1]

(1) a. All men are mortal.
Socrates is a man.
―――――――――
Socrates is mortal.

b. All men are mortal.
Socrates is mortal.
―――――――――
Socrates is a man.

If in an argument the truth of the premises guarantees the truth of the conclusion, the argument is *logical*. If the truth of the premises does not guarantee the truth of the conclusion, the argument is *not logical*. The argument (1a) is logical: independently of the facts of the real world, who Socrates is, etc., if the premises are true then the conclusion must be true. The argument (1b) is not logical: again disregarding how the world actually is, it is possible for the premises to be true but the conclusion false.

In a logical theory premises and conclusions are represented by *formulas*, and we then call an argument a *sequent*. For example, corresponding to (1) there are the sequents:

(2) a. $\forall x(Hx \to Mx), Hs \Rightarrow Ms$
b. $\forall x(Hx \to Mx), Ms \Rightarrow Hs$

[1] The research reported in the present paper was supported by DGICYT project SESAAME-BAR (TIN2008-06582-C03-01).

P. de Groote and M.-J. Nederhof (Eds.): Formal Grammar 2010/2011, LNCS 7395, pp. 146–161, 2012.

If a sequent $\Gamma \Rightarrow A$ is logical we call it a *theorem* and write $\vdash \Gamma \Rightarrow A$. If it is not logical it is not a theorem and we write $\not\vdash \Gamma \Rightarrow A$. Thus a logical theory takes the form shown in Figure 1.

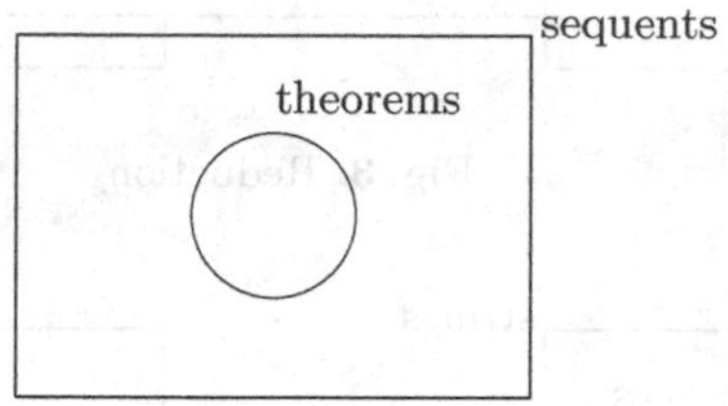

Fig. 1. Logic

A sentence comprises a string of words. Some strings of words are well-formed as sentences and we say they are grammatical, for example *John walks*; others are not well-formed as sentences and we say they are ungrammatical, for example **walks John*. Thus grammar takes the form shown in Figure 2.

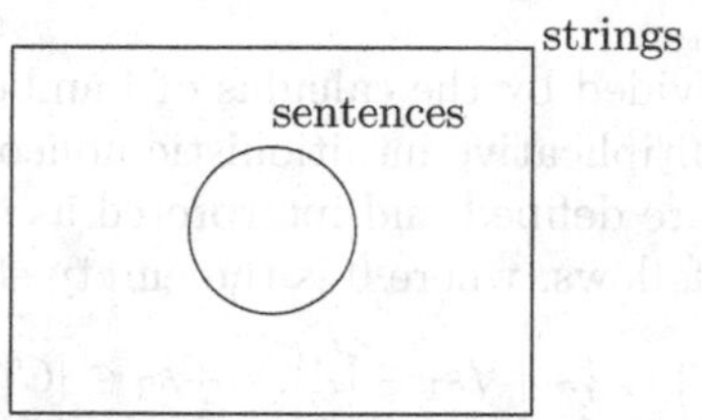

Fig. 2. Grammar

Given a subset of a domain, such as the subset of sequents that are theorems or the subset of strings that are sentences, there is the associated computational decision problem of determining whether an element of the domain belongs to the subset.

A *reduction* of one problem to another is an answer-preserving mapping from the domain of the first problem to the domain of the second problem. Thus a reduction sends members to members and nonmembers to nonmembers as shown in Figure 3. The existence of a reduction from one problem to a second means that the first problem can be solved by the composition of an algorithm for the second problem with an algorithm computing the reduction.

Logical grammar is a reduction of grammar to logic: a string is a sentence if and only if an associated sequent (or one of a set of associated sequents) is a theorem, as shown in Figure 4. Hence in logical grammar, determining grammatical properties is reduced to theorem-proving.

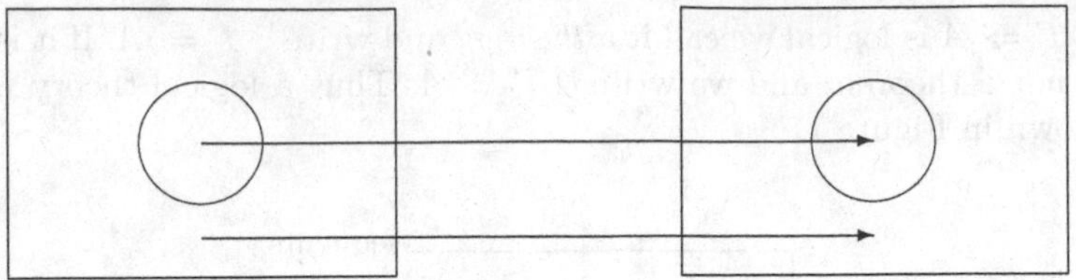

Fig. 3. Reduction

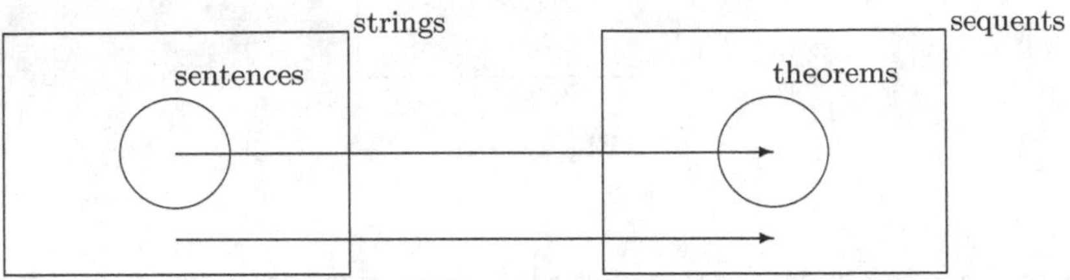

Fig. 4. Logical grammar

2 Logic of Strings: The Lambek Calculus L

Logic of strings is provided by the calculus of Lambek (1958)[4]. We consider a variant **L** which is multiplicative intuitionistic noncommutative linear logic.

The types $\mathcal{F}$ of **L** are defined and interpreted as subsets of the set of strings over a vocabulary as follows, where 0 is the empty string:

(3)
$$\begin{array}{lll} \mathcal{F} ::= \mathcal{F}\backslash\mathcal{F} & [A\backslash C] = \{s_2 |\ \forall s_1 \in [A], s_1{+}s_2 \in [C]\} & \text{under} \\ \mathcal{F} ::= \mathcal{F}/\mathcal{F} & [C/B] = \{s_1 |\ \forall s_2 \in [B], s_1{+}s_2 \in [C]\} & \text{over} \\ \mathcal{F} ::= \mathcal{F}\bullet\mathcal{F} & [A\bullet B] = \{s_1{+}s_2 |\ s_1 \in [A] \ \&\ s_2 \in [B]\} & \text{product} \\ \mathcal{F} ::= I & [I] = \{0\} & \text{product unit} \end{array}$$

The set $\mathcal{O}$ of *configurations* is defined as follows, where Λ is the empty string:[2]

(4) $\mathcal{O} ::= \Lambda \mid \mathcal{F} \mid \mathcal{O}, \mathcal{O}$

A *sequent* $\Gamma \Rightarrow A$ comprises an antecedent configuration Γ and a succedent type A. The sequent calculus of **L** is as shown in Figure 5, where $\Delta(\Gamma)$ signifies a configuration Δ with a distinguished subconfiguration Γ.

The *Cut-elimination* property of a logic is that every theorem has a Cut-free proof. Lambek (1958)[4] proved Cut-elimination for **L** without the product unit I; Lambek (1969)[5] showed that there is also Cut-elimination when the product unit is included. Cut-elimination has a series of good consequences.

Firstly, Cut-elimination means that the calculus has the *subformula property*: that every theorem has a proof containing only its subformulas. This is so because

[2] Note that this grammar is ambiguous, but that this does not matter because the product is associative.

$$\frac{}{A \Rightarrow A}\,id \qquad \frac{\Gamma \Rightarrow A \qquad \Delta(A) \Rightarrow B}{\Delta(\Gamma) \Rightarrow B}\,Cut$$

$$\frac{\Gamma \Rightarrow A \qquad \Delta(C) \Rightarrow D}{\Delta(\Gamma, A\backslash C) \Rightarrow D}\,\backslash L \qquad \frac{A, \Gamma \Rightarrow C}{\Gamma \Rightarrow A\backslash C}\,\backslash R$$

$$\frac{\Gamma \Rightarrow B \qquad \Delta(C) \Rightarrow D}{\Delta(C/B, \Gamma) \Rightarrow D}\,/L \qquad \frac{\Gamma, B \Rightarrow C}{\Gamma \Rightarrow C/B}\,/R$$

$$\frac{\Delta(A, B) \Rightarrow D}{\Delta(A{\bullet}B) \Rightarrow D}\,{\bullet}L \qquad \frac{\Gamma \Rightarrow A \qquad \Delta \Rightarrow B}{\Gamma, \Delta \Rightarrow A{\bullet}B}\,{\bullet}R$$

$$\frac{\Delta(\Lambda) \Rightarrow A}{\Delta(I) \Rightarrow A}\,IL \qquad \frac{}{\Lambda \Rightarrow I}\,IR$$

Fig. 5. Sequent calculus for **L**

every rule except Cut has the property that every type in the premises is either the same as, or is an immediate subtype of, a type in the conclusion. Thus every Cut-free proof has the subformula property, and by Cut-elimination every theorem has a Cut-free proof.

Secondly, Cut-elimination means that the calculus is *decidable*. Cut-elimination does not always have this consequence, for example full propositional linear logic enjoys Cut-elimination but is not decidable. But it follows in the present case because of the finiteness of the Cut-free search space without contraction. Every rule except Cut has the property that when a sequent is matched against the conclusions of the rule, there are only a finite number of premises from which it could have been inferred by the rule. The space of Cut-free backward chaining sequent proof search is finite. Thus, whether a sequent has a Cut-free proof can be determined in finite time, and by Cut-elimination a sequent is a theorem if and only if it has a Cut-free proof.

Thirdly, Cut-elimination means that the calculus has the *finite reading property*. Again, this does not always hold, for example intuitionistic propositional logic enjoys Cut-elimination but not the finite reading property. But here there is no contraction. Curry-Howard categorial semantics compositionally associates each proof with a derivational semantics which is its homomorphic image as an intuitionistic proof or typed lambda term. Equivalence of such semantic readings is preserved by Cut-elimination. Since the Cut-free sequent proof search space is finite, every sequent can have only a finite number of nonequivalent proofs, and hence only a finite number of semantic readings.

The Lambek calculus **L** thus has good proof-theoretic properties as a logic of strings, but as is well known, logical syntax and semantics developed on this basis does not accommodate non-peripheral discontinuities. For example, a relative pronoun type $R/(S/N)$ will produce unboundedly long-distance extraction from clause-final positions, but not clause-medial extraction such as *man who John saw today*. And a quantifier phrase type $S/(N\backslash S)$ will produce subject position quantification, and a further quantifier phrase type $(S/N)\backslash S$ will produce in addition sentence-final quantification, but neither of these types will produce sentence-medial quantification such as *John introduced everyone to Mary*.

Overall, the Lambek calculus cannot accommodate the syntax and semantics of:

(5) Discontinuous idioms (*Mary gave the man the cold shoulder*). Quantification (*John gave every book to Mary*; *Mary thinks someone left*; *Everyone loves someone*). VP ellipsis (*John slept before Mary did*; *John slept and Mary did too*). Medial extraction (*dog that Mary saw today*). Pied-piping (*mountain the painting of which by Cezanne John sold for $10,000,000*. Appositive relativization (*John, who jogs, sneezed*). Parentheticals (*Fortunately, John has perseverance*; *John, fortunately, has perseverance*; *John has, fortunately, perseverance*; *John has perseverance, fortunately*). Gapping (*John studies logic, and Charles, phonetics*). Comparative subdeletion (*John ate more donuts than Mary bought bagels*). Reflexivization (*John sent himself flowers*).

Furthermore, since the Lambek calculus is context-free in generative power (Pentus 1992)[15] it cannot generate cross-serial dependencies as in Dutch and Swiss-German (Sheiber 1985[16]).

3 Logic of Segmented Strings: The Displacement Calculus D

By *segmented strings* we mean strings over a vocabulary containing a distinguished symbol 1 which we call the *separator*. We define the *sort* of a segmented string as the number of separators it contains. Henceforth, by 'string' we shall mean 'segmented string'.

Morrill and Valentín (2010)[11] defines displacement calculus with k-ary wrapping, $k > 0$, meaning wrapping around the kth separator. Here we consider a variant **D** which is a logic of segmented strings which has continuous connectives $\{\backslash, /, \bullet\}$ for concatenation and discontinuous connectives $\{\downarrow_k, \uparrow_k, \odot_k\}_{k\in\{>,<\}}$ for left and right wrapping. The characteristic feature of this variant is that it has only a finite number of connectives. We consider also here some defined connectives for which rules are compiled.

The concatenation of a string of sort i with a string of sort j is a string of sort $i + j$. But in addition to concatenation, we define on (segmented) strings two operations of intercalation or 'wrap'. Where α and β are segmented strings and the sort of α is at least 1, we define the *left wrap* of α around β, $\alpha \times_> \beta$ as the result of replacing the leftmost separator in α by β, and we define the *right wrap* of α around β, $\alpha \times_< \beta$ as the result of replacing the rightmost separator in α by β. For example:

(6) **before+1+left+1+slept** $\times_<$ **the+man** = **before+1+left+the+man+ slept**

The types of **D** are sorted into types $\mathcal{F}_i$ of sort i interpreted as sets of strings of sort i as shown in Figure 6 where $k \in \{>, <\}$; the left hand column displays the definition of the types in Backus-Naur form, and $[A]$ where A is a type represents the natural syntactical interpretation of a type in terms of (separated) strings. The set $\mathcal{O}$ of *configurations* is defined as follows, where $[]$ is the metalinguistic

$\mathcal{F}_j ::= \mathcal{F}_i \backslash \mathcal{F}_{i+j}$	$[A\backslash C] = \{s_2 \mid \forall s_1 \in [A], s_1{+}s_2 \in [C]\}$	under
$\mathcal{F}_i ::= \mathcal{F}_{i+j} / \mathcal{F}_j$	$[C/B] = \{s_1 \mid \forall s_2 \in [B], s_1{+}s_2 \in [C]\}$	over
$\mathcal{F}_{i+j} ::= \mathcal{F}_i \bullet \mathcal{F}_j$	$[A \bullet B] = \{s_1{+}s_2 \mid s_1 \in [A] \;\&\; s_2 \in [B]\}$	product
$\mathcal{F}_0 ::= I$	$[I] = \{0\}$	product unit
$\mathcal{F}_j ::= \mathcal{F}_{i+1} \downarrow_k \mathcal{F}_{i+j}$	$[A \downarrow_k C] = \{s_2 \mid \forall s_1 \in [A], s_1 \times_k s_2 \in [C]\}$	infix
$\mathcal{F}_{i+1} ::= \mathcal{F}_{i+j} \uparrow_k \mathcal{F}_j$	$[C \uparrow_k B] = \{s_1 \mid \forall s_2 \in [B], s_1 \times_k s_2 \in [C]\}$	extract
$\mathcal{F}_{i+j} ::= \mathcal{F}_{i+1} \odot_k \mathcal{F}_j$	$[A \odot_k B] = \{s_1 \times_k s_2 \mid s_1 \in [A] \;\&\; s_2 \in [B]\}$	disc. product
$\mathcal{F}_1 ::= J$	$[J] = \{1\}$	disc. prod. unit

Fig. 6. Types of the displacement calculus **D** and their interpretation

separator:

(7) $\mathcal{O} ::= \Lambda \mid [] \mid \mathcal{F}_0 \mid \mathcal{F}_{i+1}\{\underbrace{\mathcal{O} : \ldots : \mathcal{O}}_{i+1\ \mathcal{O}'s}\} \mid \mathcal{O}, \mathcal{O}$

$A\{\Delta_1 : \ldots : \Delta_n\}$ interpreted syntactically is formed by strings $\alpha_0{+}\beta_1{+}\alpha_1{+}\cdots{+}\alpha_{n-1}{+}\beta_n{+}\alpha_n$ where $\alpha_0{+}1{+}\alpha_1{+}\cdots{+}\alpha_{n-1}{+}1{+}\alpha_n \in A$ and $\beta_1 \in \Delta_1, \ldots, \beta_n \in \Delta_n$. Where A is a type we call its sort sA. The *figure* $\overrightarrow{A}$ of a type A is defined by:

(8) $$\overrightarrow{A} = \begin{cases} A & \text{if } sA = 0 \\ A\{\underbrace{[] : \ldots : []}_{sA\ []'s}\} & \text{if } sA > 0 \end{cases}$$

The sort of a configuration is the number of metalinguistic separators it contains. Where Γ and Φ are configurations and the sort of Γ is at least 1, $\Gamma|_{>}\Phi$ signifies the configuration which is the result of replacing the leftmost separator in Γ by Φ, and $\Gamma|_{<}\Phi$ signifies the configuration which is the result of replacing the rightmost separator in Γ by Φ. Where Γ is a configuration of sort i and $\Phi_1, \ldots, \Phi_i$ are configurations, the *generalized wrap* $\Gamma \otimes \langle \Phi_1, \ldots, \Phi_i \rangle$ is the result of simultaneously replacing the successive separators in Γ by $\Phi_1, \ldots, \Phi_i$ respectively. $\Delta\langle\Gamma\rangle$ abbreviates $\Delta_0(\Gamma \otimes \langle \Delta_1, \ldots, \Delta_i \rangle)$. Thus where the usual distinguished occurrence notation $\Delta(\Gamma)$ represents a subconfiguration Γ with an *external* context Δ, our distinguished hyperconfiguration notatation $\Delta\langle\Gamma\rangle$ represents a subconfiguration Γ with *external* context Δ_0 and also *internal* contexts $\Delta_1, \ldots, \Delta_i$. A *sequent* $\Gamma \Rightarrow A$ comprises an antecedent configuration Γ of sort i and a succedent type A of sort i. The sequent calculus for the calculus of displacement **D** is as shown in Figure 7 where $k \in \{>, <\}$. Like **L**, **D** has no structural rules.

$$\frac{}{\overrightarrow{A} \Rightarrow A}\, id \qquad \frac{\Gamma \Rightarrow A \qquad \Delta\langle\overrightarrow{A}\rangle \Rightarrow B}{\Delta\langle\Gamma\rangle \Rightarrow B}\, Cut$$

$$\frac{\Gamma \Rightarrow A \qquad \Delta\langle\overrightarrow{C}\rangle \Rightarrow D}{\Delta\langle\Gamma, \overrightarrow{A\backslash C}\rangle \Rightarrow D}\, \backslash L \qquad \frac{\overrightarrow{A}, \Gamma \Rightarrow C}{\Gamma \Rightarrow A\backslash C}\, \backslash R$$

$$\frac{\Gamma \Rightarrow B \qquad \Delta\langle\overrightarrow{C}\rangle \Rightarrow D}{\Delta\langle\overrightarrow{C/B}, \Gamma\rangle \Rightarrow D}\, /L \qquad \frac{\Gamma, \overrightarrow{B} \Rightarrow C}{\Gamma \Rightarrow C/B}\, /R$$

$$\frac{\Delta\langle\overrightarrow{A}, \overrightarrow{B}\rangle \Rightarrow D}{\Delta\langle\overrightarrow{A \bullet B}\rangle \Rightarrow D}\, \bullet L \qquad \frac{\Gamma_1 \Rightarrow A \qquad \Gamma_2 \Rightarrow B}{\Gamma_1, \Gamma_2 \Rightarrow A \bullet B}\, \bullet R$$

$$\frac{\Delta\langle\Lambda\rangle \Rightarrow A}{\Delta\langle\overrightarrow{I}\rangle \Rightarrow A}\, IL \qquad \frac{}{\Lambda \Rightarrow I}\, IR$$

$$\frac{\Gamma \Rightarrow A \qquad \Delta\langle\overrightarrow{C}\rangle \Rightarrow D}{\Delta\langle\Gamma|_k \overrightarrow{A{\downarrow_k}C}\rangle \Rightarrow D}\, {\downarrow_k} L \qquad \frac{\overrightarrow{A}|_k \Gamma \Rightarrow C}{\Gamma \Rightarrow A{\downarrow_k}C}\, {\downarrow_k} R$$

$$\frac{\Gamma \Rightarrow B \qquad \Delta\langle\overrightarrow{C}\rangle \Rightarrow D}{\Delta\langle\overrightarrow{C{\uparrow_k}B}|_k \Gamma\rangle \Rightarrow D}\, {\uparrow_k} L \qquad \frac{\Gamma|_k \overrightarrow{B} \Rightarrow C}{\Gamma \Rightarrow C{\uparrow_k}B}\, {\uparrow_k} R$$

$$\frac{\Delta\langle\overrightarrow{A}|_k \overrightarrow{B}\rangle \Rightarrow D}{\Delta\langle\overrightarrow{A \odot_k B}\rangle \Rightarrow D}\, \odot_k L \qquad \frac{\Gamma_1 \Rightarrow A \qquad \Gamma_2 \Rightarrow B}{\Gamma_1|_k \Gamma_2 \Rightarrow A \odot_k B}\, \odot_k R$$

$$\frac{\Delta\langle[\,]\rangle \Rightarrow A}{\Delta\langle\overrightarrow{J}\rangle \Rightarrow A}\, JL \qquad \frac{}{[\,] \Rightarrow J}\, JR$$

Fig. 7. Sequent calculus for **D**

Morrill and Valentín (2010)[11] proves Cut-elimination for the k-ary displacement calculus, $k > 0$, and the variant **D** considered here enjoys Cut-elimination by the same reasoning since left wrap is the same as first wrap, and right wrap is k-ary wrap with k the corresponding maximal sort; see Morrill, Valentín and Fadda (forthcoming, appendix)[13]. As a consequence **D**, like **L**, enjoys in addition the subformula property, decidability, and the finite reading property. The calculus of displacement provides basic analyses of all of the phenomena itemized in (5) (Morrill and Valentín 2010[11], Morrill 2010 chapter 6[14], Morrill, Valentín and Fadda forthcoming[13]). Furthermore it analyses verb raising and cross-serial dependencies (Morrill, Valentín and Fadda 2009)[12].

4 Examples

When s is of sort 1, $s \times_> s' = s \times_< s'$ which we may write $s \times s'$. Hence, when $sA = 1$, $A{\downarrow_>}C \Leftrightarrow A{\downarrow_<}C$, which we abbreviate $A{\downarrow}C$; and when $sC - sB = 1$, $C{\uparrow_>}B \Leftrightarrow C{\uparrow_<}B$, which we may abbreviate $C{\uparrow}B$; and when $sA = 1$, $A{\odot_>}B \Leftrightarrow A{\odot_<}B$, which we may write $A{\odot}B$.

Our first example is of a discontinuous idiom, where the lexicon has to assign *give ... the cold shoulder* a non-compositional meaning 'shun':

(9) **mary+gave+the+man+the+cold+shoulder** : S

Lexical insertion yields the following sequent, which is labelled with the lexical semantics:

(10) $N : m, (N\backslash S){\uparrow}N\{N/CN : \iota, CN : man\} : shunned \;\Rightarrow\; S$

This has a proof as follows.

$$
(11)\quad \dfrac{\dfrac{\overline{CN \Rightarrow CN} \quad \overline{N \Rightarrow N}}{N/CN, CN \Rightarrow N}/L \qquad \dfrac{\overline{N \Rightarrow N} \quad \overline{S \Rightarrow S}}{N, N\backslash S \Rightarrow S}\backslash L}{N, (N\backslash S){\uparrow}N\{N/CN, CN\} \Rightarrow S}{\uparrow}L
$$

This delivers the semantics:

(12) $((shunned\ (\iota\ man))\ m)$

Consider medial extraction:

(13) **dog+that+mary+saw+today** : CN

An associated semantically annotated sequent may be as follows:

(14) $CN : dog, (CN\backslash CN)/((S{\uparrow}N){\odot}I) : \lambda A\lambda B\lambda C[(B\ C) \wedge (\pi_1 A\ C)], N : m,$
$(N\backslash S)/N : saw, (N\backslash S)\backslash(N\backslash S) : \lambda A\lambda B(today\ (A\ B)) \Rightarrow CN$

This has the sequent derivation given in Figure 8. This yields semantics:

(15) $\lambda C[(dog\ C) \wedge (today\ ((saw\ C)\ m))]$

Consider medial quantification:

(16) **john+gave+every+book+to+mary** : S

An associated semantically annotated sequent may be as follows:

(17) $N : j, (N\backslash S)/(N{\bullet}PP) : \lambda A((gave\ \pi_2 A)\ \pi_1 A), ((S{\uparrow}N){\downarrow}S)/CN :$
$\lambda A\lambda B\forall C[(A\ C) \rightarrow (B\ C)], CN : book, PP/N : \lambda AA, N : m \Rightarrow S$

This has the sequent derivation given in Figure 9. This yields semantics:

(18) $\forall C[(book\ C) \rightarrow (((gave\ m)\ C)\ j)]$

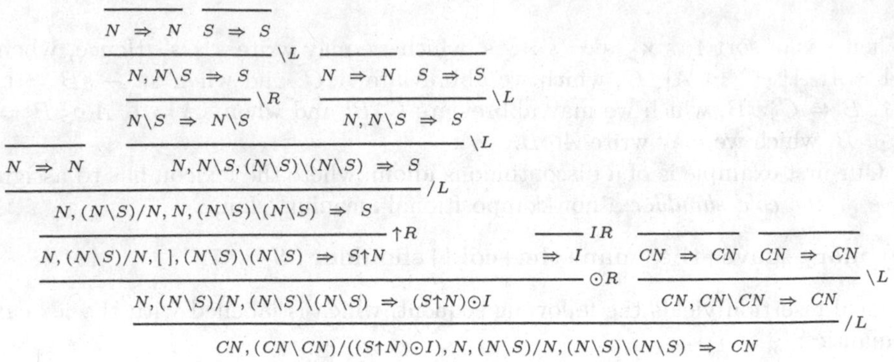

$$\dfrac{\dfrac{}{N \Rightarrow N} \quad \dfrac{\dfrac{\dfrac{\dfrac{}{N \Rightarrow N}\quad\dfrac{}{S \Rightarrow S}}{N, N\backslash S \Rightarrow S}\backslash L}{N\backslash S \Rightarrow N\backslash S}\backslash R \quad \dfrac{\dfrac{}{N \Rightarrow N}\quad\dfrac{}{S \Rightarrow S}}{N, N\backslash S \Rightarrow S}\backslash L}{N, N\backslash S, (N\backslash S)\backslash(N\backslash S) \Rightarrow S}\backslash L}{N, (N\backslash S)/N, N, (N\backslash S)\backslash(N\backslash S) \Rightarrow S}/L$$

$$\dfrac{\dfrac{\dfrac{N, (N\backslash S)/N, N, (N\backslash S)\backslash(N\backslash S) \Rightarrow S}{N, (N\backslash S)/N, [\,], (N\backslash S)\backslash(N\backslash S) \Rightarrow S\uparrow N}\uparrow R \quad \dfrac{}{\Rightarrow I}IR}{N, (N\backslash S)/N, (N\backslash S)\backslash(N\backslash S) \Rightarrow (S\uparrow N)\odot I}\odot R \quad \dfrac{\dfrac{}{CN \Rightarrow CN}\quad\dfrac{}{CN \Rightarrow CN}}{CN, CN\backslash CN \Rightarrow CN}\backslash L}{CN, (CN\backslash CN)/((S\uparrow N)\odot I), N, (N\backslash S)/N, (N\backslash S)\backslash(N\backslash S) \Rightarrow CN}/L$$

Fig. 8. Derivation of medial extraction

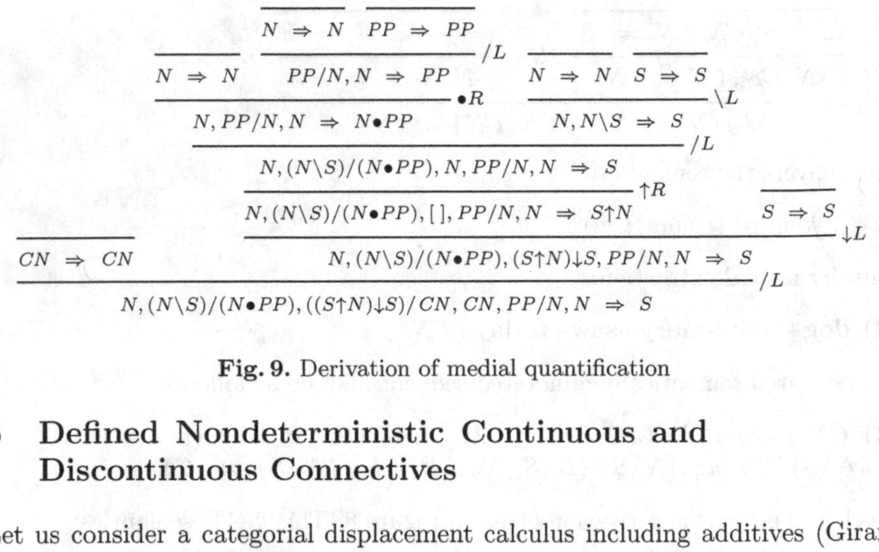

$$\dfrac{\dfrac{}{CN \Rightarrow CN} \quad \dfrac{\dfrac{\dfrac{\dfrac{\dfrac{}{N \Rightarrow N}\quad\dfrac{\dfrac{}{N \Rightarrow N}\quad\dfrac{}{PP \Rightarrow PP}}{PP/N, N \Rightarrow PP}/L}{N, PP/N, N \Rightarrow N\bullet PP}\bullet R \quad \dfrac{\dfrac{}{N \Rightarrow N}\quad\dfrac{}{S \Rightarrow S}}{N, N\backslash S \Rightarrow S}\backslash L}{N, (N\backslash S)/(N\bullet PP), N, PP/N, N \Rightarrow S}/L}{N, (N\backslash S)/(N\bullet PP), [\,], PP/N, N \Rightarrow S\uparrow N}\uparrow R \quad \dfrac{}{S \Rightarrow S}}{N, (N\backslash S)/(N\bullet PP), (S\uparrow N)\downarrow S, PP/N, N \Rightarrow S}\downarrow L}{N, (N\backslash S)/(N\bullet PP), ((S\uparrow N)\downarrow S)/CN, CN, PP/N, N \Rightarrow S}/L$$

Fig. 9. Derivation of medial quantification

5 Defined Nondeterministic Continuous and Discontinuous Connectives

Let us consider a categorial displacement calculus including additives (Girard 1987)[3] which we call displacement calculus with additives (**DA**):

(19) $\mathcal{F}_i := \mathcal{F}_i \& \mathcal{F}_i \mid \mathcal{F}_i \oplus \mathcal{F}_i$

(20) $$\dfrac{\Gamma\langle\overrightarrow{A}\rangle \Rightarrow C}{\Gamma\langle\overrightarrow{A\&B}\rangle \Rightarrow C}\&L_1 \qquad \dfrac{\Gamma\langle\overrightarrow{B}\rangle \Rightarrow C}{\Gamma\langle\overrightarrow{A\&B}\rangle \Rightarrow C}\&L_2$$

$$\dfrac{\Gamma \Rightarrow A \qquad \Gamma \Rightarrow B}{\Gamma \Rightarrow A\&B}\&R$$

$$\dfrac{\Gamma\langle\overrightarrow{A}\rangle \Rightarrow C \qquad \Gamma\langle\overrightarrow{B}\rangle \Rightarrow C}{\Gamma\langle\overrightarrow{A\oplus B}\rangle \Rightarrow C}\oplus L$$

$$\frac{\Gamma \Rightarrow A}{\Gamma \Rightarrow A \oplus B} \oplus L_1 \qquad \frac{\Gamma \Rightarrow B}{\Gamma \Rightarrow A \oplus B} \oplus L_2$$

Then we may define nondeterministic continuous and discontinuous connectives as follows, where $+(s_1, s_2, s_3)$ if and only if $s_3 = s_1{+}s_2$ or $s_3 = s_2{+}s_1$, and $\times(s_1, s_2, s_3)$ if and only if $s_3 = s_1 \times_> s_2$ or $s_3 = s_2 \times_< s_1$.

(21)
$\frac{B}{A} =_{df} (A\backslash B)\&(B/A)$ $\quad \{s \mid \forall s' \in A, s_3, +(s, s', s_3) \Rightarrow s_3 \in B\}$
nondeterministic concatenation
$A \otimes B =_{df} (A{\bullet}B)\oplus(B{\bullet}A)$ $\quad \{s_3 \mid \exists s_1 \in A, s_2 \in B, +(s_1, s_2, s_3)\}$
nondeterministic product
$A{\Downarrow}C =_{df} (A{\downarrow_>}C)\&(A{\downarrow_<}C)$ $\quad \{s_2 \mid \forall s_1 \in A, s_3, \times(s_1, s_2, s_3) \Rightarrow s_3 \in C\}$
nondeterministic infix
$C{\Uparrow}B =_{df} (C{\uparrow_>}B)\&(C{\uparrow_<}B)$ $\quad \{s_1 \mid \forall s_2 \in B, s_3, \times(s_1, s_2, s_3) \Rightarrow s_3 \in C\}$
nondeterministic extract
$A{\circledcirc}B =_{df} (A{\odot_>}B)\oplus(A{\odot_<}B)$ $\quad \{s_3 \mid \exists s_1 \in A, s_2 \in B, \times(s_1, s_2, s_3)\}$
nondeterministic disc. product

These have the derived rules shown in Figure 10 where $k \in \{>, <\}$. We call the displacement calculus extended with nondeterministic connectives the nondeterministic displacement calculus **ND**.

$$\frac{\Gamma \Rightarrow A \qquad \Delta\langle\vec{C}\rangle \Rightarrow D}{\Delta\langle\Gamma, \frac{\vec{C}}{A}\rangle \Rightarrow D} \text{-}L_1 \qquad \frac{\Gamma \Rightarrow A \qquad \Delta\langle\vec{C}\rangle \Rightarrow D}{\Delta\langle\frac{\vec{C}}{A}, \Gamma\rangle \Rightarrow D} \text{-}L_2$$

$$\frac{\vec{A}, \Gamma \Rightarrow C \qquad \Gamma, \vec{A} \Rightarrow C}{\Gamma \Rightarrow \frac{C}{A}} \text{-}R$$

$$\frac{\Delta\langle\vec{A}, \vec{B}\rangle \Rightarrow D \qquad \Delta\langle\vec{B}, \vec{A}\rangle \Rightarrow D}{\Delta\langle\overrightarrow{A \otimes B}\rangle \Rightarrow D} \otimes L$$

$$\frac{\Gamma_1 \Rightarrow A \qquad \Gamma_2 \Rightarrow B}{\Gamma_1, \Gamma_2 \Rightarrow A \otimes B} \otimes R_1 \qquad \frac{\Gamma_1 \Rightarrow B \qquad \Gamma_2 \Rightarrow A}{\Gamma_1, \Gamma_2 \Rightarrow A \otimes B} \otimes R_2$$

$$\frac{\Gamma \Rightarrow A \qquad \Delta\langle\vec{C}\rangle \Rightarrow D}{\Delta\langle\Gamma|_k \overrightarrow{A{\Downarrow}C}\rangle \Rightarrow D} {\Downarrow}L \qquad \frac{\vec{A}|_> \Gamma \Rightarrow C \qquad \vec{A}|_< \Gamma \Rightarrow C}{\Gamma \Rightarrow A{\Downarrow}C} {\Downarrow}R$$

$$\frac{\Gamma \Rightarrow B \qquad \Delta\langle\vec{C}\rangle \Rightarrow D}{\Delta\langle\overrightarrow{C{\Uparrow}B}|_k \Gamma\rangle \Rightarrow D} {\Uparrow}L \qquad \frac{\Gamma|_> \vec{B} \Rightarrow C \qquad \Gamma|_< \vec{B} \Rightarrow C}{\Gamma \Rightarrow C{\Uparrow}B} {\Uparrow}R$$

$$\frac{\Delta\langle\vec{A}|_> \vec{B}\rangle \Rightarrow D \qquad \Delta\langle\vec{A}|_< \vec{B}\rangle \Rightarrow D}{\Delta\langle\overrightarrow{A{\circledcirc}B}\rangle \Rightarrow D} {\circledcirc}L \qquad \frac{\Gamma_1 \Rightarrow A \qquad \Gamma_2 \Rightarrow B}{\Gamma_1|_k \Gamma_2 \Rightarrow A{\circledcirc}B} {\circledcirc}R$$

Fig. 10. Derived rules for the defined nondeterministic continuous and discontinuous connectives of **ND**

Concerning Cut-elimination for the nondeterministic rules, the usual Lambek-style reasoning applies. For example, using the method and definition of

Cut-degree in Morrill and Valentín (2010)[11], here we mention how the nondeterministic extract and discontinuous product behave in the Cut elimination steps. We show one case of principal Cut and one case of permutation conversion. Observe that in the last conversion the logical rule and the Cut rule are permuted by two Cuts and one logical rule, contrary to what is standard, but as required both Cut-degrees are lower.

– $\Uparrow$ principal cut case:

$$\dfrac{\dfrac{\Delta|_{>}\vec{A} \Rightarrow B \qquad \Delta|_{<}\vec{A} \Rightarrow B}{\Delta \Rightarrow B{\Uparrow}A}\,{\Uparrow}R \qquad \dfrac{\Gamma \Rightarrow A \qquad \Theta\langle\vec{B}\rangle}{\Theta\langle\overrightarrow{B{\Uparrow}A}|_k\Gamma\rangle \Rightarrow C}\,{\Uparrow}L}{\Theta\langle\Delta|_k\Gamma\rangle \Rightarrow C}\,Cut \quad \rightsquigarrow$$

$$\dfrac{\Gamma \Rightarrow A \qquad \dfrac{\Delta|_k\vec{A} \Rightarrow B \qquad \Theta\langle\vec{B}\rangle \Rightarrow C}{\Theta\langle\Delta|_k\vec{A}\rangle \Rightarrow C}\,Cut}{\Theta\langle\Delta|_k\Gamma\rangle \Rightarrow C}\,Cut$$

– $\circledcirc$ permutation conversion case:

$$\dfrac{\dfrac{\Delta\langle\vec{B}|_{>}\vec{C}\rangle \Rightarrow A \qquad \Delta\langle\vec{B}|_{<}\vec{C}\rangle \Rightarrow A}{\Delta\langle\overrightarrow{B \circledcirc C}\rangle \Rightarrow A}\,{\circledcirc}L \qquad \Theta\langle\vec{A}\rangle \Rightarrow D}{\Theta\langle\Delta\langle\overrightarrow{B \circledcirc C}\rangle\rangle \Rightarrow D}\,Cut \quad \rightsquigarrow$$

$$\dfrac{\dfrac{\Delta\langle\vec{B}|_{>}\vec{C}\rangle \Rightarrow A \qquad \Theta\langle\vec{A}\rangle \Rightarrow D}{\Theta\langle\Delta\langle\vec{B}|_{>}\vec{C}\rangle\rangle \Rightarrow D}\,Cut \qquad \dfrac{\Delta\langle\vec{B}|_{<}\vec{C}\rangle \Rightarrow A \qquad \Theta\langle\vec{A}\rangle \Rightarrow D}{\Theta\langle\Delta\langle\vec{B}|_{<}\vec{C}\rangle\rangle \Rightarrow D}\,Cut}{\Theta\langle\Delta\langle\overrightarrow{B \circledcirc C}\rangle\rangle \Rightarrow D}\,{\circledcirc}L$$

By way of linguistic applications, a functor of type $\frac{B}{A}$ can concatenate with its argument A either to the left or to the right to form a B. For example, in Catalan subjects can appear either preverbally or clause-finally (*Barcelona creix* or *Creix Barcelona* "Barcelona expands/grows"). This generalization may be captured by assigning a verb phrase such as *creix* the type $\frac{S}{N}$. And a nondeterministic concatenation product type $A \otimes B$ comprises an A concatenated with a B or a B concatenated with an A. For example, in English two prepositional complements may appear in either order (*talks to John about Mary* or *talks about Mary to John*). This generalization may be captured by assigning a verb such as *talks* the type $VP/(PP_{\mathrm{to}} \otimes PP_{\mathrm{about}})$.

5.1 Embedding Translation between ND and DA

We propose the following embedding translation $(\cdot)^{\natural} : \mathbf{ND} \longrightarrow \mathbf{DA}$ which we define recursively:[3]

[3] We assume a convention of precedence whereby the multiplicative connectives take higher priority than the additives.

$$
\begin{array}{ll}
(A)^\natural & = A \text{ for atomic types } A \\
(B{\Uparrow}A)^\natural & = B^\natural{\uparrow_>}A^\natural \& B^\natural{\uparrow_<}A^\natural \\
(A{\Downarrow}B)^\natural & = A^\natural{\downarrow_>}B^\natural \& A^\natural{\downarrow_<}B^\natural \\
(A{\circledcirc}B)^\natural & = A^\natural{\odot_>}B^\natural \oplus A^\natural{\odot_<}B^\natural \\
(\frac{B}{A})^\natural & = A^\natural\backslash B^\natural \& B^\natural/A^\natural \\
(A\otimes B)^\natural & = A^\natural \bullet B^\natural \oplus B^\natural \bullet A^\natural \\
(A\star B)^\natural & = A^\natural \star B^\natural \text{ where } \star \text{ is any other binary connective}
\end{array}
$$

We have the following interesting result:

Lemma 1. *The* $(\cdot)^\natural$ *embedding is faithful.*

Proof. ¿From **ND** to **DA**, hypersequent derivations translate without any trouble while preserving provability. Let us suppose now that we have a **DA** provable hypersequent which corresponds to the image by $(\cdot)^\natural$ of a **ND** hypersequent, i.e. $\Delta^\natural \Rightarrow A^\natural$ where Δ and A are in the language of **ND**. We want to prove that if $\Delta^\natural \Rightarrow A^\natural$ is **DA**-provable then $\Delta \Rightarrow A$ is **ND** provable. Since the Cut rule is admissible in **DA**, we can assume only **DA** Cut-free provable hypersequents $\Delta^\natural \Rightarrow A^\natural$. The proof is by induction on the length (or height) of Cut-free **DA** derivations. If the length is 0 there is nothing to prove. If the end-hypersequent is derived by a multiplicative inference there is no problem. We analyze then the cases where the last rule is an additive rule:[4]

- Left rules:

– Case where the additive active formula corresponds to $(A{\Uparrow}B)^\natural = A^\natural{\uparrow_>}B^\natural \;\&\; A^\natural{\uparrow_<}B^\natural$:[5]

$$
\frac{\Delta^\natural\langle\overrightarrow{A^\natural{\uparrow_<}B^\natural}\rangle \Rightarrow C^\natural}{\Delta^\natural\langle\overrightarrow{A^\natural{\uparrow_>}B^\natural \;\&\; A^\natural{\uparrow_<}B^\natural}\rangle \Rightarrow C^\natural}\,\&L_2
$$

By induction hypothesis (i.h.), $\Delta\langle\overrightarrow{A{\uparrow_<}B}\rangle \Rightarrow C$ is derivable in the system without additives. Since $\overrightarrow{A{\Uparrow}B} \Rightarrow A{\uparrow_<}B$ is **ND**-provable, we can apply the Cut rule as follows:

$$
\frac{\overrightarrow{A{\Uparrow}B} \Rightarrow A{\uparrow_<}B \qquad \Delta\langle\overrightarrow{A{\uparrow_<}B}\rangle \Rightarrow C}{\Delta\langle\overrightarrow{A{\Uparrow}B}\rangle \Rightarrow C}\,Cut
$$

– Case where the additive active formula corresponds to $(A{\circledcirc}B)^\natural = A^\natural \odot_> B^\natural \oplus A^\natural \odot_< B^\natural$:

$$
\frac{\Delta^\natural\langle A \odot_> B\rangle \Rightarrow C^\natural \qquad \Delta^\natural\langle A \odot_< B\rangle \Rightarrow C^\natural}{\Delta^\natural\langle A \odot_> B \oplus A \odot_< B\rangle \Rightarrow C^\natural}\,{\circledcirc}L
$$

---

[4] By way of example we only consider some cases of nondeterministic discontinuous rules: nondeterministic $\Downarrow$ and continuous connectives are similar.

[5] The other case of the & left rule, i.e. $\&L_1$, is completely similar.

By i.h. $\Delta\langle A \odot_> B\rangle \Rightarrow C$ and $\Delta\langle A \odot_< B\rangle \Rightarrow C$ are **ND**-provable. Moreover, the hypersequents $\overrightarrow{A}|_k\overrightarrow{B} \Rightarrow A \odot_k B$ for $k \in \{>, <\}$ are **ND**-provable. By Cut we have $\Delta\langle\overrightarrow{A}|_>\overrightarrow{B}\rangle \Rightarrow C$ and $\Delta\langle\overrightarrow{A}|_<\overrightarrow{B}\rangle \Rightarrow C$. Applying then the left $\circledcirc$ rule we have:

$$\frac{\Delta\langle\overrightarrow{A}|_>\overrightarrow{B}\rangle \Rightarrow C \qquad \Delta\langle\overrightarrow{A}|_<\overrightarrow{B}\rangle \Rightarrow C}{\Delta\langle\overrightarrow{A\circledcirc B}\rangle \Rightarrow C}\circledcirc L$$

- Right rules:

– Case where the additive formula corresponds to $(A{\Uparrow}B)^\natural = A^\natural{\uparrow}_>B^\natural \& A^\natural{\uparrow}_<B^\natural$:

$$\frac{\Delta^\natural \Rightarrow A^\natural{\uparrow}_>B^\natural \qquad \Delta^\natural \Rightarrow A^\natural{\uparrow}_<B^\natural}{\Delta^\natural \Rightarrow A^\natural{\uparrow}_>B^\natural \& A^\natural{\uparrow}_<B}\&R$$

By i.h. we have that the hypersequents $\Delta \Rightarrow A{\uparrow}_>B$ and $\Delta \Rightarrow A{\uparrow}_<B$ are **ND**-provable. We then apply the right $\Uparrow$ rule:

$$\frac{\Delta \Rightarrow A{\uparrow}_>B \qquad \Delta \Rightarrow A{\uparrow}_<B}{\Delta \Rightarrow A{\Uparrow}B}{\Uparrow}R$$

– Case where the additive active formula corresponds to $(A\circledcirc B)^\natural = A^\natural \odot_> B^\natural \oplus A^\natural \odot_< B^\natural$:[6]

$$\frac{\Delta^\natural \Rightarrow A^\natural \odot_< B^\natural}{\Delta^\natural \Rightarrow A^\natural \odot_> B^\natural \oplus A^\natural \odot_< B^\natural}\oplus R_2$$

By i.h. $\Delta \Rightarrow A \odot_< B$. Now it is **ND**-provable that $\overrightarrow{A \odot_< B} \Rightarrow A\circledcirc B$. Then by Cut:

$$\frac{\Delta \Rightarrow A \odot_< B \qquad \overrightarrow{A \odot_< B} \Rightarrow A\circledcirc B}{\Delta \Rightarrow A\circledcirc B}Cut$$

□

6 Defined Unary Connectives

We may define unary connectives as follows:

6 Without loss of generality we suppose that the instance of the last right $\oplus$ rule is $\oplus R_2$.

(22) $\triangleright^{-1}A =_{df} J\backslash A \quad \{s|\ 1{+}s \in A\}$
right projection
$\triangleleft^{-1}A =_{df} A/J \quad \{s|s{+}1 \in A\}$
left projection
$\triangleright A =_{df} J \bullet A \quad \{1{+}s|\ s \in A\}$
right injection
$\triangleleft A =_{df} A \bullet J \quad \{s{+}1|\ s \in A\}$
left injection
$\check{}^{>}A =_{df} A\uparrow_{>}I \ \{s|\ s \times_{>} 0 \in A\}$
first split
$\check{}^{<}A =_{df} A\uparrow_{<}I \ \{s|\ s \times_{<} 0 \in A\}$
last split
$\hat{}^{>}A =_{df} A\odot_{>}I \ \{s \times_{>} 0|\ s \in A\}$
first bridge
$\hat{}^{<}A =_{df} A\odot_{<}I \ \{s \times_{<} 0|\ s \in A\}$
last bridge

The derived rules of inference can be compiled straightforwardly. Some interdefinabilities are as follows:

(23) $\frac{B}{A} \Leftrightarrow \triangleleft^{-1}\triangleright^{-1}((B\uparrow A)\Uparrow I)$ when $sB = 1$
$A \otimes B \Leftrightarrow (\triangleleft\!\triangleright A \circledcirc I) \odot B$
$A\backslash B \Leftrightarrow \triangleright^{-1}(B\uparrow_{>} A)$
$B/A \Leftrightarrow \triangleleft^{-1}(B\uparrow_{<} A)$

When $sA = 0$, $\check{}^{>}A \Leftrightarrow \check{}^{<}A$, which we abbreviate $\check{}A$; and when $sA = 1$, $\hat{}^{>}A \Leftrightarrow \hat{}^{<}A$, which we abbreviate $\hat{}A$. By way of linguistic application, to produce particle shift (*rings up Mary* or *rings Mary up*) we may assign **rings+1+up** the type $\triangleleft^{-1}(\check{}\,VP\Uparrow N)$.

7 Discussion

The defined connectives considered in this paper facilitate more concise lexical entries, but since they are defined they do not in any way increase the expressivity of the displacement calculus (with additives). But in addition, the use of defined connectives with their derived rules can eliminate bureaucracy in sequent derivations in the case of the introduction of the additives. Consider the two following derivations which are *equal* modulo some permutations:

$$\mathcal{D}_1 \vdash \cfrac{\cfrac{\cfrac{A \Rightarrow A \qquad B\{[\,]\} \Rightarrow B}{B\uparrow_{>}A\{A:[\,]\} \Rightarrow B}\uparrow_{>}}{(B\uparrow_{>}A)\&(B\uparrow_{<}A)\{A:[\,]\} \Rightarrow B}\&L \qquad C\{[\,]\} \Rightarrow C}{C/B, (B\uparrow_{>}A)\&(B\uparrow_{<}A)\{A:[\,]\} \Rightarrow C}/L$$

$$\mathcal{D}_2 \vdash \cfrac{\cfrac{C\{[\,]\} \Rightarrow C \qquad \cfrac{A \Rightarrow A \qquad B\{[\,]\} \Rightarrow B}{B\uparrow_{>}A\{A:[\,]\} \Rightarrow B}\uparrow_{>}}{C/B, (B\uparrow_{>}A)\{A:[\,]\} \Rightarrow C}/L}{C/B, (B\uparrow_{>}A)\&(B\uparrow_{<}A)\{A:[\,]\} \Rightarrow C}\&L$$

Observe that both derivations $\mathcal{D}_1$ and $\mathcal{D}_2$ are essentially the same. The only (inessencial) difference is the permutations steps of the additive connective & and the forward slash connective /. In $\mathcal{D}_1$ the left rule $\&L$ precedes the left rule $/L$, whereas in $\mathcal{D}_2$ the left rule $/L$ precedes the left rule $\&L$. These would have the same corresponding derivation for defined connectives. It follows then that derived or compiled rules for defined connectives eliminate some undesirable bureaucracy in the derivations.

The displacement calculus has been formulated in this paper in terms of first and last wrap, as opposed to the k-ary wrap, $k > 0$, of Morrill and Valentín (2010)[11], and has a finite rather than an infinite number of connectives. This last version of displacement calculus draws together ideas spanning three decades:

(24)
- Bach (1981, 1984)[1], [2]: the idea of categorial connectives for discontinuity/wrapping; wrap, extract, infix.
- Moortgat (1988)[6]: first type logical account of extract and infix discontinuous connectives (string interpretation and sequent calculus).
- Morrill and Merenciano (1996)[10]: sorts; bridge and split.
- Morrill (2002)[8]: separators; unboundedly many points of discontinuity.
- Morrill, Fadda and Valentín (2007)[9]: nondeterministic discontinuity.
- Morrill, Valentín and Fadda (2009)[12]: projection and injection.
- Morrill and Valentín (2010)[11]: product and discontinuous product units, Cut-elimination.

This road to discontinuity has respected fully intuitionism, residuation, and natural algebraic string models. Further logical and mathematical properties of the resulting system remain to be studied, and it also remains to be seen whether it may be necessary to appeal to continuation semantics or classical (symmetric) calculi (Moortgat 2009)[7].

References

1. Bach, E.: Discontinuous constituents in generalized categorial grammars. In: Burke, V.A., Pustejovsky, J. (eds.) Proceedings of the 11th Annual Meeting of the North Eastern Linguistics Society, New York, pp. 1–12. GLSA Publications, Department of Linguistics, University of Massachussets at Amherst, Amherst, Massachussets (1981)
2. Bach, E.: Some Generalizations of Categorial Grammars. In: Landman, F., Veltman, F. (eds.) Varieties of Formal Semantics: Proceedings of the Fourth Amsterdam Colloquium, pp. 1–23. Foris, Dordrecht (1984); Reprinted in Savitch, W.J., Bach, E., Marsh, W., Safran-Naveh, G., (eds.) The Formal Complexity of Natural Language, pp. 251–279. D. Reidel, Dordrecht (1987)
3. Girard, J.-Y.: Linear logic. Theoretical Computer Science 50, 1–102 (1987)
4. Lambek, J.: The mathematics of sentence structure. American Mathematical Monthly 65, 154–170 (1958); Reprinted in Buszkowski, W., Marciszewski, W., van Benthem, J., (eds.) Categorial Grammar. Linguistic & Literary Studies in Eastern Europe, vol. 25 153–172. John Benjamins, Amsterdam (1988)

5. Lambek, J.: Deductive systems and categories, II: Standard constructions and closed categories. In: Hilton, P. (ed.) Category Theory, Homology Theory and Applications. Lecture Notes in Mathematics, vol. 86, pp. 76–122. Springer (1969)
6. Moortgat, M.: Categorial Investigations: Logical and Linguistic Aspects of the Lambek Calculus. Foris, Dordrecht, PhD thesis. Universiteit van Amsterdam (1988)
7. Moortgat, M.: Symmetric categorial grammar. Journal of Philosophical Logic (2009)
8. Morrill, G.: Towards Generalised Discontinuity. In: Jäger, G., Monachesi, P., Penn, G., Wintner, S. (eds.) Proceedings of the 7th Conference on Formal Grammar, Trento, ESSLLI, pp. 103–111 (2002)
9. Morrill, G., Fadda, M., Valentín, O.: Nondeterministic Discontinuous Lambek Calculus. In: Geertzen, J., Thijsse, E., Bunt, H., Schiffrin, A. (eds.) Proceedings of the Seventh International Workshop on Computational Semantics, IWCS 2007, pp. 129–141. Tilburg University (2007)
10. Morrill, G., Merenciano, J.-M.: Generalising discontinuity. Traitement Automatique des Langues 37(2), 119–143 (1996)
11. Morrill, G., Valentín, O.: Displacement calculus. Linguistic Analysis (forthcoming)
12. Morrill, G., Valentín, O., Fadda, M.: Dutch Grammar and Processing: A Case Study in TLG. In: Bosch, P., Gabelaia, D., Lang, J. (eds.) TbiLLC 2007. LNCS (LNAI), vol. 5422, pp. 272–286. Springer, Heidelberg (2009)
13. Morrill, G., Valentín, O., Fadda, M.: The Displacement Calculus. Journal of Logic, Language and Information (forthcoming)
14. Morrill, G.V.: Categorial Grammar: Logical Syntax, Semantics, and Processing. Oxford University Press, Oxford (2010)
15. Pentus, M.: Lambek grammars are context-free. Technical report, Dept. Math. Logic, Steklov Math. Institute, Moskow (1992); Also published as ILLC Report, University of Amsterdam, 1993, and in Proceedings Eighth Annual IEEE Symposium on Logic in Computer Science, Montreal (1993)
16. Shieber, S.: Evidence Against the Context-Freeness of Natural Language. Linguistics and Philosophy 8, 333–343 (1985); Reprinted in Savitch, W.J., Bach, E., Marsh, W., Safran-Naveh, G., (eds.) The Formal Complexity of Natural Language, pp. 320–334. D. Reidel, Dordrecht (1987)

Controlling Extraction in Abstract Categorial Grammars

Sylvain Pogodalla[1] and Florent Pompigne[2]

[1] LORIA/INRIA Nancy – Grand Est
sylvain.pogodalla@inria.fr
[2] LORIA/Nancy Université
florent.pompigne@loria.fr

Abstract. This paper proposes an approach to control extraction in the framework of Abstract Categorial Grammar (ACG). As examples, we consider embedded wh-extraction, multiple wh-extraction and tensed-clauses as scope islands. The approach relies on an extended type system for ACG that introduces dependent types and advocates for a treatment at a rather abstract (tectogrammatical) level. Then we discuss approaches that put control at the object (phenogrammatical) level, using appropriate calculi.

1 Introduction

In pursuing [2]'s program of separating the combinatorial part of grammars, the *tectogrammatical* level, from the one that realizes the operations on the surface structures, the *phenogrammatical* level, the two independently formulated frameworks of Lambda Grammar (LG) [20,21] and Abstract Categorial Grammar (ACG) [3] propose to consider the implicative fragment of linear logic as the underlying tectogrammatical calculus. While interleaving the phenogrammatical and the tectogrammatical levels as in standard Categorial Grammar and Lambek calculus (CG) [13,16] leads to using a *directed* (or non-commutative) calculus, both LG and ACG rather rely on a *non-directed* (or commutative) calculus.

As immediate outcome of this choice, extraction is easily available, in particular from medial position whereas CG permits only for peripheral extraction. So even if CG and Lambek grammars are known for their powerful treatment of extraction, LG and ACG extend these capabilities.

However, it is a common observation that extractions are not completely free in natural language in general. The power of hypothetical reasoning of Lambek calculus based grammars itself is sometimes too strong [1, p. 207]. Directionality of the calculus is not sufficient to model all kinds of *islands* to extraction, for instance with coordinate structures, and it *overgenerates*. Because of the presence of hypothetical reasoning in the LG and ACG frameworks, the question arises whether those frameworks overgenerate as well and, because they do, how to control extraction in those frameworks.

This paper aims at providing some solution to control extraction in the framework of ACG for various cases, including tensed-clauses as scope islands, embedded wh-extraction and multiple wh-extraction. The solution relies on an extended type system for ACG that Sect. 2 presents together with the ACG basics. We emphasize there

P. de Groote and M.-J. Nederhof (Eds.): Formal Grammar 2010/2011, LNCS 7395, pp. 162–177, 2012.

the compositional[1] flexibility of ACG and present a treatment at a rather abstract (tectogrammatical) level. Then Sect. 3 describes the examples and the solutions we provide. Our account focuses on using *dependent types*, both in a rather limited and in a more general setting. Section 4 compares our approach with related works. We first discuss approaches that put control at the phenogrammatical level, using appropriate calculi, then discuss other ACG models that use the same kind of architectures as the one we propose. We also discuss ways of importing solutions developed in the the CG frameworks.

2 ACG: Definitions and Properties

The ACG formalism lies within the scope of type-theoretic grammars [13,2,15,25]. In addition to relying on a small set of mathematical primitives from type-theory and λ-calculus, an important property concerns the direct control it provides over the parse structures of the grammar. This control is at the heart of the present proposal.

2.1 Definitions

The definitions we provide here follow [3] together with the type-theoretic extension of [4,7] providing the dependent product[2].

Definition 1. *The set of kinds $\mathscr{K}$, the set of types $\mathscr{T}$ and the set of terms T are defined as:*

$$\begin{aligned} \mathscr{K} &::= \mathtt{type} \mid (\mathscr{T})\mathscr{K} \\ \mathscr{T} &::= a \mid (\lambda x.\mathscr{T}) \mid (\mathscr{T}\,T) \mid (\mathscr{T} \multimap \mathscr{T}) \mid (\Pi\, x : \mathscr{T})\mathscr{T} \\ T &::= c \mid x \mid (\lambda^0 x.T) \mid (\lambda x.T) \mid (T\,T) \end{aligned}$$

where a ranges over atomic types and c over constants [3].

Assume for instance a type *Gender* and the three terms masc, fem and neut of this type. We then can define *np* with kind $(Gender)\mathtt{type}$ that derives three types: *np* masc, *np* fem and *np* neut. *np* can be seen as a feature structure whose gender value is still missing while *John* can be seen as a term of type *np* masc, *i.e.* a feature structure where the value of the *Gender* feature has been set to masc. On the other hand, an intransitive verb accepts as subject a noun phrase with any gender. So its type is typically $(\Pi x : Gender)$ $(np\, x \multimap s)$.

Definition 2 (Signature). *A* raw signature *is a sequence of declarations of the form 'a : K'or of the form 'c : α', where a ranges over atomic types, c over constants, K over kinds and α over types.*

Let Σ be a raw signature. We write A_Σ (resp. C_Σ) for the set of atomic types (resp. constants) declared in Σ and write $\mathscr{K}_\Sigma$ (resp. $\mathscr{T}_\Sigma$ and Λ_Σ) for the set of well-formed

[1] As in functional composition, not as in the compositionality principle.

[2] We don't use the record and variant types they introduced.

[3] $\lambda^0 x.T$ denotes the linear abstraction and $\lambda x.T$ the non-linear one. $(\Pi x : \alpha)$ denotes a universal quantification over variables of type α.

kinds (resp. well-kinded types and well-typed terms). In case Σ correctly introduces well-formed kinds and well-kinded types, it is said to be a well-formed signature.

We also define κ_Σ (resp. τ_Σ) the function that assigns kinds to atomic types (resp. that assigns types to constants).

There is no room here to give the typing rules detailed in [4,7], but the ones used in the next sections are quite straightforward. They all are instances of the following derivation (the sequent $\vdash_\Sigma$ (SLEEPS masc)JOHN : s is said to be *derivable*) assuming the raw signature Σ of Table 1:

$$\dfrac{\dfrac{\vdash_\Sigma \text{SLEEPS} : (\Pi x : \textit{Gender})\,(np\,x \multimap s) \qquad \vdash_\Sigma \text{masc} : \textit{Gender}}{\vdash_\Sigma \text{SLEEPS masc} : np\ \text{masc} \multimap s} \qquad \vdash_\Sigma \text{JOHN} : np\ \text{masc}}{\vdash_\Sigma (\text{SLEEPS masc})\text{JOHN} : s}$$

Table 1. Raw signature example

Σ :	*Gender* : type	masc, fem : *Gender*	JOHN : *np* masc
	np : (*Gender*)type	SLEEPS : $(\Pi x : \textit{Gender})\,(np\,x \multimap s)$	

Definition 3 (Lexicon). *A* lexicon *from Σ_A to Σ_O is a pair $\langle \eta, \theta \rangle$ where:*

- *η is a morphism form A_{Σ_A} to $\mathscr{T}_{\Sigma_O}$ (we also note η its unique extension to $\mathscr{T}_{\Sigma_A}$);*
- *θ is a morphism form C_{Σ_A} to Λ_{Σ_O} (we also note θ its unique extension to Λ_{Σ_A});*
- *for every $c \in C_{\Sigma_A}$, $\theta(c)$ is of type $\eta(\tau_{\Sigma_A}(c))$;*
- *for every $a \in A_{\Sigma_A}$, the kind of $\eta(a)$ is $\tilde{\eta}(\kappa_{\Sigma_A}(a))$ where $\tilde{\eta} : \mathscr{K}_{\Sigma_A} \to \mathscr{K}_{\Sigma_O}$ is defined by $\tilde{\eta}(\mathtt{type}) = \mathtt{type}$ and $\tilde{\eta}((\alpha)K) = (\eta(\alpha))\tilde{\eta}(K)$.*

Definition 4 (Abstract Categorial Grammar). *An* abstract categorial grammar *is a quadruple $\mathscr{G} = \langle \Sigma_A, \Sigma_O, \mathscr{L}, s \rangle$ where:*

1. *Σ_A and Σ_O are two well-formed signatures: the* abstract vocabulary *and the* object vocabulary, *respectively;*
2. *$\mathscr{L} : \Sigma_A \to \Sigma_O$ is a lexicon from the abstract vocabulary to the object vocabulary;*
3. *$s \in \mathscr{T}_{\Sigma_A}$ (in the abstract vocabulary) is the* distinguished type *of the grammar.*

While the object vocabulary specifies the surface structures of the grammars (*e.g.* strings or trees), the abstract vocabulary specifies the parse structures (*e.g.* trees, but more generally proof trees as in CG). The lexicon specifies how to map the parse structures to the surface structures.

Definition 5 (Languages). *An ACG $\mathscr{G} = \langle \Sigma_A, \Sigma_O, \mathscr{L}, s \rangle$ defines two languages:*

- *the* abstract language*: $\mathcal{A}(\mathscr{G}) = \{t \in \Lambda_{\Sigma_A} \mid \vdash_{\Sigma_A} t : s \text{ is derivable}\}$*
- *the* object language, *which is the image of the abstract language by the lexicon: $\mathcal{O}(\mathscr{G}) = \{t \in \Lambda_{\Sigma_O} \mid \exists u \in \mathcal{A}(\mathscr{G}). t = \mathscr{L}(u)\}$*

The expressive power and the complexity of ACG have been intensively studied, in particular for 2nd-order ACG. This class of ACG corresponds to a subclass of the ACG where linear implication ($\multimap$) is the unique type constructor (*core* ACG). While the parsing problem for the latter reduces to provability in the Multiplicative Exponential fragment of Linear Logic (MELL) [27], which is still unknown, parsing of 2nd-order ACG is polynomial and the generated languages correspond to mildly context-sensitive languages [5,27,10][4].

Extending the typing system with dependent products results in a Turing-complete formalism. The problem of identifying interesting and tractable fragments for this extended type system is ongoing work that we don't address in this paper. However, a signature where types only depend on finitely inhabited types (as in the former example, *np* depends on the finitely inhabited type *Gender*) can be expressed in core ACG and complexity results can be transfered. The model we propose in Sect. 3.3 has this property. For the other cases where the number of inhabitants is infinite, an actual implementation could take into account an upper bound for the number of extractions in the same spirit as [8,19] relate the processing load with the number of unresolved dependencies while processing a sentence, and could reduce these cases to the finite one.

2.2 Grammatical Architecture

Since they both are higher-order signatures, the abstract vocabulary and the object one don't show any structural difference. This property makes ACG composition a quite natural operation. Figure 1(a) exemplifies the first way to compose two ACG: the object vocabulary of the first ACG $\mathscr{G}_1$ is the abstract vocabulary of the second ACG $\mathscr{G}_2$. Its objectives are twofold:

- either a term $u \in \mathcal{A}(\mathscr{G}_2)$ has at least one antecedent by the lexicon of $\mathscr{G}_1$ in $\mathcal{A}(\mathscr{G}_1)$ (or even two or more antecedents) and $\mathscr{G}_2 \circ \mathscr{G}_1$ provides more analysis to a same object term of $\mathcal{O}(\mathscr{G}_2)$ than $\mathscr{G}_2$ does. [24,6] use this architecture to model scope ambiguity using higher-order types for quantified noun phrases at the level of Σ_{A_1} while their type remains low at the level of Σ_{A_2};
- or a term $u \in \mathcal{A}(\mathscr{G}_2)$ has no antecedent by the lexicon of $\mathscr{G}_1$ in $\mathcal{A}(\mathscr{G}_1)$. It means that $\mathscr{G}_2 \circ \mathscr{G}_1$ somehow *discards* some analysis given by $\mathscr{G}_2$ of an object term of Λ_{O_2}. We have chosen this architecture in this paper for that purpose: while some constructs are accepted by $\mathscr{G}_{\mathrm{Syn}}$ (to be defined in Sect. 3.1), an additional control at a more abstract level discard them.

Figure 1(b) illustrates the second way to compose two ACG: $\mathscr{G}_1$ and $\mathscr{G}_2$ share the same abstract vocabulary, hence define the same abstract language. This architecture arises in particular when one of the ACG specifies the syntactic structures and the other one specifies the semantic structures. The shared abstract vocabulary hence specifies the syntax-semantics interface. [23,6] precisely consider this architecture with that aim. Note that this architecture for the syntax-semantics interface corresponds to the presentation of synchronous TAG as a bi-morphic architecture [30].

[4] There are other decidable classes we don't discuss here.

(a) First composition mode (b) Second composition mode

Fig. 1. Various ways of composing ACG

Finally, mixing the two ways of composition is also possible, as Fig. 2 illustrates. Because the ACG for the semantics is linked at the highest level in Fig. 2(b), this architecture has been used in [24] and [6] to model semantic ambiguity while keeping at an intermediate level a non-ambiguous syntactic type for quantifiers. Indeed the semantics needs in that case to attach to the place where ambiguity already arised.

On the other hand, if the syntax-semantics interface takes place at an intermediate level such as in Fig. 2(a) the highest ACG can provide further control on the acceptable structures: while some syntactic constructs could be easily given a semantics, it might happen that they're forbidden in some languages. Hence the need of another control that discards those constructs. This paper uses such an architecture and we show first how to set a fairly standard syntax-semantics interface and second how to provide additional control without changing anything to this interface.

Note that in both cases, because the composition of two ACG is itself an ACG, these architectures boil down to the one of Fig. 1(b). However, keeping a multi-level architecture helps in providing some modularity for grammatical engineering, either by reusing components as in Fig. 2(a) (where the syntax-semantics interface is not affected by the supplementary control provided by the most abstract ACG) or by providing intermediate components as in Fig. 2(b) (such as the low-order type for quantifiers, contrary to CG)[5].

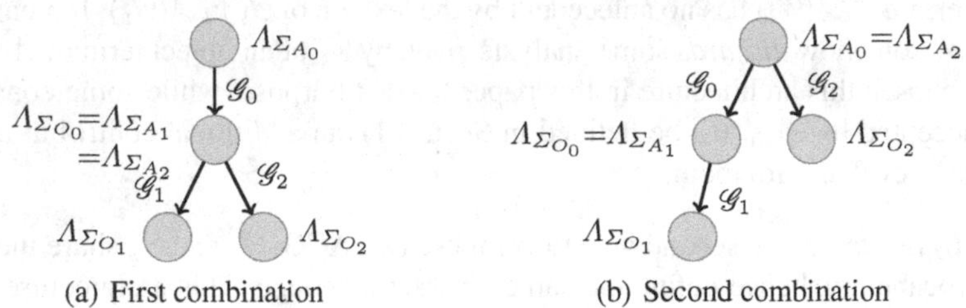

(a) First combination (b) Second combination

Fig. 2. Mixing composition modes

[5] However, for sake of simplicity, we don't use this intermediate level here and directly adopt the standard higher-order type for quantified noun-phrases.

3 Examples

3.1 The Syntax-Semantics Interface

Following the architecture presented in Sect. 2.2, we first briefly define the two ACG sharing the same abstract language defining the general syntax-semantics interface we use. Since the scope of this paper is rather the control of this interface, we don't enter the details here. It's enough to say that we basically follow standard categorial grammar approaches except that the linear non-directional implication replaces the two directional implications[6]. We define $\mathscr{G}_{\text{Syn}} = \langle \Sigma_{\text{Syn}}, \Sigma_{\text{String}}, \mathscr{L}_{\text{Syn}}, s\rangle$ the ACG that relates syntactic structures together with their surface realization. Table 2 presents Σ_{Syn} the signature for the parse structures, Σ_{String} the signature for surface realization, and $\mathscr{L}_{\text{Syn}}$ the lexicon that relates them.

Table 2. Σ_{Syn}, Σ_{String} (σ stands for the type of string, $+$ for the concatenation operation and ϵ for the empty string) and $\mathscr{L}_{\text{Syn}}$ (obvious interpretations are omitted)

Σ_{Syn} :

s, np, n : type	$C_{\text{so}}, C_{\text{ev}} : (np \multimap s) \multimap s$	$C_{\text{loves}} : np \multimap np \multimap s$
$C_{\text{Mary}}, C_{\text{John}} : np$	$C_{\text{who}} : (np \multimap s) \multimap n \multimap n$	$C_{\text{says}} : s \multimap np \multimap s$

Σ_{String} :

σ	: type
/Mary/, /John/, /someone/, ϵ, /everyone//loves/, /who/, /says/	: σ
$+$	: $\sigma \multimap \sigma \multimap \sigma$

$\mathscr{L}_{\text{Syn}}$:

$s, np, n :=_{\text{Syn}} \sigma$	$C_{\text{Mary}} :=_{\text{Syn}} /\text{Mary}/$
$C_{\text{so}} :=_{\text{Syn}} \lambda^0 p.p\, /\text{someone}/$	$C_{\text{loves}} :=_{\text{Syn}} \lambda^0 os.s + /\text{loves}/ + o$
$C_{\text{who}} :=_{\text{Syn}} \lambda^0 pn.n + /\text{who}/ + (p\,\epsilon)$	$C_{\text{says}} :=_{\text{Syn}} \lambda^0 cs.s + /\text{says}/ + c$

In situ operators such as quantifiers have the property to (semantically) take scope over complex (surface) expressions they are part of. In (1) for instance, the quantified noun phrase (QNP), while subpart of the whole sentence, has the existential quantifier of its semantic contribution taking scope over the whole proposition as in (1-a).

(1) Mary loves someone

a. $\exists x.\mathbf{love}\,\mathbf{m}\,x$

b. $C_{\text{so}}(\lambda^0 x.C_{\text{loves}}\, x\, C_{\text{Mary}})$

The way CG model these phenomena is to type QNP with the higher-order type $(np \multimap s) \multimap s$, whose first argument is a sentence missing an NP. Such an argument can be represented by a λ-term starting with an abstraction $\lambda^0 x.t$ with x occurring (free) in t that plays the role of any non quantified NP having the surface position of the QNP. So, in the previous example, t would represent the expression *Mary loves* x, and

[6] ACG manages word order at the surface level. For discussion on relations between ACG and CG, see [26].

the representation of (1) is (1-b). We leave it to the reader to check that the string representation is indeed the image by $\mathscr{L}_{\text{Syn}}$ of (1-b).

The case of wh-words where the movement is overt is dealt with in almost the same way: the first argument is a sentence missing an NP. The difference (overt *vs.* covert) rests in what is provided to this first argument to get the surface form: in the case of covert movements, there is an actual realization with the QNP form (see $\mathscr{L}_{\text{Syn}}(C_{\text{so}})$) while there is no realization of overt movements (see $\mathscr{L}_{\text{Syn}}(C_{\text{who}})$). However, in both cases, the abstract structure contains a variable that is abstracted over. In the sequel of this paper, we refer to the variable as the *extracted* variable, or as the variable *available for extraction*.

We also define $\mathscr{G}_{\text{Sem}} = \langle \Sigma_{\text{Syn}}, \Sigma_{\text{Sem}}, \mathscr{L}_{\text{Sem}}, s \rangle$ the ACG that relates syntactic structures together with their *semantic* interpretation. As expected, $\mathscr{G}_{\text{Syn}}$ and $\mathscr{G}_{\text{Sem}}$ share the abstract vocabulary Σ_{Syn} presented in Table 2. Table 3 presents Σ_{Sem} the signature for logical formulas and $\mathscr{L}_{\text{Sem}}$ the lexicon that relates them. This lexicon associates (1-b) with its meaning (1-a).

Table 3. Σ_{Sem} and $\mathscr{L}_{\text{Sem}}$

Σ_{Sem} :

e, t : `type`	$\mathbf{m}, \mathbf{j} : e$	$\forall, \exists : (e \to t) \multimap t$
$\wedge, \Rightarrow : t \multimap t \multimap t$	$\mathbf{love} : e \multimap e \multimap t$	$\mathbf{say} : t \multimap e \multimap t$

$\mathscr{L}_{\text{Sem}}$:

$s :=_{\text{Sem}} t$	$np :=_{\text{Sem}} e$
$n :=_{\text{Sem}} e \multimap t$	$C_{\text{Mary}} :=_{\text{Sem}} \mathbf{m}$
$C_{\text{so}} :=_{\text{Sem}} \lambda^o p.\forall x.p\,x$	$C_{\text{loves}} :=_{\text{Sem}} \lambda^o os.s(\lambda^o x.o(\lambda^o y.\mathbf{love}\,x\,y))$
$C_{\text{who}} :=_{\text{Sem}} \lambda^o pn.\lambda x.(n\,x) \wedge (p\,x)$	$C_{\text{says}} :=_{\text{Sem}} \lambda^o cs.\mathbf{say}\,x\,c$

Because $\mathscr{G}_{\text{Syn}}$ is a straightforward adaptation of standard treatments of quantification and relativization in CG, it overgenerates as well. Indeed, when building a term using free variables, those variables can be arbitrarily deep in the term, and can be abstracted over in any order (resulting in particular in scope ambiguity), as close of the top level as we want. However, natural languages are not completely free with that respect, and the next sections are devoted to deal with some of these cases and to show how to introduce some control.

The principle we adopt is based on the following observation: operators triggering extractions get the general pattern $(\alpha \multimap \beta) \multimap \gamma$ for their type. However, not all elements of a same type α can be extracted. For instance, if α is *np*, it is required sometimes to be nominative and sometimes to be accusative. These constraints can be accomodated adding feature structures (here dependent types) to the syntactic type.

But this is not enough since β might also express some additional constaints. For instance, if β is s, extraction is sometimes possible under the assumption that no other extraction occured. This can also be expressed using feature structures added to s.

Finally, it might happen that not all combinations for the constraints on α and β are possible, meaning that the extraction constraints are described by a *relation*, distinct from the cartesian product, between their feature structures. For instance extraction of the subject inside a clause is possible provided this is the very subject of that clause. Dependent types allows us to implement such relations. This approach shares a lot of similarities with [17]s' usage of first order linar logic where first order variables also implements some kind of relation between constituents.

3.2 Tensed Clauses as Scope Islands for Quantifiers

(2) is a first example of such a constraint. It is indeed sometimes considered that in such sentences, the QNP *everyone* should not be able to take scope over *someone*, or even *says* as in (2-b) and (2-c): the QNP *everyone* cannot take its scope outside its minimal tensed sentence[7].

(2) Someone said everyone loves Mary
 a. $C_{\text{so}}(\lambda^0 x.C_{\text{says}}\,(C_{\text{ev}}(\lambda^0 y.C_{\text{loves}}\,C_{\text{Mary}}\,y))\,x)$
  $\exists x.\mathbf{say}\,x\,(\forall y.\mathbf{love}\,y\,\mathbf{m})$
 b. $*C_{\text{so}}(\lambda^0 x.C_{\text{ev}}(\lambda^0 y.C_{\text{says}}\,(C_{\text{loves}}\,C_{\text{Mary}}\,y)\,x))$
  $*\exists x.\forall y.\mathbf{say}\,x\,(\mathbf{love}\,y\,\mathbf{m})$
 c. $*C_{\text{ev}}(\lambda^0 y.C_{\text{so}}(\lambda^0 x.C_{\text{says}}\,(C_{\text{loves}}\,C_{\text{Mary}}\,y)\,x))$
  $*\forall y.\exists x.\mathbf{say}\,x\,(\mathbf{love}\,y\,\mathbf{m})$

The fact that a QNP cannot take its scope outside its minimal tensed sentence means that whenever such a sentence is argument of a verb like *says*, it should not contain any free variable, hence any variable available for extraction, anymore. To model that, we decorate the *s* and *np* types with an integer feature that contains the actual number of free variables of type *np* occurring in it. Because any *np* introduced by the lexicon is decorated by 0, *np* with a feature strictly greater than 0 can only be introduced by hypothetical reasoning, hence by free variables. A clause without any left free variable is then of type *s* decorated with 0: this is required for the first argument of the verb *says* for instance.

In order to avoid changing the syntax-semantics interface we defined in Sect. 3.1, we implement the control using a more abstract level. This level introduces the counter feature using a new signature Σ_{Cont_1}, as Table 4 shows. The new types are very similar to the ones of Σ_{Syn} (Table 2) except that they now depend on an integer meant to denote the number of free variables occurring in the subterms. We then define $\mathscr{G}_{\text{Cont}_1} = \langle \Sigma_{\text{Cont}_1}, \Sigma_{\text{Syn}}, \mathscr{L}_{\text{Cont}_1}, s\,0\rangle$ the ACG that realizes the control over the syntactic structures. Lexicon $\mathscr{L}_{\text{Cont}_1}$ (Table 4) basically removes the dependent product and transforms Σ_{Cont_1} into Σ_{Syn}.

Having constants producing terms of type $s\,i$ like D_{loves}, where the feature indicates the number of current free variables that can be abstracted over in the subterms they are the head of, we are now in position of controlling the scope of QNP. Because the sentence argument of D_{says} is required to carry 0 free variables, all the quantified variables must have met their scope-taking operator before the resulting term is passed as argument, preventing them from escaping the scope island.

[7] This is arguable, and the tensed clauses island may be less straightforward, but this point is not ours here.

Table 4. Σ_{Cont_1} and $\mathscr{L}_{\text{Cont}_1}$

Σ_{Cont_1} :

int	$: \texttt{type}$	s, np, n	$: (int)\,\texttt{type}$
next	$: int \multimap int$	D_{loves}	$: (\Pi i, j : int)\,(np\, i \multimap np\, j \multimap s\,(i+j))$
$+$	$: int \multimap int \multimap int$	$D_{\text{so}}, D_{\text{ev}}$	$: (\Pi i : int)\,((np\, 1 \multimap s\,(\mathsf{next}\, i)) \multimap s\, i)$
$D_{\text{John}}, D_{\text{Mary}}$	$: np\, 0$	D_{says}	$: (\Pi i : int)\,(s\, 0 \multimap np\, i \multimap s\, i)$

$\mathscr{L}_{\text{Cont}_1}$:

$s :=_{\text{Cont}_1} \lambda x.\, s$	$np :=_{\text{Cont}_1} \lambda x.\, np$
$n :=_{\text{Cont}_1} \lambda x.\, n$	$D_{\text{x}} :=_{\text{Cont}_1} C_{\text{x}}$

(3) is a well-typed term (of type $s\,0$) of $\Lambda_{\Sigma_{\text{Cont}_1}}$. It has the same structure as (2-a) which, indeed, is its image by $\mathscr{L}_{\text{Cont}_1}$. On the other hand, the type $np\,0 \multimap s\,0$ of (4) (that would be the counterpart of the subterm of (2-c)) prevents it from being argument of a quantifier. Here, D_{says} requires y to be of type $np\,0$ in order to have its argument $D_{\text{love}}\,0\,0\,D_{\text{Mary}}\,y$ of type $s\,0$.

(3) $D_{\text{so}}\,0\,\underbrace{(\lambda^0 x.\, \overbrace{D_{\text{says}}\,1\,\underbrace{(D_{\text{ev}}\,0\,\underbrace{(\lambda^0 y.\, \overbrace{D_{\text{love}}\,0\,1\,D_{\text{Mary}}\,\overbrace{y}^{np\,1}}^{s\,1})}_{np\,1 \multimap s\,1})}_{s\,0}\,\overbrace{x}^{np\,1}}^{})}_{np\,1 \multimap s\,1}$ — with $D_{\text{so}}\,0\,(\ldots) : s\,0$

(4) $\lambda^0 y.\, \underbrace{D_{\text{so}}\,0\,\underbrace{(\lambda^0 x.\, \overbrace{D_{\text{says}}\,1\,(D_{\text{love}}\,0\,0\,D_{\text{Mary}}\,\overbrace{y}^{np\,0})\,\overbrace{x}^{np\,1}}^{s\,0})}_{np\,1 \multimap s\,1}}_{s\,0}$

This example could be easily adapted to other tensed clauses, as if-clauses or relative clauses. The next examples use the same principle: all types depend on a feature that expresses whether some free variables in the subterms are available for extraction. Then, wh-words put the condition on how many of them are simultaneously possible for extraction to take place while islands still require this number to be set to 0.

Note that in each case, we introduce a new feature for the particular phenomenon under study. Using record types (that np, n and s would depend on) with a proper field for each of them makes the different solutions work together without any interaction. Feature structures for each type might of course become complex, however this complexity can be dealt with in a very modular way.

3.3 Rooted and Embedded Wh-Extraction

We now focus on extractions in relative clauses, in which a distinction should be made between rooted extractions and embedded extractions: while an embedded object can be extracted by a relative pronoun, embedded subjects cannot. Only main-clause subjects (rooted subjects) can be extracted. This is illustrated in:

(5) *The man who$_1$ John said that t$_1$ loves Mary sleeps
*$C_{\text{sleep}}\,(C_{\text{the}}\,(C_{\text{who}}(\lambda^0 x.C_{\text{say that}}\,(C_{\text{love}}\,C_{\text{Mary}}\,x)\,C_{\text{John}})\,C_{\text{man}}))$

(6) The man whom$_1$ John said that Mary loves t$_1$ sleeps
$C_{\text{sleep}}\,(C_{\text{the}}\,(C_{\text{whom}}(\lambda^0 x.C_{\text{say that}}\,(C_{\text{love}}\,x\,C_{\text{Mary}})\,C_{\text{John}})\,C_{\text{man}}))$

Relative clauses are extraction islands, so we know that acceptable terms should never have more than one free variable available to extraction in the same clause. Hence we don't need an unbound counter for them and we use instead a 3-valued type that distinguishes: the absence of extraction, the existence of a rooted extraction, and the existence of an embedded one. The new abstract signature is given in Table 5 (for the sake of clarity, *who* will only refer to subject extraction and case is omitted). The corresponding ACG $\mathscr{G}_{\text{Cont}_2} = \langle \Sigma_{\text{Cont}_2}, \Sigma_{\text{Syn}}, \mathscr{L}_{\text{Cont}_2}, s\,\text{no}\rangle$ is built in the same way as the previous example.

Table 5. Σ_{Cont_2}

Σ_{Cont_2} :

value, *extraction*	: `type`	D_{sleeps}	: $(\Pi x : \textit{value})\,(np\,x \multimap s\,(f\,x\,\text{cst}))$
var, cst	: *value*	D_{loves}	: $(\Pi x, y : \textit{value})$ $(np\,x \multimap np\,y \multimap s\,(f\,x\,y))$
no, root, emb	: *extraction*	D_{the}	: $(\Pi x : \textit{value})\,(n\,x \multimap np\,x)$
np, *n*	: (*value*) `type`	$D_{\text{says that}}$	: $(\Pi x : \textit{extraction}, y : \textit{value})\,(s\,x)$ $\multimap np\,y \multimap s\,(g\,x\,y)$
s	: (*extraction*) `type`		
D_{man}	: *n* cst	D_{who}	: $(np\,\text{var} \multimap s\,\text{root}) \multimap n\,\text{cst} \multimap n\,\text{cst}$
$D_{\text{John}}, D_{\text{Mary}}$	: *np* cst	D_{whom}	: $(np\,\text{var} \multimap s\,\text{root}) \multimap n\,\text{cst} \multimap n\,\text{cst}$
		D'_{whom}	: $(np\,\text{var} \multimap s\,\text{emb}) \multimap n\,\text{cst} \multimap n\,\text{cst}$

$$\text{with } f : \begin{cases} \text{var } x \longrightarrow \text{root} \\ \text{cst var} \longrightarrow \text{root} \\ \text{cst cst} \longrightarrow \text{no} \end{cases} \qquad g : \begin{cases} \text{no var} \longrightarrow \text{root} & \text{root cst} \longrightarrow \text{emb} \\ \text{no cst} \longrightarrow \text{no} & \text{emb var} \longrightarrow \text{emb} \\ \text{root var} \longrightarrow \text{root} & \text{emb cst} \longrightarrow \text{emb} \end{cases}$$

The behavior of a transitive verb such as D_{loves} is to percolate the information that a free variable occurs in its parameters. So the resulting type depends on no only when both the subject and the object don't themselves depend on a var term. Function f in Table 5 implements it.

Verbs requiring subordinate clauses as $D_{\text{says that}}$ also needs to percolate the information as to whether a free variable occurs in the main clause and/or if a free variable occurs in the subordinate clause (in that case, the extraction is embedded). Function g in Table 5 implements these conditions.

Finally, relative pronouns need to check the type of their argument. In particular subject extractor can't accept an argument clause with type $(np\,\text{var} \multimap s\,\text{emb})$ while other pronouns can. This prevents extractions of embedded subject from being generated while extraction of embedded objects can, as shown with the abstract term (7) of type s no associated to (6).

$$(7)\ D_{\text{sleep}}\ \text{cst}\ D_{\text{the}}\ \text{cst}\ (D'_{\text{whom}}\ (\underbrace{\lambda^0 x.\ \overbrace{D_{\text{says that}}\ \text{root cst}\ \underbrace{(D_{\text{love}}\ \text{var cst}\ \overbrace{x}^{np\ \text{var}}\ D_{\text{Mary}})}_{s\ \text{root}}\ D_{\text{John}}}^{s\ \text{emb}}}_{np\ \text{var} \multimap s\ \text{emb}})\ D_{\text{man}})$$

On the other hand, $D_{\text{says that}}\ (D_{\text{love}}\ D_{\text{Mary}}\ x)\ D_{\text{John}}$ is typable only with type s emb or s no (because D_{John} is of type np cst), hence $\lambda^0 x.D_{\text{says that}}\ (D_{\text{love}}\ x\ D_{\text{Mary}})\ D_{\text{John}}$ cannot be of type np var $\multimap$ s root and cannot be an argument of D_{who}. Then (5) cannot get an antecedent by $\mathscr{L}_{\text{Cont}_2}$[8].

The same technique can be used to model the fact that a nominative interrogative pronoun can form a root question with a sentence that is missing its main clause subject as in (8) but not with one that is missing an embedded subject as in (9).

(8) Who left?

(9) *Who$_1$ Mary said that t_1 left?

3.4 Multiple Extraction

Nested-dependencies constraints, exemplified in (10) and (11), specify that only the leftmost trace can be bound (for sake of clarity, we forget here about the control verb nature of *know*).

(10) Which$_1$ problems does John know whom$_2$ to talk to t$_2$ about t$_1$?

a. $C_{\text{which?}}\ C_{\text{problems}}\ (\lambda^0 x.\ C_{\text{know}}\ (C_{\text{whom?}}\ (\lambda^0 y.\ C_{\text{to talk to about}}\ y\ x))\ C_{\text{John}})$

(11) *Whom$_1$ does John know which$_2$ problems to talk to t$_1$ about t$_2$?

a. $*C_{\text{whom?}}\ (\lambda^0 y.\ C_{\text{know}}\ (C_{\text{which?}}\ C_{\text{problems}}\ (\lambda^0 x.\ C_{\text{to talk to about}}\ y\ x))\ C_{\text{John}})$

The interrogative extraction follows a first in last out pattern. Despite the close relation of this pattern to the linear order of the sentence, we again implement control at the abstract level. As in Sect. 3.2, extractions are associated with counters that reflect the argument position in the canonical form. Table 6 describes the abstract signature for modelling these cases and $\mathscr{G}_{\text{Cont}_3} = \langle \Sigma_{\text{Cont}_3}, \Sigma_{\text{Syn}}, \mathscr{L}_{\text{Cont}_3}, s\,0\rangle$ is defined the usual way.

Basically, pronouns and their traces get the same counter value. The type of the interrogative pronouns requires sequences of them to have increasing values, greater numbers being abstracted first.

Let us consider a term $t = D_{\text{to talk to about}}\ i\ j\ y\ x$ (to be read as *to talk to* y *about* x) of type $q(h\,i\,j)$ with y of type $np\ i$ and x of type $np\ j$. We show that in order to extract both x and y (and bind them with interrogative pronouns), y has to be extracted first:

[8] The felicity of *The man who John said loves Mary sleeps*, without the complementizer, suggests a type assignment to D_{says} that does not switch the dependant product to emb the way $D_{\text{says that}}$ does.

Table 6. Σ_{Cont_3}

int	: type	np, n, s, q	: (int) type
D_{John}	: $np\,0$	$D_{\text{to talk to about}}$	: $(\Pi i, j : int)\,(np\,i \multimap np\,j \multimap q\,(h\,i\,j))$
D_{problems}	: $n\,0$	D_{know}	: $(\Pi i, j : int)\,(q\,i \multimap np\,j \multimap q\,(h\,i\,j))$
next	: $int \multimap int$	$D_{\mathrm{whom?}}$	: $(\Pi i : int)\,((np\,(\mathsf{next}\,i) \multimap q\,(\mathsf{next}\,i)) \multimap q\,i)$
		$D_{\mathrm{which?}}$	: $(\Pi i : int)\,(n\,0 \multimap (np\,(\mathsf{next}\,i) \multimap q\,(\mathsf{next}\,i)) \multimap q\,i)$

$$h : \begin{cases} i & 0 \longrightarrow i \\ 0 & j \longrightarrow j \\ \mathsf{next}\,i\,j \longrightarrow \mathsf{next}\,i \end{cases}$$

- let's assume x is extracted first. The type of the result is $np\,j \multimap q\,i$. Making it a suitable argument of an interrogative pronoun requires $i = j$. But the application results in a term of type $q\,(i-1)$. Then an abstraction of y would result in a term of type $np\,i \multimap q\,(i-1)$ that cannot be argument of another interrogative pronoun. Hence (11-a) can't have an antecedent by $\mathscr{L}_{\mathrm{Cont}_3}$;
- let's now assume that y is extracted first. The type of the result is $np\,i \multimap q\,i$, and when argument of an interrogative pronoun, it results in a term of type $q\,(i-1)$. The result of abstracting then over x is a term of type $np\,j \multimap q\,(i-1)$. To have the latter a suitable argument for an interrogative pronoun requires that $j = i - 1$, or $i = \mathsf{next}\,j$.
  Then, provided $i \geq 2$,

$$\begin{array}{l} D_{\mathrm{which?}}\,(i-2)\,D_{\mathrm{problems}} \\ \quad (\lambda^0 x.D_{\mathrm{know}}\,(i-1)\,0\,(D_{\mathrm{whom?}}\,(i-1)\,(\lambda^0 y.D_{\text{to talk to about}}\,i\,(i-1)\,y\,x))\,D_{\mathrm{John}}) \end{array}$$

  is typable (of type $q\,(i-2)$) and is an antecedent of (10-a) by $\mathscr{L}_{\mathrm{Cont}_3}$.

4 Related Approaches

4.1 Parallel Architectures

In this section, we wish to contrast our approach that modifies the abstract level with approaches in which control comes from a specific calculus at the object level. One of this approach specifically relates to the LG framework [22] and aims at introducing Multimodal Categorial Grammar (MMCG) [16] analysis at the phenogrammatical level. The other approach [12] also builds on MMCG analysis. It can actually bee seen as a parallel framework where the both the tectogrammatical level and the phenogrammatical level are MMCG. What is of interest to us is the proposal permitting phonological changes at the phenogrammatical level while the tectogrammatical one is unchanged.

In order to compare the three approaches, it is convenient to introduce the following notations:

Definition 6 (Signs and languages). *A* sign $s = \langle a, o, m\rangle$ *is a triple where:*

- *a is a term belonging to the tectogrammatical level*
- *o is a term belonging to the phenogrammatical level describing the* surface *form associated to a*
- *m is a term belonging to the phenogrammatical level describing the* logical *form associated to a*

In the case of LG and ACG, a is a linear λ-term whereas it is a MMCG proof term in [12].

In all frameworks, a sign $s = \langle a, o, m\rangle$ belong to the language whenever a is of a distinguished type s. Following [22], we call it a generated *sign.*

It is easy to see that in ACG and the approach we developed, o is a λ-term, possibly using the string concatenation operation.

On the other hand, [22] makes o be a multimodal logical formula build from constants and (unary and binary) logical connectives. It not only includes a special binary connective $\circ$ basically representing concatenation, but also any other required connective, in particular families of $\Diamond_i$ and $\Box_i$ operators. Then, the phenogrammatical level can be provided with a consequence relation $\sqsubseteq$ and also, as is standard in MMCG, with proper axioms, or *postulates*. It can then inherit all models of this framework such as [18]'s one for controlling extraction.

Hence, for any sign $s = \langle a, o, m\rangle$, it is possible to define a notion of derivability:

Definition 7 (Derivable and string-meaning signs). *Let $s = \langle a, o, m\rangle$ be a generated sign and o' a logical formula such that $o \sqsubseteq o'$. Then $s' = \langle a, o', m\rangle$ is called a* derivable *sign.*

Let $s = \langle a, o, m\rangle$ be a sign such that o is made only from constants and $\circ$. Then o is said to be readable[9] *and s is said to be a* string-meaning *sign.*

From that perspective, what is now of interest is not the generated signs as such but rather the string-meaning signs. In particular, if $s = \langle a, o, m\rangle$ is a generated sign, the interesting question is whether there exist some o' with $o \sqsubseteq o'$ and o' *readable*. If such an o' exists, then s is expressible, otherwise it is not.

[22, example (35)] is very similar to Example (2). Its analysis is as follows: (2-a), (2-b) and (2-c) are all possible abstract terms so that $s_a = \langle\text{(2-a)}, o_a, m_a\rangle$, $s_b = \langle\text{(2-b)}, o_b, m_b\rangle$ and $s_c = \langle\text{(2-c)}, o_c, m_c\rangle$ are all generated signs. However, there is no readable o such that $o_b \sqsubseteq o$ or $o_c \sqsubseteq o$ because o_b and o_c make use of different kinds of modalities that don't interact through postulates. Hence s_b and s_c can be generated but don't have any readable (or pronounceable) form and only s_a gives rise to a string-meaning sign and is expressible. The approach of [12] is very similar except that the phenogrammatical level is an algebra with a preorder whose maximal elements are the only pronounceable ones.

4.2 Continuation Semantics

In order to take into account constraints on scope related to scope ambiguity and polar sensitivity, [29] uses control operators, in particular delimited continuations with **shift** and **reset** operators in the semantic calculus.

[9] [12] defines *pronounceable* because it deals with phonology rather than with strings.

Parallel architecture such as LG or ACG could also make use of such operators in the syntactic calculus, achieving some of the effects we described. However, applying the continuation-passing style (CPS) transform to those constructs results in a significant increase of the order of types. The impact on the parsing complexity should then be studied carefully in order to get tractable fragments.

4.3 TAG and Lambek Grammars in ACG

We also wish to relate our proposal with similar architectures that have been proposed to model other grammatical formalisms, namely Minimalist Grammars (MG) [31], Tree Adjoining Grammar (TAG) [9], and non-associative Lambek grammars (NL) [14] .

In order to study MG from a logical point of view, [28] studies MG derivations in the ACG framework. Derivations are described at an abstract level (using **move** and **merge** operations) and are further interpreted to get the syntactic representation and the semantic representation at object levels. But rather than giving a direct translation, it is possible to add an intermediate level that corresponds to what is shared between syntax and semantics, but that contains much more than only MG derivations. This is reminiscent of the architecture of Fig. 2(a).

An other example where such an architecture takes place is given in [11] where a first abstract level specifies a syntax-semantics interface for TAG. However, this interface is not constrained enough and accept more than just TAG derivations. Then more abstract levels are added to control the derivations and accept only TAG, local MCTAG and non-local MCTAG.

The encoding of NL into ACG [26] also involves such an architecture. It defines a syntax-semantics interface very close to the one proposed here, and a more abstract level controls in turn this interface in order to discard derivations that are not NL derivations. This last result gives another interesting link to MMCG at a tectogrammatical level rather than at a phenogrammatical one as described in Sect. 4.1, in particular in the case of extraction because of the relation between NL and the calculus with the bracket operator of [18] to deal with islands.

5 Conclusion

Studying constraints related to extraction phenomena, we propose to use dependent types to implement them at an abstract level in the ACG framework. Using dependent types allows us to get finer control on derivations and to discard overgenerating ones. The same methodology has been used to model constraints related to bounding scope displacement, wh-extraction and multiple wh-extraction. This approach, where what appears as constraints at the surface level are rendered at an abstract level, contrasts with other approaches where a derivability notion on surface forms is introduced, and where some of the surface forms get the special status of *readable*.

Interestingly, these two ways to introduce or relax control on derivations are completely orthogonal, hence they could be used together. This gives rise to the question of determining the most appropriate approach given one particular phenomena. Answers could come both from linguistic considerations and from tractability issues of the underlying calculi. Another question is whether the relational semantics behind MMCG could be used, together with the dependent types, to model MMCG derivations within the ACG framework.

Acknowledgments. We would like to thank Carl Pollard and Philippe de Groote for fruitful discussions on earlier versions of this paper.

References

1. Carpenter, B.: Type-Logical Semantics. The MIT Press (1997)
2. Curry, H.B.: Some logical aspects of grammatical structure. In: Jakobson, R. (ed.) Structure of Language and its Mathematical Aspects: Proceedings of the Twelfth Symposium in Applied Mathematics, pp. 56–68. American Mathematical Society (1961)
3. de Groote, P.: Towards Abstract Categorial Grammars. In: 39th Annual Meeting and 10th Conference of the European Chapter, Proceedings of the Conference on Association for Computational Linguistics, pp. 148–155 (2001)
4. de Groote, P., Maarek, S.: Type-theoretic extensions of Abstract Categorial Grammars. In: Proceedings of New Directions in Type-Theoretic Grammars, pp. 18–30 (2007), `http://let.uvt.nl/general/people/rmuskens/ndttg/ndttg2007.pdf`
5. de Groote, P., Pogodalla, S.: On the expressive power of abstract categorial grammars: Representing context-free formalisms. Journal of Logic, Language and Information 13(4), 421–438 (2004), `http://hal.inria.fr/inria-00112956/fr/`
6. de Groote, P., Pogodalla, S., Pollard, C.: On the Syntax-Semantics Interface: From Convergent Grammar to Abstract Categorial Grammar. In: Ono, H., Kanazawa, M., de Queiroz, R. (eds.) WoLLIC 2009. LNCS, vol. 5514, pp. 182–196. Springer, Heidelberg (2009), `http://hal.inria.fr/inria-00390490/en/`
7. de Groote, P., Maarek, S., Yoshinaka, R.: On Two Extensions of Abstract Categorial Grammars. In: Dershowitz, N., Voronkov, A. (eds.) LPAR 2007. LNCS (LNAI), vol. 4790, pp. 273–287. Springer, Heidelberg (2007)
8. Johnson, M.: Proof nets and the complexity of processing center embedded constructions. Journal of Logic, Language and Information 7(4) (1998)
9. Joshi, A.K., Schabes, Y.: Tree-adjoining grammars. In: Rozenberg, G., Salomaa, A. (eds.) Handbook of Formal Languages, ch. 2. Springer (1997)
10. Kanazawa, M.: Parsing and generation as datalog queries. In: Proceedings of the 45th Annual Meeting of the Association of Computational Linguistics (ACL), pp. 176–183. Association for Computational Linguistics, Prague (2007), `http://www.aclweb.org/anthology/P/P07/P07-1023`
11. Kanazawa, M., Pogodalla, S.: Advances in Abstract Categorial Grammars: Language theory and linguistic modelling. In: ESSLLI 2009, Bordeaux, France. Lecture Notes (2009), `http://www.loria.fr/equipes/calligramme/acg/publications/esslli-09/2009-esslli-acg-week-2-part-2.pdf`
12. Kubota, Y., Pollard, C.: Phonological Interpretation into Preordered Algebrasa. In: Ebert, C., Jäger, G., Michaelis, J. (eds.) MOL 10. LNCS, vol. 6149, pp. 200–209. Springer, Heidelberg (2010)
13. Lambek, J.: The mathematics of sentence structure. American Mathematical Monthly 65(3), 154–170 (1958)
14. Lambek, J.: On the calculus of syntactic types. In: Jacobsen, R. (ed.) Structure of Language and its Mathematical Aspects. Proceedings of Symposia in Applied Mathematics, vol. XII. American Mathematical Society (1961)

15. Montague, R.: The proper treatment of quantification in ordinary english. In: Formal Philosophy: Selected Papers of Richard Montague. Yale University Press (1974); re-edited in Formal Semantics: The Essential Readings, Portner, P., Partee, B.H., (eds.) Blackwell Publishers (2002)
16. Moortgat, M.: Categorial type logics. In: van Benthem, J., ter Meulen, A. (eds.) Handbook of Logic and Language, pp. 93–177. Elsevier Science Publishers, Amsterdam (1996)
17. Moot, R., Piazza, M.: Linguistic applications of first order intuitionistic linear logic. Journal of Logic, Language and Information 10, 211–232 (2001)
18. Morrill, G.: Categorial formalisation of relativisation: Islands, extraction sites and pied piping. Tech. Rep. LSI-92-23-R, Departament de Llenguatges i Sistemes Informàtics, Universitat Politècnica de Catalunya (1992)
19. Morrill, G.V.: Incremental processing and acceptability. Computational Linguistics 26(3), 319–338 (2000)
20. Muskens, R.: Lambda Grammars and the Syntax-Semantics Interface. In: van Rooy, R., Stokhof, M. (eds.) Proceedings of the Thirteenth Amsterdam Colloquium, Amsterdam, pp. 150–155 (2001)
21. Muskens, R.: Lambdas, Language, and Logic. In: Kruijff, G.J., Oehrle, R. (eds.) Resource Sensitivity in Binding and Anaphora. Studies in Linguistics and Philosophy, pp. 23–54. Kluwer (2003)
22. Muskens, R.: Separating syntax and combinatorics in categorial grammar. Research on Language and Computation 5(3), 267–285 (2007)
23. Pogodalla, S.: Computing semantic representation: Towards ACG abstract terms as derivation trees. In: Proceedings of the Seventh International Workshop on Tree Adjoining Grammar and Related Formalisms (TAG+7), pp. 64–71 (May 2004), `http://www.cs.rutgers.edu/TAG+7/papers/pogodalla.pdf`
24. Pogodalla, S.: Generalizing a proof-theoretic account of scope ambiguity. In: Geertzen, J., Thijsse, E., Bunt, H., Schiffrin, A. (eds.) Proceedings of the 7th International Workshop on Computational Semantics - IWCS 2007, pp. 154–165. Tilburg University, Deparment of Communication and Information Sciences (2007), `http://hal.inria.fr/inria-00112898`
25. Ranta, A.: Type Theoretical Grammar. Oxford University Press (1994)
26. Retoré, C., Salvati, S.: A faithful representation of non-associative lambek grammars in abstract categorial grammars. Journal of Logic, Language and Information 19(2), 185–200 (2010), `http://www.springerlink.com/content/f48544n414594gw4/`
27. Salvati, S.: Problèmes de filtrage et problèmes d'analyse pour les grammaires catégorielles abstraites. Ph.D. thesis. Institut National Polytechnique de Lorraine (2005)
28. Salvati, S.: Minimalist Grammars in the Light of Logic. In: Pogodalla, S., Quatrini, M., Retoré, C. (eds.) Logic and Grammar. LNCS, vol. 6700, pp. 81–117. Springer, Heidelberg (2011)
29. Shan, C.C.: Delimited continuations in natural language: Quantification and polarity sensitivity. In: Thielecke, H. (ed.) Proceedings of the 4th continuations workshop, pp. 55–64. School of Computer Science, University of Birmingham (2004)
30. Shieber, S.M.: Unifying synchronous tree-adjoining grammars and tree transducers via bimorphisms. In: Proceedings of the 11th Conference of the European Chapter of the Association for Computational Linguistics (EACL 2006), Trento, Italy, April 3-7 (2006), `http://www.aclweb.org/anthology-new/E/E06/E06-1048.pdf`
31. Stabler, E.: Derivational minimalism. In: Retoré, C. (ed.) LACL 1996. LNCS (LNAI), vol. 1328, pp. 68–95. Springer, Heidelberg (1997)

Plural Quantifications and Generalized Quantifiers

Byeong-Uk Yi

Department of Philosophy, University of Toronto, 170 St George St, Toronto, ON M5R 2M8, Canada
b.yi@utoronto.ca

Abstract. This paper discusses two important results about expressive limitations of elementary languages due to David Kaplan, and clarifies how they relate to the expressive power of plural constructions of natural languages. Kaplan proved that such plural quantifications as the following cannot be paraphrased into elementary languages:

Most things are funny. (1)

Some critics admire only one another. (2)

The proof that (1) cannot be paraphrased into elementary languages is often taken to support the generalized quantifier approach to natural languages, and the proof that (2) cannot be so paraphrased is usually taken to mean that (2) is a second-order sentence. The paper presents an alternative interpretation: Kaplan's results provide important steps toward clarifying the expressive power of plural constructions of natural languages vis-à-vis their singular cousins. In doing so, the paper compares and contrasts (regimented) plural languages with generalized quantifier languages, and plural logic with second-order logic.

Keywords: semantics, natural language, plural construction, plural logic, Geach-Kaplan sentence, generalized quantifier theory, Rescher quantifier, plural quantifier, the semantics of 'most'.

1 Introduction

Plural constructions (in short, *plurals*) are as prevalent in natural languages as singular constructions (in short, *singulars*). This contrasts natural languages with the usual symbolic languages, e.g., elementary languages or their higher-order extensions.[1] These are singular languages, languages with no counterparts of natural language plurals, because they result from regimenting singular fragments of natural languages (e.g., Greek, German, or English).[2] But it is commonly thought that the lack of plurals

[1] What I call *elementary languages* are often called *first-order languages*. I avoid this terminology, because it suggests contrasts only with higher-order languages.

[2] Some natural languages (e.g., Chinese, Japanese, or Korean) have neither singulars nor plurals, because they have no grammatical number system. This does not mean that those languages, like the usual symbolic languages, have no counterparts of plurals or that they have no expressions for talking about many things (as such).

P. de Groote and M.-J. Nederhof (Eds.): Formal Grammar 2010/2011, LNCS 7395, pp. 178–191, 2012.

in the usual symbolic languages result in no deficiency in their expressive power. There is no need to add counterparts of natural language plurals to symbolic languages, some might hold, because plurals are more or less devices for abbreviating their singular cousins: 'Ali and Baba are funny' and 'All boys are funny', for example, are more or less abbreviations of 'Ali is funny and Baba is funny' and 'Every boy is funny', respectively. But there are more recalcitrant plurals, plurals that cannot be considered abbreviations of singulars, e.g., 'The scientists who discovered the structure of DNA cooperated' or 'All the boys cooperated to lift a piano.' So I reject the traditional view of plurals as abbreviation devices, and propose an alternative view that departs radically from the tradition that one can trace back to Aristotle through Gottlob Frege. Plurals, on my view, are not redundant devices, but fundamental linguistic devices that enrich our expressive power, and help to extend the limits of our thoughts. They belong to *basic linguistic categories* that complement the categories to which their singular cousins belong, and have a *distinct semantic function*: plurals are by and large devices for talking about *many* things (as such), whereas singulars are more or less devices for talking about *one* thing ('at a time').[3] It has been a few decades since David Kaplan established two important limitations of elementary languages by considering natural language plurals. In this paper, I ruminate on his results to clarify the expressive power of plurals.

2 Expressive Limitations of Elementary Languages

Elementary languages can be taken to have five kinds of primitive expressions:

(a) *Singular Constants*: 'a', 'b', etc.
(b) *Singular Variables*: 'x', 'y', 'z', etc.
(c) *Predicates*
  (i) 1-place predicates: 'B^1', 'C^1', 'F^1', etc.
  (ii) 2-place predicates: '$=$', 'A^2', etc.
  (iii) 3-place predicates: 'G^3', etc.
  Etc.
(d) *Boolean Sentential Connectives*: '$\neg$', '$\wedge$', etc.
(e) *Elementary Quantifiers*: the singular existential '$\exists$', and the singular universal '$\forall$'

The constants amount to proper names of natural languages, e.g., 'Ali' or 'Baba'; variables to singular pronouns, e.g., 'he', 'she', or 'it', as used anaphorically (as in 'A boy loves a girl, and *she* is happy'); the predicates to verbs (or verb phrases) in the singular form, e.g., 'is a boy', 'is identical with', 'admires', or 'gives ... to —'; and the quantifiers to 'something' and 'everything'. 'Something is a funny boy', for example, can be paraphrased by the elementary language sentence '$\exists x[B(x) \wedge F(x)]$', where '$B$' and '$F$' are counterparts of 'is a boy' and 'is funny', respectively.[4] Elementary

[3] See, e.g., Yi [17]–[20] for an account of plurals based on the view sketched in this paragraph.

[4] The superscript of a predicate is omitted, if the number of its argument places is clear from the context.

languages, note, have no quantifier that directly amounts to the determiner 'every' in, e.g., 'Every boy is funny', but this determiner can be defined in elementary languages, as is well-known. 'Every boy is funny' can be paraphrased by the universal conditional '$\forall x[B(x) \rightarrow F(x)]$', which amounts to 'Everything is such that if it is a boy, then it is funny.'

Now, say that natural language sentences (e.g., 'Something is a funny boy', 'Every boy is funny') are *elementary*, if they can be paraphrased into elementary languages. Kaplan showed that the following plural constructions are *not* elementary:

Most are funny. (1a)

Most boys are funny. (1b)

Some critics admire only one another.[5] (2)

He showed that these cannot be paraphrased into elementary languages with predicates amounting to 'is a boy' and 'is funny' (e.g., 'B' and 'F'). An important feature of elementary languages that one can use to obtain such a result is that their logic, elementary logic, is *compact*, that is,

> *Compactness of elementary logic*:
> If some sentences of an elementary language $\mathcal{L}$ logically imply a sentence of $\mathcal{L}$, then there are finitely many sentences among the former that logically imply the latter.

So if a sentence is logically implied by infinitely many elementary sentences, but not by any finitely many sentences among them, this means that the sentence is not elementary.

Now, consider (1a) and (1b). Rescher [12] considers two quantifiers that roughly amount to the two uses of 'most' in these sentences. (1a) might be considered an abbreviation of a sentence in which a noun (in the plural form) follows 'most' (e.g., (1b) or 'Most things are funny'). But one might regard 'most' in (1a) as a *unary* quantifier, one that, like 'everything', can combine with one (1-place) predicate (in the plural form) to form a sentence. Rescher proposes to add to elementary languages a new quantifier, 'M', that corresponds to 'most' in (1a), so construed. Using the quantifier, which came to be known as (Rescher's) *plurality quantifier*,[6] he puts (1a) as follows:

$$MxF(x) \qquad (1a^*)$$

---

[5] (2) can be paraphrased by 'There are some critics each one of whom is a critic, and admires something only if it is one of them but is not identical with him- or herself.'

[6] Rescher says that sentences with the quantifier 'M' involve "the new mode of *plurality-quantification*", but calls the quantifier itself "M-quantifier" ([13], 373). Kaplan [6]–[7] calls it "the plurality quantifier".

where 'F' is an elementary language predicate amounting to 'is funny'. He takes the quantification $Mx\varphi(x)$ to say that "the set of individuals for which φ is true has a greater cardinality than the set for which it is false" ([12], 373). And he states that the quantifier 'M' cannot be defined in elementary languages (*ibid*., 374), which means that (1a) and the like are not elementary,[7] but without indication of how to prove it. The statement was proved by Kaplan [6].[8] Here is a simple proof: '$\neg MxF(x) \wedge Mx[F(x) \vee x=a]$' is satisfiable in any finite model, but in no infinite model; but elementary languages, whose logic is compact, have no such sentence.[9]

In response to Kaplan's result, some might argue that the quantifier 'most' (or 'M') is not definable in elementary languages because it is not a logical but a mathematical expression. They might hold that (1a), for example, is a statement about numbers or sets: it means that the *number* of funny things is greater than the *number* of non-funny things (or that the *cardinality* of the *set* of funny things is greater than the *cardinality* of the *set* of non-funny things). If so, they might conclude, Kaplan's result merely confirms the well-known limitations of elementary languages in expressing mathematical truths, e.g., the mathematical induction principle, which helps to give a categorical characterization of arithmetical truths.

Although it might seem plausible that 'most' has an implicit reference to numbers or sets, it does not seem plausible to hold the same about 'some' in (2). This sentence does not seem to pertain to mathematics at all. So Boolos says that (2) and the like, unlike (1), "*look* as if they 'ought to' be symbolizable" in elementary languages ([3], 433; original italics). But Kaplan proved that they are not.[10] (2), which came to be known as the *Geach-Kaplan sentence*, cannot be paraphrased into elementary languages with counterparts of 'is a critic' and 'admires' (e.g., 'C^1' and 'A^2').

[7] To relate the undefinability of 'M' to the natural language quantifier 'most', it is necessary to assume that the former (as Rescher explains it) captures the latter. This is a controversial assumption; I think 'most' is usually used to mean *nearly all*, rather than *more than half* or *a majority* (*of*). (Westerståhl [15] holds that it is an ambiguous expression with two readings, which I doubt.) But we can take Rescher's plurality quantifier to correspond to 'more than half' or 'a majority', and the Rescher-Kaplan result to pertain to *this* quantifier.

[8] Kaplan [6] gives sketches of proofs of this and other interesting facts about languages that contain 'M'. It is straightforward to define 'most' in (1a) in terms of the binary determiner 'most' in (1b); (1a) is equivalent to 'Most things that are identical with themselves are funny.' So Kaplan's proof of the undefinability of the former extends to the latter.

[9] Rescher ([13], 374) does not introduce a symbolic counterpart of the binary determiner 'most' in (1b), but he states that it cannot be defined in elementary languages or even their extensions that result from adding 'M'. Barwise & Cooper ([2], 214f) prove this, and note that Kaplan proved it in 1965, but did not publish the proof. (Barwise & Cooper's proof yields a stronger result.)

[10] Kaplan communicated his proof to Quine. Quine ([11], 238f) states Kaplan's result, and argues that (2) is a sentence about sets or classes of critics. Kaplan's proof is reproduced in Boolos ([3], 432f). See also Almog [1].

His proof begins by paraphrasing (2) by a second-order sentence:

$$\exists^2 X\{\exists x X(x) \wedge \forall x[X(x) \rightarrow C(x)] \wedge \forall x\forall y[X(x) \wedge A(x, y) \rightarrow x{\neq}y \wedge X(y)]\} \quad (2a)$$

where 'X' is a second-order variable, and '$\exists^2$' the second-order existential quantifier. By replacing '$C(x)$' and '$A(x, y)$' in (2a) with '$N(x) \wedge x{\neq}0$' and '$S(x, y)$', where 'N' and 'S' amount to 'is a natural number' and 'is a successor of',[11] respectively, we can get the following:[12]

$$\exists^2 X\{\exists x X(x) \wedge \forall x[X(x) \rightarrow N(x) \wedge x{\neq}0] \wedge \forall x\forall y[X(x) \wedge S(x, y) \rightarrow x{\neq}y \wedge X(y)]\} \quad (2a^*)$$

which amounts to 'Some non-zero natural numbers are successors only of one another.' So if (2a) can be paraphrased into elementary languages with 'C' and 'A', so can (2a*) into those with 'N' and 'S'. But (2a*) cannot; its negation is equivalent to the second-order mathematical induction principle, which cannot be expressed in them.[13]

Most of those who discuss Kaplan's proof take it to show that (2) also turns out to be a covert statement of a mathematical fact. It is commonly held that (2) *is* a second-order sentence, comparable to (2a), which has the second-order existential quantifier '$\exists^2$', and it is usual to take second-order quantifiers to range over sets (or classes), e.g., sets of critics.

But Kaplan's proof does not support the conclusion that (2) is a second-order sentence or a sentence that implies the existence of a non-empty set (or class) of critics. To see this, consider the following sentences:

Ezra is a critic, Thomas is a critic, Ezra is not Thomas, Ezra admires only Thomas, and Thomas admires only Ezra. (2.1)

Ezra and Thomas are critics who admire only one another. (2.2)

We can intuitively see that (2.1) logically implies (2.2), and that (2.2) logically implies (2). So (2.1) must logically imply (2). If so, (2) cannot imply the existence of a set because (2.1), which has straightforward elementary language counterparts, does not do so.[14] And we can prove that (2) is not elementary without assuming that it can be paraphrased by a second-order sentence (e.g., (2a)). Consider the following series of infinitely many elementary language sentences:

---

[11] A natural number x is said to *be a successor of* a natural number y, if $x = y + 1$.

[12] (2a*) is slightly different from the arithmetical sentence used in Kaplan's proof. See the sentence (C) in Boolos (1984, 432). But it is straightforward to see that they are logically equivalent.

[13] The mathematical induction principle, added to the other Dedekind-Peano axioms, which are elementary, yields a categorical characterization of arithmetical truths. But one cannot give a categorical characterization of arithmetical truths in an elementary language, which can be proved using the compactness of elementary logic.

[14] For an elaboration of this argument, see Yi ([18], Ch. 1). See also Yi [17] & [19].

c_1 is a critic who admires only c_2. (3.1)

c_2 is a critic who admires only c_3. (3.2)

. . . .

c_n is a critic who admires only c_{n+1}. (3.n)

. . . .

where 'c_1', 'c_2', 'c_3', etc. are different proper names. We can intuitively see that these sentences, taken together, logically imply (2): if they hold, then c_1, c_2, c_3, etc. are critics who admire only one another. But no finitely many sentences among them logically imply (2): (3.1)-(3.n), for example, do not logically imply that c_{n+1} is a critic who admires nothing but c_1, c_2, . . ., c_{n+1}. So (2) cannot be paraphrased into elementary languages, whose logic is compact.[15]

Now, some might think that (2) is non-elementary only because it concerns cases involving infinitely many things. Then those who think, plausibly or not, that we are not concerned with any such cases outside mathematics might argue that elementary languages are powerful enough as long as we do not engage in higher mathematical enterprise, and restrict our domain of discourse to finite domains.[16] It would be wrong to do so. There is no elementary language sentence that agrees with (2) *even on all finite domains*. We can show this by applying basic results of Finite Model Theory.[17]

3 From Singular Languages to Plural Languages

The Geach-Kaplan sentence (2), we have seen, cannot be paraphrased into elementary languages. This is usually taken to show that it is a second-order sentence with an implicit quantification over sets of critics. But there is no good reason to take it to be a second-order sentence by taking the plural quantifier 'some' in the sentence as a second-order quantifier. Although Kaplan's proof of its non-elementary character proceeds by paraphrasing it by its second-order analogue, there are alternative, more direct proofs that do not rest on the paraphrase as we have seen. By contrast, there is a clear contrast between the plural constructions involved in (2) and the singular constructions involved in, e.g., (2.1)–(2.2) and (3.1)–(3.n), which have straightforward elementary language counterparts. So there is a good reason to take Kaplan's result on (2) to show the limitations of singulars (especially those incorporated into elementary

[15] See Yi [21], which argues that logic is not axiomatizable, for an elaboration of this argument. See also Yi ([20], 262).

[16] Sol Feferman once made this response.

[17] See Appendix, where it is also proved that no elementary language sentence agrees with '$MxF(x)$' on all finite models.

languages) vis-à-vis plurals.[18] If so, it would be useful to develop *plural extensions* of elementary languages, symbolic languages that have refinements of natural language plurals while containing elementary languages as their singular fragments, to give a theory of the *logical relations pertaining to plurals*, e.g., those that relate (2) to (2.1)-(2.2) or those that relate (2) to the sentences in (3). I have presented such languages by regimenting basic plural constructions of natural languages, and characterized their logic in some other publications.[19] Let me explain the basics of those symbolic languages, called *(first-order) plural languages*,[20] to show that they have natural paraphrases of basic plural constructions of natural languages.

Plural languages extend elementary languages by including plural cousins of singular variables, predicates, and quantifiers of elementary languages:

(b*) *plural variables*: '*xs*', '*ys*', '*zs*', etc.[21]
(c*) *plural predicates*: '$\boldsymbol{C}^1$' (for '*to* cooperate'), '$\mathbf{H}^2$' (for 'is one of'), '$\boldsymbol{D}^2$' (for '*to* discover'), '$\boldsymbol{L}^2$' (for '*to* lift'), '$\boldsymbol{W}^2$' (for '*to* write'), etc.
(e*) *plural quantifiers*: the existential '$\boldsymbol{\Sigma}$', and the universal '$\boldsymbol{\Pi}$'

Plural variables are refinements of the plural pronoun 'they' as used anaphorically (as in 'Some scientists worked in Britain, and *they* discovered the structure of DNA', where 'they' takes 'some scientists' as the antecedent). Plural quantifiers, which bind plural variables, are refinements of 'some things' and 'any things'. And plural predicates are refinements of usual natural language predicates (e.g., '*to* discover'). That is, they can combine with plural terms (e.g., 'they') as in the above-mentioned sentence, and have one or more argument places that admit a plural term: the only argument place of '$\boldsymbol{C}^1$', the first argument place of '$\boldsymbol{D}^2$', the second argument place of '$\mathbf{H}^2$', etc.[22] Elementary language predicates, by contrast, are refinements of the *singular forms* of natural language predicates (e.g., 'is funny' or 'admires'), and have no argument place that admits plural terms; so they can combine only with singular terms (i.e., singular constants or variables).

One of the plural predicates, '$\mathbf{H}^2$', which amounts to 'is one of', has a special logical significance. Like the elementary language predicate '=', it is a logical predicate. And we can use it to define complex plural predicates that result from 'expanding' singular predicates. We can define the *plural* (or *neutral*) *expansion* π^N of π as follows:

---

[18] Kaplan's results do not suffice to show that we cannot accommodate natural language plurals into the usual, singular higher-order languages. But we can add to Kaplan's results other results that show this. See, e.g., Yi ([17], 172-4) and ([19], 472-6).

[19] See, e.g., Yi [17]–[20].

[20] The logic of plural languages is called *plural logic*.

[21] I add '*s*' to a lower-case letter of English alphabet to write a plural variable, but plural variables, like singular variables, are simple expressions with no components of semantic significance.

[22] Such argument places are called *plural argument places*. The plural argument places (of plural language predicates) are *neutral* ones, that is, they admit singular terms as well.

Def. 1 (neutral expansion):
$\pi^{N}(xs) \equiv_{df} \forall y[\mathbf{H}(y, xs) \to \pi(y)]$ (to use the λ notation, $\pi^{N} =_{df} \lambda xs \forall y[\mathbf{H}(y, xs) \to \pi(y)]$).

Then the neural expansion 'C^{N}' of the elementary language counterpart '*C*' of 'is a critic', for example, amounts to the predicate 'are critics' (or, more precisely, '*to* be critics), which can be taken to paraphrase '*to* be such that any one of them *is a critic*'.

Now, we can paraphrase (2) into plural languages with '*C*' and '*A*' as follows:

$$\Sigma xs\{C^{N}(xs) \wedge \forall x \forall y[\mathbf{H}(x, xs) \wedge A(x, y) \to x \neq y \wedge \mathbf{H}(y, xs)]\}. \quad (2^*)$$[23]

This amounts to an English sentences that we can see is a natural paraphrase of (2): 'Some things are such that they are critics (i.e., any one of them is a critic), and any one of them admires something only if the latter is not the former and is one of them.'[24]

We can now consider sentences with 'most', such as those mentioned above:

Most are funny. (1a)

Most boys are funny. (1b)

These sentences involve plurals, as much as (2) does.[25] But Rescher's symbolic counterpart of (1a), '$MxF(x)$', results from reducing (1a) into a singular construction. The so-called plurality quantifier '*M*', like the elementary quantifiers '∃' and '∀', is a *singular quantifier*, one that can combine with singular variables (e.g., '*x*'), but not with plural variables (e.g., '*xs*').[26] Rescher was working in the framework of the incipient generalized quantifier theory,[27] which adds quantifiers to elementary languages without changing their underlying singular character. Similarly, the generalized quantifier theory introduces a quantifier that corresponds to the use of 'most' in (1b), 'Q^{most}', as a binary quantifier that is on a par with the elementary language quantifiers except that it takes two elementary language predicates. (This is called the Rescher quantifier.) One can then paraphrase (1b) as follows:

$$Q^{most}x(B(x), F(x)). \quad (1b^*)$$

---

[23] That is, $\Sigma xs\{\forall y[\mathbf{H}(y, xs) \to C(y)] \wedge \forall x \forall y[\mathbf{H}(x, xs) \wedge A(x, y) \to x \neq y \wedge \mathbf{H}(y, xs)]\}$.

[24] Using this paraphrase of (2) while analyzing the logic of the expressions involved in (2*), we can explain that (2.1) implies (2) while (2) does not imply the existence of a set of critics.

[25] Note that 'most' is used as the superlative of 'many', rather than of 'much', in (1a)–(1b). The quantifier as so used combines only with plurals.

[26] As a result, Rescher's English reading of '$Ma\varphi(a)$' (or '$Mx\varphi(x)$') is incoherent and vacillating. He reads it sometimes as "For *most* <u>individuals</u> *a* . . . φ*a*" ([12], 373; original italics, my underline), and sometimes as "For most *x's* (of the empty domain D) φ*x*" ([12], 374). In the latter sentence, it seems that he spontaneously uses '*x's*' as a plural variable. Plurals die hard!

[27] Rescher mentions Mostowski [10], who introduces generalized quantifiers and consider them (singular) second-order predicates that take elementary language predicates as arguments.

Because it ignores the plural character of 'most', the generalized quantifier theory cannot give a proper treatment of siblings of (1a)–(1b) that involve predicates whose analogues cannot be found in elementary languages, such as the following (where 'Bob' refers to a huge piano):

Most of the boys lifted Bob. (1c)

Most of the boys who surrounded Bob lifted Bob. (1d)

One cannot paraphrase these into generalized quantifier languages (where 'b', 'L', and 'S' amount to 'Bob', 'lifted', and 'surrounded', respectively) as follows:

$$Q^{most}x(B(x), L(x, b)) . \tag{1c*}$$

$$Q^{most}x(S(x, b), L(x, b)) . \tag{1d*}$$

(1c*) amounts to 'Most of the boys individually lifted Bob' (or 'Every one of some things that are most of the boys lifted Bob'). But this is not equivalent to (1c), which is true if most of the boys (none of whom can lift Bob) cooperated to lift Bob. Similarly, (1d*) cannot be taken to paraphrase (1d), because it amounts to 'Most of the boys who *each* surrounds Bob *individually* lifted Bob.'

To deal with this problem, advocates of the generalized quantifier theory might consider building up their languages on plural languages.[28] But those who embrace plurals as peers of singulars need not revert to the generalized quantifier approach to accommodate 'most' and the like. They can take a natural approach unconstrained by its bias for singulars.

In plural languages, we can introduce a one-place plural predicate, '$\mathbf{Most}^1$', and a two-place one, '$\mathbf{Most}^2$', that amount to 'most' in (1a) and (1b), respectively. Then '$\mathbf{Most}^1(xs)$' amounts to 'They *are most* (of all the things)', and '$\mathbf{Most}^2(xs, ys)$' to 'The former *are most of* the latter' (where 'the former' and 'the latter' are used as anaphoric pronouns).[29] And we can paraphrase (1a) into plural languages with '$\mathbf{Most}^1$' and the singular predicate 'F' as follows:

$$\mathbf{\Sigma} xs[\mathbf{Most}^1(xs) \,!\, F^N(xs)] \tag{1*a}$$

where 'F^N' is the neutral expansion of 'F'. This amounts to 'There are some things that *are most* (of all the things), and they (each) are funny', which (1a) can be taken to paraphrase. Similarly, we can paraphrase (1b) as follows:

---

[28] McKay ([9], Ch. 5) takes this approach.

[29] The 1-place predicate '$\mathbf{Most}^1$' can be defined in terms of the 2-place '$\mathbf{Most}^2$':

$$\mathbf{Most}^1(xs) \equiv_{df} \mathbf{\Sigma} ys[\forall z\mathbf{H}(z, ys) \wedge \mathbf{Most}^2(xs, ys)] .$$

$$\Sigma xs\{\forall z[\mathbf{H}(z, xs) \leftrightarrow B(z)] \wedge \Sigma ys[\mathbf{Most}^2(ys, xs) \wedge F^{\mathrm{N}}(xs)]\} \quad (1^*\mathrm{b})$$

which amounts to 'There are some *things of which anything is one if and only if it is a boy*, and there are some things that are most of the former, and are funny', which (1b) can be taken to paraphrase.

Moreover, we can give a parallel paraphrase of (1c) into plural languages as follows:

$$\Sigma xs\{\forall z[\mathbf{H}(z, xs) \leftrightarrow B(z)] \wedge \Sigma ys[\mathbf{Most}^2(ys, xs) \wedge \boldsymbol{L}(ys, b)]\} \quad (1^*\mathrm{c})$$

where '**L**' is a two-place plural predicate that amounts to 'surrounded', as used in 'They surrounded Bob'. (1*c) differs from (1*b) in an important respect: while (1*c) has this plural predicate, (1*b) has no *undefined* plural predicate except '$\mathbf{Most}^2$', of which '$\mathrm{Q}^{\mathrm{most}}$' is an analogue.[30] So (1c) (or, equivalently, (1*c)) cannot be paraphrased into the usual generalized quantifier languages while (1b) (or, equivalently, (1*b)) can, just as 'Some boys lifted Bob' cannot be paraphrased into elementary languages while 'Some boys are funny' can.

Note that (1b*) and (1c*) amount to 'Most of *the boys* lifted Bob' and 'Most of *the boys* are funny', respectively, because 'They are *(all) the boys*' can be taken to paraphrase 'They are some *things of which anything is one if and only if it is a boy.*' So it is useful to introduce symbolic counterparts of the plural definite description 'the boys' and the like. Let '$<x: B(x)>$' be the symbolic counterpart of 'the boys'. Then we can use it to abbreviate '$\forall z[\mathbf{H}(z, xs) \leftrightarrow B(z)]$' in (1*b) and (1*c) as follows:[31]

$$\Sigma ys[\mathbf{Most}^2(ys, <x: Bx>) \wedge F^{\mathrm{N}}(xs)] \text{ (or, } \lambda xs\{\Sigma ys[\mathbf{Most}^2(ys, xs) \wedge F^{\mathrm{N}}(xs)]\}(<x: Bx>)) . \quad (1^*\mathrm{b}^*)$$

$$\Sigma ys[\mathbf{Most}^2(ys, <x: Bx>) \wedge \boldsymbol{L}(ys, b)] \text{ (or, } \lambda xs\{\Sigma ys[\mathbf{Most}^2(ys, xs) \wedge \boldsymbol{L}(ys, b)]\}(<x: Bx>)) . \quad (1^*\mathrm{c}^*)$$

Now, some things are, e.g., the boys if and only if they are *things of which anything is one if and only if it is a boy*. So we can give a contextual definition of the definite description '$<x: B(x)>$' and the like in plural languages as follows:

Def. 2 (*Plural definite descriptions of the first kind*)
$\pi(<x: \varphi(x)>) \equiv_{\mathrm{df}} \Sigma xs\{\forall z[\mathbf{H}(z, xs) \leftrightarrow \varphi(z)] \wedge \pi(xs)\}$, where π is a predicate.

Applying this to (1*b*) and (1*c*) yields (1*b) and (1*c), respectively.[32, 33]

---

30 'F^{N}' is a plural predicate, but is introduced as defined in terms of the singular predicate 'F' (with logical expressions of plural languages).

31 In both (1*b*) and (1*c*), '$<x: Bx>$' takes the widest scope.

32 Using the definite description, we can give a simple formulation of (1*a): '$\mathbf{Most}^1(<x: F(x)>)$.' Applying *Def.* 2 (together with *Def.* 1) to this yields a straightforward logical equivalent of (1*a).

The plural definite description in (1d), 'the boys who surrounded Bob', requires a different treatment. It cannot be analyzed in the same way as, e.g., 'the boys'. Compare the following:

Something is one of the boys if and only if it is a boy. (a)

Something is one of the boys who surrounded Bob if and only if it is a boy who surrounded Bob. (b)

Although (a) is a logical truth, which provides the basis for *Def.* 2, (2) might be false. So it is necessary to give a different analysis of (1d) to paraphrase it into plural languages. Now, we can introduce into plural languages plural definite descriptions of another kind, those that amount to 'the boys who surrounded Bob' and the like. Let '$(\text{I}zs)\boldsymbol{S}(zs, b)$', for example, be a definite description that amounts to this. Using this, we can paraphrase (1d) as follows:

$$\boldsymbol{\Sigma}ys[\mathbf{Most}^2(ys, (\text{I}xs)\boldsymbol{S}(xs, b)) \wedge \boldsymbol{L}(ys, b)] \text{ (or, } \lambda xs\{\boldsymbol{\Sigma}ys[\mathbf{Most}^2(ys, xs) \wedge \boldsymbol{L}(ys, b)]\}((\text{I}xs)\boldsymbol{S}(xs, b))\text{)} . \qquad (1^*\text{d})$$

This, which results from replacing '$<x: Bx>$' in (1*c*) with '$(\text{I}xs)\boldsymbol{S}(xs, b)$', amounts to 'There are some things that are most of *the boys who surrounded Bob*, and they lifted Bob.' And one can give a contextual definite of plural definite descriptions of the second kind as well in plural languages. Here is the definition as applied to '$(\text{I}xs)\boldsymbol{S}(xs, b)$':[34]

[33] (1*b) (or, equivalently, (1*b*)) implies the existence of a boy. So 'Every boy is funny', which does not imply the existence of a boy, does not imply 'Most boys are funny' on my analysis. The generalized quantifier analysis of 'most' yields the same result. Moreover, the usual generalized quantifier account consider 'all' in 'All boys are funny' or 'All the boys are funny' a mere variant of 'every' to take these sentences to be equivalents of 'Every boy is funny' that fail to imply (1b). But we can treat 'all' (which combines only with plural forms of count nouns) like 'most'. In plural languages, we can introduce a two-place plural predicate, '$\mathbf{All}^2$', that amounts to 'are *all* of' in, e.g., 'They *are all of* my friends.' We can then take 'All boys are funny' to be equivalent to 'All the boys are funny', and paraphrase it by '$\boldsymbol{\Sigma}xs[\mathbf{All}(xs, <x: Bx>) \wedge F^N(xs)]$'. And we can define '$\mathbf{All}^2$' using only logical expressions of plural languages as follows:

$$\mathbf{All}^2(xs, ys) \equiv_{\text{df}} \forall z[\mathbf{H}(z, xs) \leftrightarrow \mathbf{H}(z, ys)] \text{ (or } \mathbf{All}^2 =_{\text{df}} \lambda xs,ys\forall z[\mathbf{H}(z, xs) \leftrightarrow \mathbf{H}(z, ys)]\text{)}$$

For some things are *all* of, e.g., my friends if and only if any one of them is one of my friends, and *vice versa*. We can then show that '$\boldsymbol{\Sigma}xs[\mathbf{All}(xs, <x: Bx>) \wedge F^N(xs)]$' is logically equivalent to '$\exists xB(x) \wedge \forall x[B(x) \rightarrow F(x)]$', and implies (1*b) (for '$\boldsymbol{\Pi}xs.\mathbf{Most}(xs, xs)$' or 'Any things are most of themselves' is a logical truth).

[34] See Yi ([20], *sc.* 4) for the definition and general discussions of plural definite descriptions.

Def. 3 (*Plural definite descriptions of the second kind* [an example])
$\pi((\mathrm{I}xs)\boldsymbol{S}(xs, b)) \equiv_{\mathrm{df}} \boldsymbol{\Sigma}xs\{\boldsymbol{\Pi}zs[\boldsymbol{S}(zs, b) \leftrightarrow zs \approx xs)] \wedge \pi(xs)\}$, where π is a predicate.

Here '$zs \approx xs$' abbreviates '$\forall y[\mathbf{H}(y, zs) \leftrightarrow \mathbf{H}(y, xs)]$' ('$\approx$' can be used to paraphrase '*to be the same things as*'). Applying the definition to (1*d) yield the following:

$$\boldsymbol{\Sigma}xs\{\forall zs[\boldsymbol{S}(zs, b) \leftrightarrow zs \approx xs)] \wedge \boldsymbol{\Sigma}ys[\mathbf{Most}^2(ys, xs) \wedge \boldsymbol{L}(ys, b)]\} . \qquad (1^*d^*)$$

This amounts to, roughly, 'There are some things that lifted Bob, and they are most of *the things that are the same as some things if and only if these surrounded Bob*.'

Now, note that the plural language analysis of 'most' *explains* the fact that 'most' is a so-called *conservative* quantifier: (1b) is equivalent to 'Most of the boys *are boys and* are funny', because any things that are most of the boys must be boys. Although the generalized quantifier theory attributes conservativity to 'most', however, its conservativity has an important limitation in scope: (1d) is not equivalent to 'Most of the boys who surrounded Bob *are boys who surrounded Bob and* lifted Bob.' The plural language analysis of 'most' explains this limitation as well: those who are most of *the boys who surrounded Bob* might not themselves suffice to *surround it*. Proponents of the generalized quantifier theory fail to note the limitation, let alone explain it, because they in effect begin by placing, e.g., (1d) beyond the scope of the theory because it cannot be put in their favorite languages. But it would be wrong to hold that 'most' is used differently in (1d) than in (1b) and (1c). My plural language analysis of the quantifier explains the logical disparity as arising from the difference in the logical character between two kinds of definite descriptions.

Acknowledgments. I would like to thank Yiannis Moschovakis for suggesting the possibility of applying finite model theory to the Geach-Kaplan sentence, and Tom McKay and anonymous referees for *Formal Grammar 10* for comments on earlier versions of this paper. The work for this paper was supported in part by an ACLS Fellowship that I held in 2009-10. The support of the fellowship is hereby gratefully acknowledged.

I wish to dedicate this paper to David Kaplan.

References

1. Almog, J.: The complexity of marketplace logic. Linguistics and Philosophy 20, 545–569 (1997)
2. Barwise, J., Cooper, R.: Generalized quantifiers and natural language. Linguistics and Philosophy 4, 159–219 (1981)
3. Boolos, G.: To be is to be a value of a variable (or to be some values of some variables). Journal of Philosophy 81, 430–449 (1984)
4. Ebbinghaus, H.-D., Flum, J.: Finite Model Theory, 2nd revised edn. Springer, Berlin (1999)
5. Fagin, R.: Finite-model theory – a personal perspective. Theoretical Computer Science 116, 3–31 (1993)

6. Kaplan, D.: Rescher's plurality-quantification (Abstract). Journal of Symbolic Logic 31, 153–154 (1966)
7. Kaplan, D.: Generalized plurality quantification (Abstract). Journal of Symbolic Logic 31, 154–155 (1966)
8. Keenan, E.L., Stavi, J.: A semantic characterization of natural language determiners. Linguistics and Philosophy 9, 253–326 (1986)
9. McKay, T.: Plural Predication. Oxford University Press, Oxford (2006)
10. Mostowski, A.: On a generalization of quantifiers. Fundamenta Mathematicae 44, 12–36 (1957)
11. Quine, W.V.: Methods of Logic, 3rd edn. London, RKP (1974)
12. Rescher, N.: Plurality-quantification (Abstract). Journal of Symbolic Logic 27, 373–374 (1962)
13. Rescher, N.: Plurality quantification revisited. Philosophical Inquiry 26, 1–6 (2004)
14. Väänänen, J.: A Short Course on Finite Model Theory. Manuscript Based on Lectures in 1993-1994 (1994), `http://www.math.helsinki.fi/logic/people/jouko.vaananen/shortcourse.pdf`
15. Westerståhl, D.: Logical constants in quantifier languages. Linguistics and Philosophy 8, 387–413 (1985)
16. Westerståhl, D.: Quantifiers in formal and natural languages. In: Gabbay, D., Guethner, F. (eds.) The Handbook of Philosophical Logic, vol. IV, pp. 1–131. Reidel, Dordrecht (1989)
17. Yi, B.-U.: Is two a property? Journal of Philosophy 95, 163–190 (1999)
18. Yi, B.-U.: Understanding the Many. Routledge, New York & London (2002)
19. Yi, B.-U.: The logic and meaning of plurals. Part I. Journal of Philosophical Logic 34, 459–506 (2005)
20. Yi, B.-U.: The logic and meaning of plurals. Part II. Journal of Philosophical Logic 35, 239–288 (2006)
21. Yi, B.-U.: Is logic axiomatizable? unpublished manuscript

Appendix

We can use some basic results of finite model theory to show the following:

(A) There is no elementary language sentence that agrees with Rescher's "plurality" quantification '$MxF(x)$' on all finite models.

(B) There is no elementary sentence that agrees with the Geach-Kaplan sentence (2), 'Some critics admire only one another', on all finite domains.

Let $\mathcal{L}_F$ be the elementary language whose only non-logical expression is 'F^1', and $\mathcal{L}_A$ the elementary language whose only non-logical expression is 'A^2'. Then a model **M** of $\mathcal{L}_F$ is a pair $<\mathbf{D}^{\mathbf{M}}, F^{\mathbf{M}}>$ such that $\mathbf{D}^{\mathbf{M}}$ is a non-empty set and $F^{\mathbf{M}}$ a subset of $\mathbf{D}^{\mathbf{M}}$; and a model **M** of $\mathcal{L}_A$ a pair $<\mathbf{D}^{\mathbf{M}}, A^{\mathbf{M}}>$ such that $\mathbf{D}^{\mathbf{M}}$ is a non-empty set and $A^{\mathbf{M}}$ a subset of

$\mathbf{D}^{\mathbf{M}} \times \mathbf{D}^{\mathbf{M}}$. Say that a model **M** is *finite*, if its domain, $\mathbf{D}^{\mathbf{M}}$, is a finite set. Then we can show the following:[35]

(A*) There is no sentence φ of $\mathcal{L}_F$ such that any finite model **M** of $\mathcal{L}_F$ satisfies φ if and only if $|\mathbf{D}^{\mathbf{M}} \backslash F^{\mathbf{M}}| = |F^{\mathbf{M}}|$.

But '$\neg MxF(x) \wedge \neg Mx\neg F(x)$' is such a sentence. So (A) holds. To state a theorem that we can use to show (B), it is useful to use the following notions:

Definitions: Let **M** $(=\langle \mathbf{D}^{\mathbf{M}}, A^{\mathbf{M}} \rangle)$ be a model of $\mathcal{L}_A$. Then

1. **M** is a *graph*, if for any members a and b of $\mathbf{D}^{\mathbf{M}}$, not $A^{\mathbf{M}}(a, a)$, and if $A^{\mathbf{M}}(a, b)$, then $A^{\mathbf{M}}(b, a)$.
2. There is *a path* between a and b, if a and b are members of $\mathbf{D}^{\mathbf{M}}$ and either $A^{\mathbf{M}}(a, b)$ or there are finitely many members $x_1, x_2, \ldots, x_{n-1}, x_n$ of $\mathbf{D}^{\mathbf{M}}$ such that $A^{\mathbf{M}}(a, x_1)$, $A^{\mathbf{M}}(x_1, x_2)$, ..., $A^{\mathbf{M}}(x_{n-1}, x_n)$, $A^{\mathbf{M}}(x_n, y)$.
3. A graph is *connected*, if there is a path between any two members of $\mathbf{D}^{\mathbf{M}}$.

Then the following is a theorem of finite model theory:[36]

(B*) There is no sentence φ of $\mathcal{L}_A$ such that any finite model **M** of $\mathcal{L}_A$ satisfies φ if and only if **M** is a connected graph.

Now, if there is an elementary language sentence that agrees with (2) on all finite models, then there is an elementary language sentence that agrees with the following on all finite models:

> There is something such that there are some things that are not identical with it that admire only one another.

(And we may assume such a sentence is a sentence of $\mathcal{L}_A$.) So let φ be a sentence of $\mathcal{L}_A$ that agrees with the above sentence on all finite models. Then let φ* be $[\forall x \neg A(x, x) \wedge \forall x \forall y (A(x, y) \leftrightarrow A(y, x)) \wedge \varphi]$. Then a finite model **M** of $\mathcal{L}_A$ satisfies φ* if and only if **M** is a connected graph, which violates (B*). So (B) holds.

[35] This is a variant of the theorem of undefinability of the class of even-numbered models (among finite models). For a proof of the theorem, see, e.g., Väänänen ([14], 6). We can prove it using compactness of elementary logic.

[36] See, e.g., Ebbinghaus & Flum ([4], 22f) or Väänänen ([14], 9) for a proof.

Polynomial Time Learning of Some Multiple Context-Free Languages with a Minimally Adequate Teacher

Ryo Yoshinaka[1,*] and Alexander Clark[2]

[1] MINATO Discrete Structure Manipulation System Project, ERATO, Japan Science and Technology Agency
ryoshinaka@erato.ist.hokudai.ac.jp

[2] Department of Computer Science, Royal Holloway, University of London
alexc@cs.rhul.ac.uk

Abstract. We present an algorithm for the inference of some Multiple Context-Free Grammars from Membership and Equivalence Queries, using the Minimally Adequate Teacher model of Angluin. This is an extension of the congruence based methods for learning some Context-Free Grammars proposed by Clark (ICGI 2010). We define the natural extension of the syntactic congruence to tuples of strings, and demonstrate we can efficiently learn the class of Multiple Context-Free Grammars where the non-terminals correspond to the congruence classes under this relation.

1 Introduction

In this paper we look at efficient algorithms for the inference of multiple context free grammars (MCFGs). MCFGs are a natural extension of context free grammars (CFGs), where the non-terminal symbols derive tuples of strings, which are then combined with each other in a limited range of ways. Since some natural language phenomena were found not to be context-free, the notion of *mildly context-sensitive languages* was proposed and studied for better describing natural languages, while keeping tractability [1]. MCFGs are regarded as a representative mildly context-sensitive formalism, which has many equivalent formalisms such as linear context-free rewriting systems [2], multicomponent tree adjoining grammars [3,4], minimalist grammars [5], hyperedge replacement grammars [6], etc.

In recent years, there has been rapid progress in the field of context-free grammatical inference using techniques of distributional learning. The first result along these lines was given by Clark and Eyraud [7]. They showed a polynomial

* He is concurrently working in Graduate School of Information Science and Technology, Hokkaido University. This work was supported in part by Grant-in-Aid for Young Scientists (B-20700124) from the Ministry of Education, Culture, Sports, Science and Technology of Japan.

P. de Groote and M.-J. Nederhof (Eds.): Formal Grammar 2010/2011, LNCS 7395, pp. 192–207, 2012.

identification in the limit result from positive data alone for a class of languages called the substitutable context free languages.

We assume we have a finite non-empty alphabet Σ and an extra hole or gap symbol, $\Box$, which is not in Σ. A context is a string with one hole in, written as $l\Box r$, and the distribution of a string u in a language L is defined as $L/u = \{ l\Box r \mid lur \in L\}$. There is a natural congruence relation called the syntactic congruence which is defined as $u \equiv_L v$ iff $L/u = L/v$.

The learning approach of Clark and Eyraud is based on the following observations:

Lemma 1. *If* $u \equiv_L u'$ *then* $uw \equiv_L u'w$. *If* $u \equiv_L u'$ *and* $v \equiv_L v'$ *then* $uv \equiv_L u'v'$.

This means that it is trivial to construct a CFG where the non-terminals correspond to congruence classes under this relation, as we can construct rules of the form $[uv] \rightarrow [u], [v]$. All rules of this form will be valid since $[u][v] \subseteq [uv]$ by Lemma 1.

Let us consider now the language $L_c = \{ wcwc \mid w \in \{a, b\}^* \}$. This is a slight variant of the classic copy language which is not context-free. It is formally similar to the constructions in Swiss German which established that the class of natural languages is not included in the class of context-free languages and is thus a classic test case for linguistic representation [8, 9].

We will extend the approach by considering relationships not between strings, but between tuples of strings. Here for concreteness we will consider the case of tuples of dimension 2, i.e. pairs; but in the remainder of the paper we will define this for tuples of arbitrary dimension.

Let $\langle u, v\rangle$ be an ordered pair of strings. We define $L/\langle u, v\rangle = \{ l\Box m\Box r \mid lumvr \in L \}$ and we then define an equivalence relation between pairs of strings which is analogous to the syntactic congruence: $\langle u, v\rangle \equiv_L \langle u', v'\rangle$ iff $L/\langle u, v\rangle = L/\langle u', v'\rangle$.

Note that when we consider L_c it is easy to see that $\langle c, c\rangle \equiv_{L_c} \langle ac, ac\rangle \equiv_{L_c} \langle bc, bc\rangle$, but that $\langle a, a\rangle$ is not congruent to $\langle b, b\rangle$, since for example $\Box\Box caac$ is in $L/\langle a, a\rangle$ but not $L/\langle b, b\rangle$.

Just as with Lemma 1 above, this congruence has interesting properties.

Lemma 2. *If* $\langle u, v\rangle \equiv_L \langle u', v'\rangle$ *and* $\langle x, y\rangle \equiv_L \langle x', y'\rangle$, *then* $\langle ux, vy\rangle \equiv_L \langle u'x', v'y'\rangle$.

Note that there is now more than one way of combining these two elements. If they are single strings, then there are only two ways - uv and vu. When there are multiple strings we can combine them in many different ways. For example, it is also the case that $\langle uxy, v\rangle \equiv_L \langle u'x'y', v'\rangle$, but in general, it is not the case that $\langle uyx, v\rangle \equiv_L \langle u'y'x', v'\rangle$.

In the specific case of Lemma 2 we can consider the concatenation operation as a function $f : (\Sigma^* \times \Sigma^*)^2 \rightarrow \Sigma^* \times \Sigma^*$ defined as $f(\langle u, v\rangle, \langle x, y\rangle) = \langle ux, vy\rangle$. If we use the notation $\boldsymbol{w}$ for tuples then the lemma says that if $\boldsymbol{u} \equiv_L \boldsymbol{u'}$ and $\boldsymbol{v} \equiv_L \boldsymbol{v'}$, then $f(\boldsymbol{u}, \boldsymbol{v}) \equiv_L f(\boldsymbol{u'}, \boldsymbol{v'})$.

We can then generalise Lemma 2 to all functions f that satisfy a certain set of conditions; as we discuss formally later on, these must be linear regular and non-permuting.

Our learning algorithm will work by constructing an MCFG where each non-terminal generates a congruence class of tuples of strings.

For our learning model, we will use in this paper the Minimally Adequate Teacher (MAT) model introduced by Angluin in [10]. This model assumes the existence of a teacher who can answer two sorts of queries: first, the learner can ask Membership Queries (MQs), where the learner queries whether a given string is in the target language or not, and secondly the learner can ask Equivalence Queries (EQs), where the learner queries whether a particular hypothesis is correct or not; if the hypothesis is not correct, then the teacher will give a counter-example to the learner. The learner is required to use only a limited amount of computation before terminating – in particular there must be a polynomial p such that the total amount of computation used is less than $p(n, \ell)$, where n is the size of the target representation, and ℓ is the length of the longest counter-example returned by the teacher.

This model is rather abstract; but it has a number of important advantages that make it appropriate for this paper. First of all, it is a very restrictive model. There are strong negative results [11] that show that certain natural classes – regular grammars, CFGs are not learnable under this MAT paradigm. Secondly, it is also permissive – it allows reasonably large classes of languages to be learned. Angluin's famous LSTAR algorithm for the inference of regular languages is a classic and well-studied algorithm, and one of the great success stories of grammatical inference.

Finally, if we look at the history of regular and context-free grammatical inference it appears to work well as a substitute for more realistic probabilistic models. Probabilistic DFAs turn out to be PAC-learnable under a very rigorous model [12], albeit stratified by certain parameters. Thus though this model is unrealistic it seems a good intermediate step, and it is technically easy to work with.

The main result of this paper is a polynomial MAT learning algorithm for a class of MCFGs. In Section 2 we will introduce our notation. Our learning algorithm is discussed in Section 3. We conclude this paper in Section 4.

2 Preliminaries

2.1 Basic Definitions and Notations

The set of non-negative integers is denoted by $\mathbb{N}$ and this paper will consider only numbers in $\mathbb{N}$. The cardinality of a set S is denoted by $|S|$. If w is a string over an alphabet Σ, $|w|$ denotes its length. $\varnothing$ is the empty set and λ is the empty string. Σ^* denotes the set of all strings over Σ. We write $\Sigma^+ = \Sigma^* - \{\lambda\}$, $\Sigma^k = \{\, w \in \Sigma^* \mid |w| = k \,\}$ and $\Sigma^{\leq k} = \{\, w \in \Sigma^* \mid |w| \leq k \,\}$. Any subset of Σ^* is called a *language (over Σ)*. If L is a finite language, its size is defined as $\|L\| = |L| + \sum_{w \in L} |w|$. An *m-word* is an m-tuple of strings and we denote the set of m-words by $(\Sigma^*)^{\langle m\rangle}$. Similarly we define $(\cdot)^{\langle *\rangle}$, $(\cdot)^{\langle +\rangle}$, $(\cdot)^{\langle \leq m\rangle}$. Any m-word is called a *multiword.* Thus $(\Sigma^*)^{\langle *\rangle}$ denotes the set of all multiwords. For $\boldsymbol{w} = \langle w_1, \dots, w_m\rangle \in (\Sigma^*)^{\langle m\rangle}$, $|\boldsymbol{w}|$ denotes its length m and $\|\boldsymbol{w}\|$ denotes its *size* $m + \sum_{1\leq i\leq m} |w_i|$. We use the symbol $\Box$, assuming that $\Box \notin \Sigma$, for representing a "hole", which is supposed to be replaced by another string. We write $\Sigma_\Box$ for $\Sigma \cup \{\Box\}$. A string $\mathbf{x}$ over $\Sigma_\Box$ is called an *m-context* if $\mathbf{x}$ contains m occurrences of $\Box$. m-contexts are also called *multicontexts.* For an m-context $\mathbf{x} = x_0\Box x_1\Box \dots \Box x_m$ with $x_0, \dots, x_m \in \Sigma^*$ and an m-word $\boldsymbol{y} = \langle y_1, \dots, y_m\rangle \in (\Sigma_\Box^*)^{\langle m\rangle}$, we define

$$\mathbf{x} \odot \boldsymbol{y} = x_0 y_1 x_1 \dots y_n x_n$$

and say that $\boldsymbol{y}$ is a *sub-multiword* of $\mathbf{x} \odot \boldsymbol{y}$. Note that $\Box$ is the empty context and we have $\Box \odot \langle y\rangle = y$ for any $y \in \Sigma^*$. For $L \subseteq \Sigma^*$ and $p \geq 1$, we define

$$\mathcal{S}^{\leq p}(L) = \{\, \boldsymbol{y} \in (\Sigma^+)^{\langle\leq p\rangle} \mid \mathbf{x} \odot \boldsymbol{y} \in L \text{ for some } \mathbf{x} \in \Sigma_\Box^* \,\},$$
$$\mathcal{C}^{\leq p}(L) = \{\, \mathbf{x} \in \Sigma_\Box^* \mid \mathbf{x} \odot \boldsymbol{y} \in L \text{ for some } \boldsymbol{y} \in (\Sigma^+)^{\langle\leq p\rangle} \,\},$$

and for $\boldsymbol{y} \in (\Sigma^*)^{\langle *\rangle}$, we define

$$L/\boldsymbol{y} = \{\, \mathbf{x} \in \Sigma_\Box^* \mid \mathbf{x} \odot \boldsymbol{y} \in L \,\}.$$

Obviously computation of $\mathcal{S}^{\leq p}(L)$ can be done in $O(\|L\|^{2p})$ time if L is finite.

2.2 Linear Regular Functions

Let us suppose a countably infinite set Z of *variables* disjoint from Σ. A function f from $(\Sigma^*)^{\langle m_1\rangle} \times \dots \times (\Sigma^*)^{\langle m_n\rangle}$ to $(\Sigma^*)^{\langle m\rangle}$ is said to be *linear*, if there is $\langle \alpha_1, \dots, \alpha_m\rangle \in ((\Sigma \cup \{\, z_{ij} \in Z \mid 1 \leq i \leq n,\ 1 \leq j \leq m_i \,\})^*)^{\langle m\rangle}$ such that each variable z_{ij} occurs at most once in $\langle \alpha_1, \dots, \alpha_m\rangle$ and

$$f(\boldsymbol{w}_1, \dots, \boldsymbol{w}_n) = \langle \alpha_1[\boldsymbol{z} := \boldsymbol{w}], \dots, \alpha_m[\boldsymbol{z} := \boldsymbol{w}]\rangle$$

for any $\boldsymbol{w}_i = \langle w_{i1}, \dots, w_{im_i}\rangle \in (\Sigma^*)^{\langle m_i\rangle}$ with $1 \leq i \leq n$, where $\alpha_k[\boldsymbol{z} := \boldsymbol{w}]$ denotes the string obtained by replacing each variable z_{ij} with the string w_{ij}. We call f *linear regular* if every variable z_{ij} occurs in $\langle \alpha_1, \dots, \alpha_m\rangle$. We say that f is *λ-free* when no α_k from $\langle \alpha_1, \dots, \alpha_m\rangle$ is λ. A linear regular function f is said to be *non-permuting*, if z_{ij} always occurs to the left of $z_{i(j+1)}$ in $\alpha_1 \dots \alpha_m$ for $1 \leq i \leq n$ and $1 \leq j < m_i$. The *rank* $\mathrm{rank}(f)$ of f is defined to be n and the *size* $\mathrm{size}(f)$ of f is $\|\langle \alpha_1, \dots, \alpha_m\rangle\|$.

Example 1. Among the functions defined below, where $a, b \in \Sigma$, f is not linear, while g and h are linear regular. Moreover h is λ-free and non-permuting.

$$f(\langle z_{11}, z_{12}\rangle, \langle z_{21}\rangle) = \langle z_{11}z_{12}, z_{11}z_{21}z_{11}\rangle,$$
$$g(\langle z_{11}, z_{12}\rangle, \langle z_{21}\rangle) = \langle z_{12}, z_{11}bz_{21}, \lambda\rangle,$$
$$h(\langle z_{11}, z_{12}\rangle, \langle z_{21}\rangle) = \langle a, z_{11}bz_{21}, z_{12}\rangle.$$

The following lemma is the generalization of Lemmas 1 and 2 mentioned in the introduction, on which our approach is based.

Lemma 3. *For any language $L \subseteq \Sigma^*$ and any $\boldsymbol{u}_1, \dots, \boldsymbol{u}_n, \boldsymbol{v}_1, \dots, \boldsymbol{v}_n \in (\Sigma^*)^{\langle * \rangle}$ such that $|\boldsymbol{u}_i| = |\boldsymbol{v}_i|$ and $L/\boldsymbol{u}_i = L/\boldsymbol{v}_i$ for all i, we have $L/f(\boldsymbol{u}_1, \dots, \boldsymbol{u}_n) = L/f(\boldsymbol{v}_1, \dots, \boldsymbol{v}_n)$ for any non-permuting linear regular function f.*

Proof. Let $m_i = |\boldsymbol{u}_i| = |\boldsymbol{v}_i|$. Suppose that $\mathbf{x} \in L/f(\boldsymbol{u}_1, \dots, \boldsymbol{u}_n)$, i.e., $\mathbf{x} \odot f(\boldsymbol{u}_1, \dots, \boldsymbol{u}_n) \in L$. The following inference is allowed:

$$\begin{aligned} \mathbf{x} \odot f(\Box^{\langle m_1 \rangle}, \boldsymbol{u}_2, \dots, \boldsymbol{u}_n) \in L/\boldsymbol{u}_1 = L/\boldsymbol{v}_1 &\implies \mathbf{x} \odot f(\boldsymbol{v}_1, \boldsymbol{u}_2, \dots, \boldsymbol{u}_n) \in L \\ \implies \mathbf{x} \odot f(\boldsymbol{v}_1, \Box^{\langle m_2 \rangle}, \boldsymbol{u}_3, \dots, \boldsymbol{u}_n) &\in L/\boldsymbol{u}_2 = L/\boldsymbol{v}_2 \implies \\ \mathbf{x} \odot f(\boldsymbol{v}_1, \boldsymbol{v}_2, \boldsymbol{u}_3, \dots, \boldsymbol{u}_n) \in L \implies \dots &\implies \mathbf{x} \odot f(\boldsymbol{v}_1, \dots, \boldsymbol{v}_n) \in L. \end{aligned}$$

Hence $\mathbf{x} \in L/f(\boldsymbol{v}_1, \dots, \boldsymbol{v}_n)$. □

2.3 Multiple Context-Free Grammars

A *multiple context-free grammar (*MCFG*)* is a tuple $G = \langle \Sigma, V_{\text{dim}}, F, P, I \rangle$, where

- Σ is a finite set of *terminal symbols,*
- $V_{\text{dim}} = \langle V, \text{dim} \rangle$ is the pair of a finite set V of *nonterminal symbols* and a function dim giving a positive integer, called a *dimension,* to each element of V,
- F is a finite set of *linear functions,*[1]
- P is a finite set of *rules* of the form $A \to f(B_1, \dots, B_n)$ where $A, B_1, \dots, B_n \in V$ and $f \in F$ maps $(\Sigma^*)^{\langle \text{dim}(B_1) \rangle} \times \dots \times (\Sigma^*)^{\langle \text{dim}(B_n) \rangle}$ to $(\Sigma^*)^{\langle \text{dim}(A) \rangle}$,
- I is a subset of V and all elements of I have dimension 1. Elements of I are called *initial symbols.*

We note that our definition of MCFGs is slightly different from the original [13], where grammars have exactly one initial symbol, but this change does not affect the generative capacity of MCFGs.

We will simply write V for V_{dim} if no confusion occurs. If a rule has a function f, then its right hand side must have rank(f) occurrences of nonterminals by definition. If rank$(f) = 0$ and $f() = \boldsymbol{v}$, we may write $A \to \boldsymbol{v}$ instead of $A \to f()$. If rank$(f) = 1$ and f is the identity, we may write $A \to B$ instead of $A \to f(B)$, where $\text{dim}(A) = \text{dim}(B)$. The *size* $\|G\|$ of G is defined as $\|G\| = |P| + \sum_{\rho \in P} \text{size}(\rho)$ where $\text{size}(A \to f(B_1, \dots, B_n)) = \text{size}(f) + n + 1$.

For $A \in V$, *A-derivation trees* are recursively defined as follows:

- If a rule $\pi \in P$ has the form $A \to \boldsymbol{w}$, then π is an A-derivation tree for $\boldsymbol{w}$. We call $\boldsymbol{w}$ the *yield* of π.
- If a rule $\pi \in P$ has the form $A \to f(B_1, \dots, B_n)$ and t_i is a B_i-derivation tree for $\boldsymbol{w}_i$ for all $i = 1, \dots, n$, then the tree whose root is π with immediate subtrees $t_1, \dots, t_n$ from left to right, which we write as $\pi(t_1, \dots, t_n)$, is an A-derivation tree for $f(\boldsymbol{w}_1, \dots, \boldsymbol{w}_n)$, which is called the *yield* of $\pi(t_1, \dots, t_n)$.

[1] We identify a function with its name for convenience.

– nothing else is an A-derivation tree.

For all $A \in V$ we define

$$\mathcal{L}(G, A) = \{ \boldsymbol{w} \in (\Sigma^*)^{\langle \dim(A) \rangle} \mid \text{there is an } A\text{-derivation tree for } \boldsymbol{w} \}.$$

The *language* $\mathcal{L}(G)$ *generated by* G means the set $\{ w \in \Sigma^* \mid \langle w \rangle \in \mathcal{L}(G, S)$ with $S \in I \}$, which is called a *multiple context-free language (MCFL)*. If $S \in I$, an S-derivation tree for $\langle w \rangle$ is simply called a derivation tree for $w \in \mathcal{L}(G)$. Two grammars G and G' are *equivalent* if $\mathcal{L}(G) = \mathcal{L}(G')$.

This paper assumes that all linear functions in F are linear regular, λ-free and non-permuting. In fact those assumptions do not affect their generative capacity modulo λ [13, 14].

We denote by $\mathbb{G}(p, r)$ the collection of MCFGs G whose nonterminals are assigned a dimension at most p and whose functions have a rank at most r. Then we define $\mathbb{L}(p, r) = \{ \mathcal{L}(G) \mid G \in \mathbb{G}(p, r) \}$, We also write $\mathbb{G}(p, *) = \bigcup_{r \in \mathbb{N}} \mathbb{G}(p, r)$ and $\mathbb{L}(p, *) = \bigcup_{r \in \mathbb{N}} \mathbb{L}(p, r)$. The class of context-free grammars (CFGs) is identified with $\mathbb{G}(1, *)$ and all CFGs in Chomsky normal form are in $\mathbb{G}(1, 2)$. Thus $\mathbb{L}(1, 2) = \mathbb{L}(1, *)$.

Example 2. Let G be the MCFG $\langle \Sigma, V, F, P, \{S\} \rangle$ over $\Sigma = \{a, b, c, d\}$ whose rules are

$$\begin{aligned}
&\pi_1 : S \to f(A, B) \text{ with } f(\langle z_{11}, z_{12} \rangle, \langle z_{21}, z_{22} \rangle) = \langle z_{11} z_{21} z_{12} z_{22} \rangle, \\
&\pi_2 : A \to g(A) \text{ with } g(\langle z_1, z_2 \rangle) = \langle a z_1, c z_2 \rangle, \qquad \pi_3 : A \to \langle a, c \rangle, \\
&\pi_4 : B \to h(B) \text{ with } h(\langle z_1, z_2 \rangle) = \langle z_1 b, z_2 d \rangle, \qquad \pi_5 : B \to \langle b, d \rangle,
\end{aligned}$$

where $V = \{S, A, B\}$ with $\dim(S) = 1$, $\dim(A) = \dim(B) = 2$, and F consists of f, g, h and the constant functions appearing in the rules π_3 and π_5. One can see, for example, $aabccd \in \mathcal{L}(G)$ thanks to the derivation tree $\pi_1(\pi_2(\pi_3), \pi_5)$: $\langle a, c \rangle \in \mathcal{L}(G, A)$ by π_3, $\langle aa, cc \rangle \in \mathcal{L}(G, A)$ by π_2, $\langle b, d \rangle \in \mathcal{L}(G, B)$ by π_5 and $\langle aabccd \rangle \in \mathcal{L}(G, S)$ by π_1. We have $\mathcal{L}(G) = \{ a^m b^n c^m d^n \mid m, n \geq 1 \}$.

Seki et al. [13] and Rambow and Satta [15] have investigated the hierarchy of MCFLs.

Proposition 1 (Seki et al. [13], Rambow and Satta [15]).
For $p \geq 1$, $\mathbb{L}(p, *) \subsetneq \mathbb{L}(p + 1, *)$.
For $p \geq 2$, $r \geq 1$, $\mathbb{L}(p, r) \subsetneq \mathbb{L}(p, r + 1)$ *except for* $\mathbb{L}(2, 2) = \mathbb{L}(2, 3)$.
For $p \geq 1$, $r \geq 3$ *and* $1 \leq k \leq r - 2$, $\mathbb{L}(p, r) \subseteq \mathbb{L}((k + 1)p, r - k)$.

Theorem 1 (Seki et al. [13], Kaji et al. [16]). *Let* p *and* r *be fixed. It is decidable in* $O(\|G\|^2 |w|^{p(r+1)})$ *time whether* $w \in \mathcal{L}(G)$ *for any* MCFG $G \in \mathbb{G}(p, r)$ *and* $w \in \Sigma^*$.

2.4 Congruential Multiple Context-Free Grammars

Now we introduce the languages of our learning target.

Definition 1. We say that an MCFG $G \in \mathbb{G}(p,r)$ is *p-congruential* if for every nonterminal A and any $\boldsymbol{u},\boldsymbol{v} \in \mathcal{L}(G,A)$, it holds that $\mathcal{L}(G)/\boldsymbol{u} = \mathcal{L}(G)/\boldsymbol{v}$.

In a p-congruential MCFG (p-CMCFG or CMCFG for short), one can merge two nonterminals A and B without changing the language when $\mathcal{L}(G)/\boldsymbol{u} = \mathcal{L}(G)/\boldsymbol{v}$ for $\boldsymbol{u} \in \mathcal{L}(G,A)$ and $\boldsymbol{v} \in \mathcal{L}(G,B)$.

We let $\mathbb{CG}(p,r)$ denote the class of p-CMCFGs from $\mathbb{G}(p,r)$ and $\mathbb{CL}(p,r)$ the corresponding class of languages. Our learning target is $\mathbb{CL}(p,r)$ for each $p,r \geq 1$. The class $\mathbb{CL}(1,2)$ corresponds to congruential CFLs introduced by Clark [17], which include all regular languages.

The grammar G in Example 2 is 2-congruential. It is easy to see that for any $\langle a^m, c^m \rangle \in \mathcal{L}(G,A)$, we have

$$\mathcal{L}(G)/\langle a^m, c^m \rangle = \{\, a^i \Box a^j b^n c^k \Box c^l d^n \mid i+j = k+l \geq 0 \text{ and } n \geq 1 \,\},$$

which is independent of m. Similarly one sees that all elements of $\mathcal{L}(G,B)$ share the same set of multicontexts. Obviously $\mathcal{L}(G)/\langle a^m b^n c^m d^n \rangle = \{\Box\}$ for any $\langle a^m b^n c^m d^n \rangle \in \mathcal{L}(G,S)$.

Another example of a CMCFL is $L_c = \{\, wcwc \mid w \in \{a,b\}^* \,\} \in \mathbb{CL}(2,1)$. On the other hand, neither $L_1 = \{\, a^m b^n \mid 1 \leq m \leq n \,\}$ nor $L_2 = \{\, a^n b^n \mid n \geq 1 \,\} \cup \{\, a^n b^{2n} \mid n \geq 1 \,\}$ is p-congruential for any p. We now explain why $L_2 \notin \mathbb{CL}(*,*)$. A similar discussion is applicable to L_1. If an MCFG generates L_2, it must have a nonterminal that derives multiwords $\boldsymbol{u}$ and $\boldsymbol{v}$ that have different number of occurrences of a. Then it is not hard to see that $\boldsymbol{u}$ and $\boldsymbol{v}$ do not share the same set of multicontexts. However $\{\, a^n b^n c \mid n \geq 1 \,\} \cup \{\, a^n b^{2n} d \mid n \geq 1 \,\} \in \mathbb{L}(1,1) \cap \mathbb{CL}(2,1) - \mathbb{CL}(1,*)$.

3 Learning of Congruential Multiple Context-Free Grammars with a Minimally Adequate Teacher

3.1 Minimally Adequate Teacher

Our learning model is based on Angluin's MAT learning [18]. A learner has a teacher who answers two kinds of queries from the learner: *membership queries* (MQs) and *equivalence queries* (EQs). An instance of an MQ is a string w over Σ and the teacher answers "YES" if $w \in L_*$ and otherwise "NO", where we use L_* to refer to the learning target. An instance of an EQ is a grammar $\hat{G}$ and the teacher answers "YES" if $\mathcal{L}(\hat{G}) = L_*$ and otherwise returns a counter-example $w \in (L_* - \mathcal{L}(\hat{G})) \cup (\mathcal{L}(\hat{G}) - L_*)$. If $w \in L_* - \mathcal{L}(\hat{G})$, then it is called a *positive counter-example*, and if $w \in \mathcal{L}(\hat{G}) - L_*$, it is called a *negative counter-example*. The teacher is supposed to answer every query in constant time. The learning process finishes when the teacher answers "YES" to an EQ. In this learning scheme, we fix a class $\mathbb{G}$ of grammars representing our learning targets and require a learner to output a correct grammar in polynomial time in $\|G_*\|$ and ℓ where G_* is a smallest grammar in $\mathbb{G}$ such that $\mathcal{L}(G_*) = L_*$ and ℓ is the length of the longest counter-example given by the teacher.

We remark that we do not restrict instances of EQs to grammars in the class $\mathbb{G}$. Queries of this type are often called extended EQs to emphasize the difference from the restricted type of EQs.

3.2 Hypotheses

Hereafter we arbitrarily fix two natural numbers $p \geq 1$ and $r \geq 1$. Let $L_* \subseteq \Sigma^*$ be the target language from $\mathbb{CL}(p, r)$. Our learning algorithm computes grammars in $\mathbb{G}(p, r)$ from three parameters $K \subseteq \mathcal{S}^{\leq p}(L_*)$, $X \subseteq \mathcal{C}^{\leq p}(L_*)$ and L_* where K and X are always finite. Of course we cannot take L_* as a part of the input, but in fact a finite number of MQs is enough to construct the following MCFG $\mathcal{G}^r(K, X, L_*) = \langle \Sigma, V, F, P, I \rangle$. The set of nonterminal symbols is $V = K$ and we will write $[\![\boldsymbol{v}]\!]$ instead of $\boldsymbol{v}$ for clarifying that it means a nonterminal symbol (indexed with $\boldsymbol{v}$). The dimension $\dim([\![\boldsymbol{v}]\!])$ is $|\boldsymbol{v}|$. The set of initial symbols is

$$I = \{ [\![\langle w \rangle]\!] \mid \langle w \rangle \in K \text{ and } w \in L_* \},$$

where every element of I is of dimension 1. The set F of functions consists of all the λ-free and non-permuting functions that appear in the definition of P. The rules of P are divided into the following two types:

- (Type I) $[\![\boldsymbol{v}]\!] \to f([\![\boldsymbol{v}_1]\!], \dots, [\![\boldsymbol{v}_n]\!])$, if $0 \leq n \leq r$, $\boldsymbol{v}, \boldsymbol{v}_1, \dots, \boldsymbol{v}_n \in K$ and $\boldsymbol{v} = f(\boldsymbol{v}_1, \dots, \boldsymbol{v}_n)$ for f λ-free and non-permuting;
- (Type II) $[\![\boldsymbol{u}]\!] \to [\![\boldsymbol{v}]\!]$, if $L_*/\boldsymbol{u} \cap X = L_*/\boldsymbol{v} \cap X$ and $\boldsymbol{u}, \boldsymbol{v} \in K$,

where rules of the form $[\![\boldsymbol{v}]\!] \to [\![\boldsymbol{v}]\!]$ are of Type I and Type II at the same time, but they are anyway superfluous. Obviously rules of Type II form an equivalence relation in V. One can merge nonterminals in each equivalence class to compact the grammar, but the results are slightly easier to present in this non-compact form.

We want each nonterminal symbol $[\![\boldsymbol{v}]\!]$ to derive $\boldsymbol{v}$. Here the construction of I appears to be trivial: initial symbols derive elements of K if and only if they are in the language L_*. This property is realized by the rules of Type I. For example, for $p = r = 2$ and $K = \mathcal{S}^{\leq 2}(\{ab\})$, one has the following rules $\pi_1, \dots, \pi_5$ of Type I that have $[\![\langle a, b \rangle]\!]$ on their left hand side:

$$\begin{aligned}
\pi_1 &: [\![\langle a, b \rangle]\!] \to \langle a, b \rangle, \\
\pi_2 &: [\![\langle a, b \rangle]\!] \to f_a([\![\langle b \rangle]\!]) \quad \text{with} \quad f_a(\langle z \rangle) = \langle a, z \rangle, \\
\pi_3 &: [\![\langle a, b \rangle]\!] \to f_b([\![\langle a \rangle]\!]) \quad \text{with} \quad f_b(\langle z \rangle) = \langle z, b \rangle, \\
\pi_4 &: [\![\langle a, b \rangle]\!] \to g([\![\langle a \rangle]\!], [\![\langle b \rangle]\!]) \text{ with } g(\langle z_1 \rangle, \langle z_2 \rangle) = \langle z_1, z_2 \rangle, \\
\pi_5 &: [\![\langle a, b \rangle]\!] \to [\![\langle a, b \rangle]\!],
\end{aligned}$$

where π_1 indeed derives $\langle a, b \rangle$, while π_5 is superfluous. Instead of deriving $\langle a, b \rangle$ directly by π_1, one can derive it by two steps with π_3 and $\pi_6 : [\![\langle a \rangle]\!] \to \langle a \rangle$ (or π_2 and $\pi_7 : [\![\langle b \rangle]\!] \to \langle b \rangle$), or by three steps by π_4, π_6 and π_7. One may regard application of rules of Type I as a decomposition of the multiword that appears on its left hand side. It is easy to see that there are finitely many rules of Type I, because K is finite and nonterminals on the right hand side of a rule are all λ-free sub-multiwords of that on the left hand side. If the grammar had only rules of Type I, then it should derive all and only elements of I.

The intuition behind rules of Type II is explained as follows. If $L_*/\boldsymbol{u} = L_*/\boldsymbol{v}$, then L_* is closed under exchanging occurrences of $\boldsymbol{v}$ and $\boldsymbol{u}$ in any strings, and such an exchange is realized by the two symmetric rules $[\![\boldsymbol{u}]\!] \to [\![\boldsymbol{v}]\!]$ and $[\![\boldsymbol{v}]\!] \to [\![\boldsymbol{u}]\!]$ of Type II. The algorithm cannot check whether $L_*/\boldsymbol{u} = L_*/\boldsymbol{v}$ in finitely many steps, but it can approximate this relation by $L_*/\boldsymbol{u} \cap X = L_*/\boldsymbol{v} \cap X$ which we can check using MQs, because X is finite. Clearly $L_*/\boldsymbol{u} = L_*/\boldsymbol{v}$ implies that $L_*/\boldsymbol{u} \cap X = L_*/\boldsymbol{v} \cap X$, but the inverse is not true. We say that a rule $[\![\boldsymbol{u}]\!] \to [\![\boldsymbol{v}]\!]$ of Type II is *incorrect (with respect to L_*)* if $L_*/\boldsymbol{u} \neq L_*/\boldsymbol{v}$.

In this construction of $\mathcal{G}^r(K, X, L_*)$, the initial symbols are determined by K and L_* and the rules of Type I are constructed solely by K, while X is used only for determining rules of Type II. Our algorithm monotonically increases K which will monotonically increase the set of rules, and monotonically increases X which will decrease the set of rules of Type II.

3.3 Observation Tables

The process of computing rules of Type II can be handled by a collection of matrices, called *observation tables.* For each dimension $m \leq p$, we have an observation table T_m. Let K_m and X_m be the sets of m-words from K and m-contexts from X, respectively. The rows of the table T_m are indexed with the elements of K_m and the columns are indexed with the elements of X_m. For each pair $\boldsymbol{u}, \boldsymbol{v} \in K_m$, to compare the sets $L_*/\boldsymbol{u} \cap X$ and $L_*/\boldsymbol{v} \cap X$, one needs to know whether $\mathbf{x} \odot \boldsymbol{u} \in L_*$ or not for all of $\mathbf{x} \in X_m$. The membership of $\mathbf{x} \odot \boldsymbol{u}$ is recorded in the corresponding entry of the observation table with the aid of an MQ. By comparing the entries of the rows indexed with $\boldsymbol{u}$ and $\boldsymbol{v}$, one can determine whether the grammar should have the rule $[\![\boldsymbol{u}]\!] \to [\![\boldsymbol{v}]\!]$.

Example 3. Let $p = 2$, $r = 1$ and

$$L_* = \{\, a^m b^n c^m d^n \mid m + n \geq 1 \,\} \in \mathbb{CL}(2, 1).$$

Indeed L_* is generated by the 2-CMCFG $G_* \in \mathbb{CG}(2, 1)$ whose rules are

$$\begin{aligned} S &\to f(E) \text{ with } f(\langle z_1, z_2\rangle) = \langle z_1 z_2\rangle, \\ E &\to g_a(E) \text{ with } g_a(\langle z_1, z_2\rangle) = \langle a z_1, c z_2\rangle, \quad E \to \langle a, c\rangle, \\ E &\to g_b(E) \text{ with } g_b(\langle z_1, z_2\rangle) = \langle z_1 b, z_2 d\rangle, \quad E \to \langle b, d\rangle \end{aligned}$$

and whose initial symbol is S only.

Let $\hat{G} = \mathcal{G}^1(K, X, L_*)$ for

$$\begin{aligned} K &= \{\, \langle abcd\rangle,\ \langle a, c\rangle,\ \langle b, d\rangle,\ \langle ab, cd\rangle,\ \langle aab, ccd\rangle \,\}, \\ X &= \{\, \Box,\ a\Box bc\Box d,\ ab\Box cd\Box \,\}. \end{aligned}$$

We have the following rules of Type I:

$$\pi_1 : [\![\langle abcd\rangle]\!] \to \langle abcd\rangle, \quad \pi_2 : [\![\langle abcd\rangle]\!] \to f([\![\langle ab, cd\rangle]\!]),$$
$$\pi_3 : [\![\langle abcd\rangle]\!] \to f \circ g_b([\![\langle a, c\rangle]\!]), \quad \pi_4 : [\![\langle abcd\rangle]\!] \to f \circ g_a([\![\langle b, d\rangle]\!]),$$
$$\pi_5 : [\![\langle a, c\rangle]\!] \to \langle a, c\rangle, \quad \pi_6 : [\![\langle b, d\rangle]\!] \to \langle b, d\rangle, \quad \pi_7 : [\![\langle ab, cd\rangle]\!] \to \langle ab, cd\rangle,$$
$$\pi_8 : [\![\langle ab, cd\rangle]\!] \to g_b([\![\langle a, c\rangle]\!]), \quad \pi_9 : [\![\langle ab, cd\rangle]\!] \to g_a([\![\langle b, d\rangle]\!]),$$
$$\pi_{10} : [\![\langle aab, ccd\rangle]\!] \to \langle aab, ccd\rangle, \quad \pi_{11} : [\![\langle aab, ccd\rangle]\!] \to g_a([\![\langle ab, cd\rangle]\!]),$$
$$\pi_{12} : [\![\langle aab, ccd\rangle]\!] \to g_{aa}([\![\langle b, d\rangle]\!]) \quad \text{with} \quad g_{aa}(\langle z_1, z_2\rangle) = \langle aaz_1, ccz_2\rangle,$$
$$\pi_{13} : [\![\langle aab, ccd\rangle]\!] \to h_1([\![\langle a, c\rangle]\!]) \quad \text{with} \quad h_1(\langle z_1, z_2\rangle) = \langle z_1ab, z_2cd\rangle,$$
$$\pi_{14} : [\![\langle aab, ccd\rangle]\!] \to h_2([\![\langle a, c\rangle]\!]) \quad \text{with} \quad h_2(\langle z_1, z_2\rangle) = \langle z_1ab, cz_2d\rangle,$$
$$\pi_{15} : [\![\langle aab, ccd\rangle]\!] \to h_3([\![\langle a, c\rangle]\!]) \quad \text{with} \quad h_3(\langle z_1, z_2\rangle) = \langle az_1b, z_2cd\rangle,$$
$$\pi_{16} : [\![\langle aab, ccd\rangle]\!] \to h_4([\![\langle a, c\rangle]\!]) \quad \text{with} \quad h_4(\langle z_1, z_2\rangle) = \langle az_1b, cz_2d\rangle,$$

where superfluous rules of the form $[\![v]\!] \to [\![v]\!]$ are suppressed. On the other hand we have the following observation tables:

T_1	$\square$
$\langle abcd\rangle$	1

T_2	$a\square bc\square d$	$ab\square cd\square$
$\langle a, c\rangle$	1	0
$\langle b, d\rangle$	1	1
$\langle ab, cd\rangle$	1	0
$\langle aab, ccd\rangle$	1	0

Thus the three nonterminals $[\![\langle a, c\rangle]\!]$, $[\![\langle ab, cd\rangle]\!]$ and $[\![\langle aab, ccd\rangle]\!]$ can be identified thanks to the corresponding rules of Type II.

$$\rho_1 : [\![\langle a, c\rangle]\!] \to [\![\langle ab, cd\rangle]\!], \ \rho_2 : [\![\langle a, c\rangle]\!] \to [\![\langle aab, ccd\rangle]\!], \ \rho_3 : [\![\langle ab, cd\rangle]\!] \to [\![\langle aab, ccd\rangle]\!],$$
$$\rho_4 : [\![\langle ab, cd\rangle]\!] \to [\![\langle a, c\rangle]\!], \ \rho_5 : [\![\langle aab, ccd\rangle]\!] \to [\![\langle a, c\rangle]\!], \ \rho_6 : [\![\langle aab, ccd\rangle]\!] \to [\![\langle ab, cd\rangle]\!].$$

The unique initial symbol of $\hat{G}$ is $[\![\langle abcd\rangle]\!]$.

Let us see some derivations of $\hat{G}$. Derivation trees of the form

$$\overbrace{\rho_3(\pi_{11}(\dots(\rho_3(\pi_{11}}^{(m-1)\text{-times}}(\rho_4(\pi_5))))\dots))$$

give us $\langle a^m, c^m\rangle \in \mathcal{L}([\![\langle ab, cd\rangle]\!])$ for all $m \geq 1$. Similarly

$$\overbrace{\pi_8(\rho_1(\dots(\pi_8(\rho_1}^{n\text{-times}}(\overbrace{\rho_3(\pi_{11}(\dots(\rho_3(\pi_{11}}^{(m-1)\text{-times}}(\rho_4(\pi_5))))\dots)))))\dots))$$

give us $\langle a^m b^n, c^m d^n\rangle \in \mathcal{L}([\![\langle ab, cd\rangle]\!])$ for all $n \geq 0$. Finally by applying π_2, we see $a^m b^n c^m d^n \in \mathcal{L}(\hat{G})$.

However the rules $\rho_1, \rho_2, \rho_4, \rho_5$ of Type II, which involve $[\![\langle a, c\rangle]\!]$, are all incorrect, because $\square ab\square cd \in L_*/\langle a, c\rangle - L_*/\langle ab, cd\rangle = L_*/\langle a, c\rangle - L_*/\langle aab, ccd\rangle$. In fact the derivation tree $\pi_2(\rho_3(\pi_{13}(\rho_1(\pi_7))))$, for instance, yields $ababcdcd \in \mathcal{L}(\hat{G}) - L_*$.

Lemma 4. *One can compute $\hat{G} = \mathcal{G}^r(K, X, L_*)$ in polynomial time in $\|K\|$ and $\|X\|$.*

Proof. We first estimate the number of rules of Type I that have a fixed $[\![\boldsymbol{v}]\!]$ on the left hand side, which are of the form $[\![\boldsymbol{v}]\!] \to f([\![\boldsymbol{v}_1]\!], \dots, [\![\boldsymbol{v}_n]\!])$. Roughly speaking, this is the number of ways to decompose $\boldsymbol{v}$ into sub-multiwords $\boldsymbol{v}_1, \dots, \boldsymbol{v}_n$ with the aid of a linear regular function f. Once one has fixed where the occurrence of each component from $\boldsymbol{v}_1, \dots, \boldsymbol{v}_n$ starts and ends in $\boldsymbol{v}$, the function f is uniquely determined. We have at most pr components in $\boldsymbol{v}_1, \dots, \boldsymbol{v}_n$, hence the number of ways to determine such starting and ending points are at most $\|\boldsymbol{v}\|^{2pr}$. Thus, the number of rules of Type I is at most $O(|K|\ell^{2pr})$ where ℓ is the maximal size of elements of K. Clearly the description size of each rule is at most $O(\ell)$.

One can construct the observation tables by at most $|K||X|$ MQs. Then one can determine initial symbols and rules of Type II in polynomial time. □

The next subsection discusses how our learner determines K and X.

3.4 Undergeneralization

The problems that we have to deal with are undergeneralization and overgeneralization. We first show when the data are sufficient to avoid undergeneralization.

Lemma 5. *Let $G_* \in \mathbb{CG}(p, r)$ be such that $\mathcal{L}(G_*) = L_*$. If a finite set $D \subseteq L_*$ is such that every rule of G_* occurs in a derivation tree for some $w \in D$, then for any X such that $\Box \in X$, we have $L_* \subseteq \mathcal{L}(\hat{G})$ where $\hat{G} = \mathcal{G}^r(\mathcal{S}^{\leq p}(D), X, L_*)$.*

Proof. Let $G_* = \langle \Sigma, V_*, F_*, P_*, I_* \rangle$ and $K = \mathcal{S}^{\leq p}(D)$. By induction we show that if $\boldsymbol{w} \in \mathcal{L}(G_*, A)$, then $\boldsymbol{w} \in \mathcal{L}(\hat{G}, [\![\boldsymbol{v}]\!])$ for some $\boldsymbol{v} \in \mathcal{L}(G_*, A) \cap K$. In particular when $A \in I_*$, $[\![\boldsymbol{v}]\!]$ will be an initial symbol of $\hat{G}$, so this proves the lemma.

Suppose that $\pi(t_1, \dots, t_n)$ is an A-derivation tree for $\boldsymbol{w}$ of G_* where π is of the form $A \to f(B_1, \dots, B_n)$ and t_i is a B_i-derivation tree for $\boldsymbol{w}_i$ for all $i = 1, \dots, n$. By induction hypothesis, there are $\boldsymbol{u}_i \in \mathcal{L}(G_*, B_i) \cap K$ such that $\boldsymbol{w}_i \in \mathcal{L}(\hat{G}, [\![\boldsymbol{u}_i]\!])$ for all i. On the other hand, by the assumption, we have $w \in D$ that is derived by using π, which can be represented as

$$w = \mathbf{x} \odot \boldsymbol{v} = \mathbf{x} \odot f(\boldsymbol{v}_1, \dots, \boldsymbol{v}_n)$$

where $\boldsymbol{v}_i \in \mathcal{L}(G_*, B_i) \cap K$ for all $i = 1, \dots, n$ and $f(\boldsymbol{v}_1, \dots, \boldsymbol{v}_n) \in \mathcal{L}(G_*, A) \cap K$. Let $\boldsymbol{v} = f(\boldsymbol{v}_1, \dots, \boldsymbol{v}_n)$. Then $\hat{G}$ has the rule

$$[\![\boldsymbol{v}]\!] \to f([\![\boldsymbol{v}_1]\!], \dots, [\![\boldsymbol{v}_n]\!])$$

of Type I and moreover rules $[\![\boldsymbol{v}_i]\!] \to [\![\boldsymbol{u}_i]\!]$ of Type II for all $i = 1, \dots, n$ whatever X is, since G_* is a CMCFG. By applying those rules of Types I and II to $\boldsymbol{w}_i \in \mathcal{L}(\hat{G}, [\![\boldsymbol{u}_i]\!])$ for $i = 1, \dots, n$, we obtain $\boldsymbol{w} \in \mathcal{L}(\hat{G}, [\![\boldsymbol{v}]\!])$ with $\boldsymbol{v} \in \mathcal{L}(G_*, A) \cap K$. □

When our algorithm gets a positive counter-example $w \in L_* - \mathcal{L}(\hat{G})$, it adds all multiwords in $\mathcal{S}^{\leq p}(\{w\})$ to K.

3.5 Overgeneralization

Overgeneralization is a little more difficult to treat. Because overgeneralization is caused by incorrect rules of Type II, we must add appropriate multicontexts to X in order to remove them.

Suppose that our conjecture $\hat{G}$ is a correct CMCFG for the target L_*. Then it must satisfy that

– for any $[\![\boldsymbol{v}]\!] \in V$ and any $\boldsymbol{u} \in \mathcal{L}(\hat{G}, [\![\boldsymbol{v}]\!])$, it holds that $L_*/\boldsymbol{u} = L_*/\boldsymbol{v}$.

When we get a negative counter-example w from the teacher, this means that we have $\langle w \rangle \in \mathcal{L}(\hat{G}, [\![\langle v \rangle]\!])$ for some initial symbol $[\![\langle v \rangle]\!]$, but $\square \in L_*/\langle v \rangle - L_*/\langle w \rangle$. The triple $([\![\langle v \rangle]\!], \langle w \rangle, \square)$ is a witness for the fact that $\hat{G}$ does not satisfy the above desired property. It is not hard to see by Lemma 3 that in such a case $\hat{G}$ must have an incorrect rule of Type II which is used for deriving w. In order to find such an incorrect rule, we first parse the string w with $\hat{G}$ and get a derivation tree t for w. We then look for an incorrect rule in t that causes this violation by recursively searching t in a topdown manner as described below, where the initial value of $\mathbf{x}$ is $\square$.

Suppose that we have a pair $(t, \mathbf{x})$ such that t is a $[\![\boldsymbol{v}]\!]$-derivation tree for $\boldsymbol{u}$ and $\mathbf{x} \in L_*/\boldsymbol{v} - L_*/\boldsymbol{u}$. We have two cases.

Case 1. Suppose that $t = \pi(t_1, \dots, t_n)$ for some rule π of Type I. Then, there are $[\![\boldsymbol{v}_1]\!], \dots, [\![\boldsymbol{v}_n]\!] \in V$, $\boldsymbol{u}_1, \dots, \boldsymbol{u}_n \in (\Sigma^+)^{\langle * \rangle}$ and f such that

- π is of the form $[\![\boldsymbol{v}]\!] \to f([\![\boldsymbol{v}_1]\!], \dots, [\![\boldsymbol{v}_n]\!])$, i.e., $\boldsymbol{v} = f(\boldsymbol{v}_1, \dots, \boldsymbol{v}_n)$,
- $\boldsymbol{u} = f(\boldsymbol{u}_1, \dots, \boldsymbol{u}_n)$,
- t_i is a $[\![\boldsymbol{v}_i]\!]$-derivation tree for $\boldsymbol{u}_i$ for $i = 1, \dots, n$.

By the assumption we have

$$\mathbf{x} \odot \boldsymbol{v} = \mathbf{x} \odot f(\boldsymbol{v}_1, \dots, \boldsymbol{v}_n) \in L_*,$$
$$\mathbf{x} \odot \boldsymbol{u} = \mathbf{x} \odot f(\boldsymbol{u}_1, \dots, \boldsymbol{u}_n) \notin L_*.$$

This means that n cannot be 0. One can find k such that

$$\mathbf{x} \odot f(\boldsymbol{v}_1, \dots, \boldsymbol{v}_{k-1}, \boldsymbol{v}_k, \boldsymbol{u}_{k+1}, \dots, \boldsymbol{u}_n) \in L_*,$$
$$\mathbf{x} \odot f(\boldsymbol{v}_1, \dots, \boldsymbol{v}_{k-1}, \boldsymbol{u}_k, \boldsymbol{u}_{k+1}, \dots, \boldsymbol{u}_n) \notin L_*.$$

That is, for

$$\mathbf{x}' = \mathbf{x} \odot f(\boldsymbol{v}_1, \dots, \boldsymbol{v}_{k-1}, \square^{\langle |\boldsymbol{v}_k| \rangle}, \boldsymbol{u}_{k+1}, \dots, \boldsymbol{u}_n),$$

we have $\mathbf{x}' \in L/\boldsymbol{v}_k - L/\boldsymbol{u}_k$. We then recurse with $(t_k, \mathbf{x}')$.

Case 2. Suppose that $t = \pi(t')$ where $\pi : [\![\boldsymbol{v}]\!] \to [\![\boldsymbol{v}']\!]$ is a rule of Type II. Then t' is a $[\![\boldsymbol{v}']\!]$-derivation tree for $\boldsymbol{u}$. If $\mathbf{x} \odot \boldsymbol{v}' \notin L_*$, this means that $[\![\boldsymbol{v}]\!] \to [\![\boldsymbol{v}']\!]$ is an incorrect rule, because $\mathbf{x} \in L_*/\boldsymbol{v} - L_*/\boldsymbol{v}'$. We then add $\mathbf{x}$ to X for removing the incorrect rules $[\![\boldsymbol{v}]\!] \to [\![\boldsymbol{v}']\!]$ and $[\![\boldsymbol{v}']\!] \to [\![\boldsymbol{v}]\!]$. Otherwise, we know that $\mathbf{x} \in L_*/\boldsymbol{v}' - L_*/\boldsymbol{u}$. We recurse with $(t', \mathbf{x})$.

We use FindContext$(\hat{G}, w)$ to refer to this procedure for computing a multicontext that witnesses incorrect rules from a negative counter-example w. The size of the output $\mathbf{x}$ is bounded by $|w|\ell$ for $\ell = \max\{\, \|\boldsymbol{v}\| \mid \boldsymbol{v} \in K\}$, because $\mathbf{x}$ is obtained from w by replacing some occurrences of nonempty sub-multiwords $\boldsymbol{u}_i$ by $\boldsymbol{v}_i \in K$ and an occurrence of a sub-multiword by a hole $\square$. The procedure FindContext contains parsing as a subroutine. We can use any of polynomial-time parsing algorithms for MCFGs proposed so far, e.g., [13, 19].

Lemma 6. FindContext$(\hat{G}, w)$ *runs in polynomial time in* $|w|$ *and* $\|\hat{G}\|$.

3.6 Algorithm

We are now ready to describe the overall structure of our algorithm. The pseudocode is presented in Algorithm 1.

Algorithm 1. Learn $\mathbb{CL}(p, r)$

```
let K := ∅; X := {□}; Ĝ = G^r(K, X, L*);
while the teacher does not answer "YES" to the EQ on Ĝ do
   let w be the counter-example from the teacher;
   if w ∈ L* − L(Ĝ) then
      let K := K ∪ S^{≤p}({w});
   else
      let X := X ∪ {FindContext(Ĝ, w)};
   end if
   let Ĝ = G^r(K, X, L*);
end while
output Ĝ;
```

Let $G_* = \langle \Sigma, V_*, F_*, P_*, I_* \rangle \in \mathbb{CG}(p, r)$ represent the learning target L_*.

Lemma 7. *The algorithm receives a positive counter-example at most* $|P_*|$ *times. The cardinality of* K *is bounded by* $O(|P_*|\ell^{2p})$*, where* ℓ *is the length of a longest positive counter-example given so far.*

Proof. The proof of Lemma 5 in fact claims that any of the rules of G_* that are used for deriving positive counter-examples can be simulated by $\hat{G}$. That is, whenever the learner gets a positive counter-example w from the teacher, there is a rule of G_* that is used for deriving w but not used for any of previously presented positive counter-examples. Therefore the learner receives a positive counter-example at most $|P_*|$ times.

It is easy to see that $|\mathcal{S}^{\leq p}(\{w\})| \in O(|w|^{2p})$. Hence $|K| \in O(|P_*|\ell^{2p})$ where ℓ is the length of a longest positive counter-example given so far. $\square$

Lemma 8. *The number of times that the algorithm receives a negative counter-example and the cardinality of* X *are both bounded by* $O(|P_*|\ell^{2p})$.

Proof. Each time the learner receives a negative counter-example, one element is added to X and some incorrect rules of Type II are removed. That is, the number of equivalence classes induced by rules of Type II is increased. Because there can be at most $|K|$ equivalence classes in V, the learner expands X at most $|K|$ times, that is, $|X| \in O(|P_*|\ell^{2p})$ by Lemma 7. □

Theorem 2. *Our algorithm outputs a grammar representing the target language L_* and halts in polynomial time in $\|G_*\|$ and ℓ where $G_* \in \mathbb{CG}(p,r)$ is a grammar representing L_* and ℓ is the length of a longest counter-example from the teacher.*

Proof. By Lemmas 4 and 6, each time the learner receives a counter-example, it computes the next conjecture in polynomial time in $\|G_*\|$ and ℓ. By Lemmas 7 and 8, it asks EQs at most polynomial times in $\|G_*\|$ and ℓ. All in all, it runs in polynomial time in $\|G_*\|$ and ℓ. The correctness of the output of the algorithm is guaranteed by the teacher. □

3.7 Slight Enhancement

Let us say that $L \in \mathbb{L}(p,r)$ is *almost p-congruential* if $\$L\$ \in \mathbb{CL}(p,r)$ where \$ is a new marker not in Σ. While every language in $\mathbb{CL}(p,r)$ is almost p-congruential, the converse does not hold. An example in the difference is $L_{\mathrm{copy}} = \{\, w\overline{w} \in \{a,b,\bar{a},\bar{b}\}^* \mid w \in \{a,b\}^* \,\}$, where $\bar{\ }$ denotes the homomorphism that maps a,b to $\bar{a},\bar{b}$, respectively. One can easily modify our learning algorithm so that it learns all almost p-congruential languages in $\mathbb{L}(p,r)$ just by assuming the marker \$ on the head and the tail on each counter-example.

4 Conclusion

This work is based on a combination of the MAT algorithm for learning congruential CFGs presented in [17] and the algorithm for learning some sorts of MCFGs from positive data and MQs in [20].

The conjecture $\hat{G}$ of our algorithm is not always consistent with the observation tables in the sense that it might be the case $\mathcal{L}(\hat{G}) \cap (X \odot K) \neq L_* \cap (X \odot K)$. For $\mathbf{x} \in X$ and $\boldsymbol{y} \in K$, one can regard $\mathbf{x} \odot \boldsymbol{y} \in L_* - \mathcal{L}(\hat{G})$ as a positive counter-example and $\mathbf{x} \odot \boldsymbol{y} \in \mathcal{L}(\hat{G}) - L_*$ as a negative counter-example. One can modify our algorithm so that the conjecture is always consistent by substituting this self-diagnosis method for EQs to the teacher. EQs will be used only when the learner does not find any inconsistency. Verification of the consistency can be done in polynomial time, however we do not employ this idea in our algorithm, because each time the learner extracts a counter-example from the observation tables, they will be recursively expanded and this might cause exponential-time computation, when the size of a counter-example computed by the learner is not considered to be a parameter of the input size.

Lemma 3 offers another method for finding incorrect rules in some cases without getting a negative counter-example from the teacher. That is, if there are rules of Type I $[\![\boldsymbol{u}]\!] \to f([\![\boldsymbol{u}_1]\!], \dots, [\![\boldsymbol{u}_n]\!])$ and $[\![\boldsymbol{v}]\!] \to f([\![\boldsymbol{v}_1]\!], \dots, [\![\boldsymbol{v}_n]\!])$ such that

$L_*/\boldsymbol{u}_i \cap X = L_*/\boldsymbol{v}_i \cap X$ for all i and $L_*/\boldsymbol{u} \cap X \neq L_*/\boldsymbol{v} \cap X$, then one knows that $[\![\boldsymbol{u}_k]\!] \to [\![\boldsymbol{v}_k]\!]$ is an incorrect rule for some k. Yet we do not take this idea into our algorithm for the same reason discussed in the previous paragraph.

In Clark et al. [21] the approach uses representational primitives that correspond not to congruence classes but to elements of the form:

$$\{\, v \mid L/v \supseteq L/u \,\} \tag{1}$$

An advantage of using congruence classes is that any element of the class will be as good as any other element; if we use the idea of (1), we cannot be sure that we will get the right elements, and it is difficult to get a pure MAT result.

Distributional lattice grammars (DLGs) are another interesting development in the field of rich language classes that are also learnable [22]. While these models can represent some non-context-free languages, it is not yet clear whether they are rich enough to account for the sorts of cross-serial dependency that occur in natural languages; they are basically context free grammars, with an additional operation, somewhat analogous to that of Conjunctive Grammars [23].

We suggest two directions for future research: a PAC-learning result along the lines of the PAC result for NTS languages in [24] seems like a natural extension, since MCFGs have a natural stochastic variant. Secondly, combining DLGs and MCFGs might be possible – we would need to replace the single concatenation operation in DLGs with a more complex range of operations that reflect the variety of ways that tuples of strings can be combined.

We also note that given the well-known relationship between synchronous CFGs and MCFGs [25] this algorithm gives as a special case the first learning algorithm for learning synchronous context free grammars, and therefore of syntax-directed translation schemes. Given the current interest in syntax-based statistical machine translation, we intend to explore further this particular class of algorithms.

Acknowledgement. The authors are very grateful to Anna Kasprzik and the anonymous reviewers for valuable comments and suggestions on a draft of this paper.

References

1. Joshi, A.K.: Tree adjoining grammars: how much context-sensitivity is required to provide reasonable structural descriptions? In: Dowty, D.R., Karttunen, L., Zwicky, A. (eds.) Natural Language Parsing, pp. 206–250. Cambridge University Press, Cambridge (1985)
2. Vijay-Shanker, K., Weir, D.J., Joshi, A.K.: Characterizing structural descriptions produced by various grammatical formalisms. In: Proceedings of the 25th Annual Meeting of Association for Computational Linguistics, Stanford, pp. 104–111 (1987)
3. Joshi, A.K., Levy, L.S., Takahashi, M.: Tree adjunct grammars. Journal of Computer and System Sciences 10(1), 136–163 (1975)

4. Joshi, A.K.: An introduction to tree adjoining grammars. In: Manaster-Ramer, A. (ed.) Mathematics of Languge. John Benjamins (1987)
5. Stabler, E.P.: Derivational Minimalism. In: Retoré, C. (ed.) LACL 1996. LNCS (LNAI), vol. 1328, pp. 68–95. Springer, Heidelberg (1997)
6. Engelfriet, J., Heyker, L.: The string generating power of context-free hypergraph grammars. Journal of Computer and System Sciences 43(2), 328–360 (1991)
7. Clark, A., Eyraud, R.: Polynomial identification in the limit of substitutable context-free languages. Journal of Machine Learning Research 8, 1725–1745 (2007)
8. Huybrechts, R.A.C.: The weak inadequacy of context-free phrase structure grammars. In: de Haan, G., Trommelen, M., Zonneveld, W. (eds.) Van Periferie naar Kern, Foris, Dordrecht, Holland (1984)
9. Shieber, S.M.: Evidence against the context-freeness of natural language. Linguistics and Philosophy 8, 333–343 (1985)
10. Angluin, D.: Learning regular sets from queries and counterexamples. Information and Computation 75(2), 87–106 (1987)
11. Angluin, D., Kharitonov, M.: When won't membership queries help? Journal of Computer and System Sciences 50(2), 336–355 (1995)
12. Clark, A., Thollard, F.: PAC-learnability of probabilistic deterministic finite state automata. Journal of Machine Learning Research 5, 473–497 (2004)
13. Seki, H., Matsumura, T., Fujii, M., Kasami, T.: On multiple context-free grammars. Theoretical Computer Science 88(2), 191–229 (1991)
14. Kracht, M.: The Mathematics of Language. Studies in Generative Grammar, vol. 63, pp. 408–409. Mouton de Gruyter (2003)
15. Rambow, O., Satta, G.: Independent parallelism in finite copying parallel rewriting systems. Theoretical Computer Science 223(1-2), 87–120 (1999)
16. Kaji, Y., Nakanishi, R., Seki, H., Kasami, T.: The universal recognition problems for parallel multiple context-free grammars and for their subclasses. IEICE Transaction on Information and Systems E75-D(7), 499–508 (1992)
17. Clark, A.: Distributional learning of some context-free languages with a minimally adequate teacher. In: Proceedings of the ICGI, Valencia, Spain (September 2010)
18. Angluin, D.: Learning regular sets from queries and counterexamples. Information and Computation 75(2), 87–106 (1987)
19. Kanazawa, M.: A prefix-correct earley recognizer for multiple context-free grammars. In: Proceedings of the Ninth International Workshop on Tree Adjoining Grammars and Related Formalisms, pp. 49–56 (2008)
20. Yoshinaka, R.: Polynomial-Time Identification of Multiple Context-Free Languages from Positive Data and Membership Queries. In: Sempere, J.M., García, P. (eds.) ICGI 2010. LNCS, vol. 6339, pp. 230–244. Springer, Heidelberg (2010)
21. Clark, A., Eyraud, R., Habrard, A.: A Polynomial Algorithm for the Inference of Context Free Languages. In: Clark, A., Coste, F., Miclet, L. (eds.) ICGI 2008. LNCS (LNAI), vol. 5278, pp. 29–42. Springer, Heidelberg (2008)
22. Clark, A.: A learnable representation for syntax using residuated lattices. In: Proceedings of the 14th Conference on Formal Grammar, Bordeaux, France (2009)
23. Okhotin, A.: Conjunctive grammars. Journal of Automata, Languages and Combinatorics 6(4), 519–535 (2001)
24. Clark, A.: PAC-Learning Unambiguous NTS Languages. In: Sakakibara, Y., Kobayashi, S., Sato, K., Nishino, T., Tomita, E. (eds.) ICGI 2006. LNCS (LNAI), vol. 4201, pp. 59–71. Springer, Heidelberg (2006)
25. Melamed, I.D.: Multitext grammars and synchronous parsers. In: Proceedings of NAACL/HLT, pp. 79–86 (2003)

Locality and the Complexity of Minimalist Derivation Tree Languages

Thomas Graf

Department of Linguistics
University of California, Los Angeles
tgraf@ucla.edu
http://tgraf.bol.ucla.edu

Abstract. Minimalist grammars provide a formalization of Minimalist syntax which allows us to study how the components of said theory affect its expressivity. A central concern of Minimalist syntax is the locality of the displacement operation Move. In Minimalist grammars, however, Move is unbounded. This paper is a study of the repercussions of limiting movement with respect to the number of *slices* a moved constituent is allowed to cross, where a slice is the derivation tree equivalent of the phrase projected by a lexical item in the derived tree. I show that this locality condition 1) has no effect on weak generative capacity 2) has no effect on a Minimalist derivation tree language's recognizability by top-down automata 3) renders Minimalist derivation tree languages strictly locally testable, whereas their unrestricted counterparts aren't even locally threshold testable.

Keywords: Minimalist grammars, locality, subregular tree languages, first-order logic, top-down tree automata.

Introduction

Even though Minimalist grammars (MGs) weren't introduced by Stabler [16] with the sole intent of scrutinizing the merits of ideas put forward by syntacticians in the wake of Chomsky's Minimalist Program [2], a lot of work on MGs certainly focuses on this aspect [cf. 17]. Recently, considerable attention has also been directed towards the role played by derivation trees in MGs [4, 7, 8]. It is now known that every MG's derivation tree language is regular and "almost" closed under intersection with regular tree languages (some refinement of category labels is usually required), but it is still an open question which class of tree languages approximates them reasonably well. This paper combines both research strands by taking the linguistically motivated restriction to local movement as its vantage point for an examination of the structural complexity of Minimalist derivation tree languages (MDTLs). The main result is that while bounding the distance of movement leaves weak generative capacity unaffected, the complexity of MDTLs is lowered to a degree where they become strictly locally testable. Since MGs are fully characterized by their MDTLs, lowering the

P. de Groote and M.-J. Nederhof (Eds.): Formal Grammar 2010/2011, LNCS 7395, pp. 208–227, 2012.

upper bound from regular to strictly locally testable may prove useful for various practical applications operating directly on the derivation trees, in particular parsing [9, 18, 21].

The paper is laid out as follows: After some general technical remarks, Sec. 2 introduces MGs, their derivation trees, and the important concept of slices. Building on these notions, I define movement-free and k-local MGs, and I prove that every MG can be converted into a k-local one. The complexity of these derivation trees is then studied with respect to several subregular languages classes in Sec. 3. I first show that MDTLs can be recognized by l-r-deterministic top-down tree automata, but not by sensing tree automata, which entails non-recognizability by deterministic top-down tree automata. Furthermore, every k-local MG G has a strictly locally κ-testable derivation tree language, with the value of κ depending on several parameters of G. This result is followed by a demonstration that unrestricted MDTLs are not locally threshold testable. However, they are definable in first-order logic with predicates for left child, right child, proper dominance, and equivalence (Sec. 4).

1 Preliminaries and Notation

As usual, $\mathbb{N}$ denotes the set of non-negative integers. A *tree domain* is a finite subset D of $\mathbb{N}^*$ such that, for $w \in \mathbb{N}^*$ and $j \in \mathbb{N}$, $wj \in D$ implies both $w \in D$ and $wi \in D$ for all $i < j$. Every $n \in D$ is called a *node*. Given nodes $m, n \in D$, *m immediately dominates n* iff $n = mi$, $i \in \mathbb{N}$. In this case we also say m is the mother of n, or conversely, n is a daughter of m. The transitive closure of the immediate dominance relation is called *(proper) dominance*. A node that does not dominate any other nodes is a *leaf*, and the unique node that isn't dominated by any nodes is called the *root*.

Now let Σ be a *ranked* alphabet, i.e. every $\sigma \in \Sigma$ has a unique non-negative *rank* (*arity*); $\Sigma^{(n)}$ is the set of all n-ary symbols in Σ. A Σ-tree is a pair $T := \langle D, \ell \rangle$, where D is a tree domain and $\ell : D \to \Sigma$ is a function assigning each node n a *label* drawn from Σ such that $\ell(n) \in \Sigma^{(d)}$ iff n has d daughters. Usually the alphabet will not be indicated in writing when it is irrelevant or can be inferred from the context. Sometimes trees will be given in functional notation such that $f(t_1, \dots, t_n)$ is the tree whose root node is labeled f and immediately dominates trees $t_1, \dots, t_n$. I denote by T_Σ the set of all trees such that for $n \geq 0$, $f(t_1, \dots, t_n)$ is in T_Σ iff $f \in \Sigma^{(n)}$ and $t_i \in T_\Sigma$, $1 \leq i \leq n$. A *tree language* is some subset of T_Σ.

A *context* C is a $\Sigma \cup \{\Box\}$-tree, where $\Box$ is a new symbol that appears on exactly one leaf of C, designating it as the *port* of C. A context C with $\Box$ occurring in the configuration $c := \sigma(t_1, \dots, \Box, \dots, t_n)$, $\sigma \in \Sigma^{(n)}$ and each t_i a Σ-tree, can be composed with context C' (written $C \cdot C'$) by replacing c with $\sigma(t_1, \dots, C', \dots, t_n)$. This extends naturally to all cases where $C = \Box$ and C' is a tree rather than a context. Given a Σ-tree $t := \langle D, \ell \rangle$ and some node u of t, $t|_u := \langle D|_u, \ell \rangle$ denotes the subtree rooted by u in t, such that $D|_u := \{n \in D \mid u = n \text{ or } u \text{ dominates } n\}$ and dominance and the labeling function are preserved. For any tree t with nodes m and n of t such that either $m = n$ or m dominates n, $C_t[m, n)$ is the context obtained from $t|_m$ by replacing $t|_n$ by a port. If s and t are trees, r the root of s and u some node of s, then $s[u \leftarrow t] := C_s[r, u) \cdot t$.

Let m and n be nodes of some tree t. A *path* from m to n is a sequence of node $\langle i_0, \ldots, i_k \rangle$ such that $i_0 = m$, $i_k = n$, and for all $j < k$, i_j is the mother or the daughter of i_{j+1}. A path containing k nodes is of *length* $k - 1$. The *distance* between nodes m and n is the length of the shortest path from m to n. The *depth* of a tree t is identical to the greatest distance between the root of t and one of its leafs.

We now move on to defining the strictly locally testable languages, following the exposition in [21]. For each Σ-tree and choice of $k \geq 1$, we define its *k-factors*, or more precisely, its *k-prefixes*, *k-forks* and *k-suffixes* as follows:

$$p_k(\sigma(t_1, \ldots, t_n)) := \begin{cases} \sigma & \text{if } k = 1 \text{ or } \sigma \text{ has no children} \\ \sigma(p_{k-1}(t_1), \ldots, p_{k-1}(t_n)) & \text{otherwise} \end{cases}$$

$$f_k(\sigma(t_1, \ldots, t_n)) := \begin{cases} \emptyset & \text{if } \sigma(t_1, \ldots, t_n) \text{ is of depth } d < k - 1 \\ \{p_k(\sigma(t_1, \ldots, t_n))\} \cup \bigcup_{i=1}^{n} f_k(t_i) & \text{otherwise} \end{cases}$$

$$s_k(\sigma(t_1, \ldots, t_n)) := \begin{cases} \{\sigma(t_1, \ldots, t_n)\} \cup \bigcup_{i=1}^{n} s_k(t_i) & \text{if } \sigma(t_1, \ldots, t_n) \text{ is of depth } d < k - 1 \\ \bigcup_{i=1}^{m} s_k(t_i) & \text{otherwise} \end{cases}$$

A tree language $L \subseteq T_\Sigma$ is *strictly locally k-testable* (in SL_k) iff there exist three finite subsets R, F, and S, such that $t \in L$ iff $p_{k-1}(t) \in R$, $f_k(t) \subseteq F$, and $s_{k-1} \subseteq S$. A language is *local* (in LOC) iff it is in SL_2. It is locally k-threshold testable (in LTT_k) iff furthermore each k-factor must appear a specific number of times, counting up to some fixed threshold. When the threshold is set to 1, L is locally k-testable (in LT_k). We say that L belongs to one of these classes iff there is some k such that L is k-testable in the intended sense. Finally, L is *regular* iff it is the range of a transduction computed by some linear tree transducer with domain T_Σ (the reader is referred to [5] for further details).

Definition 1. *A* linear tree transducer *is a 5-tuple* $A := \langle \Sigma, \Omega, Q, Q', \Delta \rangle$, *where* Σ *and* Ω *are finite ranked alphabets,* Q *is a finite set of states,* $Q' \subseteq Q$ *the set of final states, and* Δ *is a set of productions of the form* $f(q_1(x_1), \ldots, q_n(x_n)) \rightarrow q(t)$, *where* $f \in \Sigma^{(n)}$, $q_1, \ldots, q_n, q \in Q$, $t \in T_{\Omega \cup \{x_1, \ldots, x_n\}}$, *and each* x_i *occurs at most once in* t.

It is a well-known fact that LOC $\subset$ SL $\subset$ LT $\subset$ LTT $\subset$ REG.

2 Minimalist Grammars

2.1 Introduction and Examples

MGs are a highly lexicalized formalism. Every lexical item (LI) comes equipped with a linear sequence of features that have to be "checked", or equivalently, "erased" in the right order. Features come in two varieties that can only be checked by the operations Merge and Move, respectively. Merge conjoins trees, while Move displaces subtrees. A very simple MG, for example, is instantiated by the following lexicon.

man :: n	the :: $=$n d	ε :: $=$v $+$nom t
John :: d	the :: $=$n d $-$nom	ε :: $=$t c
John :: d $-$nom	the :: $=$n d $-$top	ε :: $=$t $+$top c
John :: d $-$top	killed :: $=$d$=$d v	

The first component of an LI denotes its phonetic exponent, the second one its feature string. Features without a prefix represent categories (n for noun, d for determiner, and so on). A category feature, say n, of LI l is checked whenever Merge combines l with another LI l' such that the respective first unchecked features of l and l' are n and the matching selector feature $=$n. This is also the only feature configuration in which Merge may apply. Hence Merge could combine *the* and *man*, yielding *the man*, which in turn can be merged with *killed*, but not *the* and *John* (no compatible features at all), or *killed* and *the* (the first unchecked feature of *the* is $=$n, which is incompatible with the $=$d on *killed*). The feature combinatorics of the Move operation are essentially the same, with the only difference being that Move applies to features prefixed with $+$ and $-$ (licensor and licensee, respectively). Note that Merge introduces new material into the derivation, whereas Move merely displaces old material — intuitively, the subtree headed by an LI l with some licensee feature $-f$ as its first unchecked feature is moved into the specifier of the closest LI l' that c-commands l and has $+f$ as its first unchecked feature.

An utterance is well-formed if it can be assigned a derivation in which all features were checked except for the category feature of the last LI to be merged, which must be a so-called final category (usually c). The MG above generates the following eight sentences, and only those (assuming that c is the only final category):

(1) a. John/The man killed John/the man.
    b. John/The man, John/the man killed.

A derivation tree for one of the sentences with topicalization is given in Fig. 1.

Despite its simplicity, the feature calculus controlling Move allows for a dazzling array of movement configurations, in particular *remnant movement*, in which some XP is extracted from some YP via Move before YP itself is moved to a higher position. Remnant movement allows for an elegant reanalysis of cases where apparently non-phrasal constituents end up in positions reserved for phrases, such as in the German example below.

(2) $[_{\text{CP}}$ Geküsst$_i$ hat$_j$ $[_{\text{TP}}$ der Hans die Maria t_j t_i.$]]$
    kissed has the John the Mary.
    'John kissed Mary.'

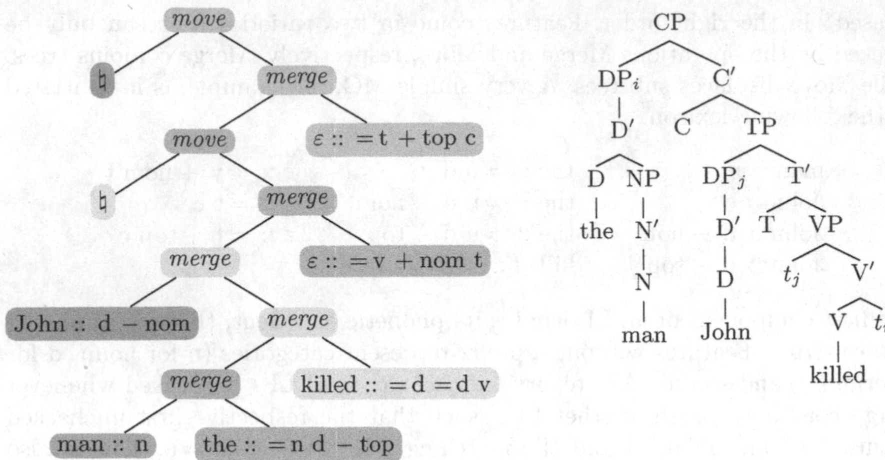

Fig. 1. Left: derivation tree of *The man, John killed,* depicted in the final format adopted in this paper and with slices indicated by color; Right: Corresponding X′-tree

Instead of having just the V-head *geküsst* move into SpecCP (which is at odds with standard assumptions about phrase structure), one can fall back to a remnant movement analysis: the object moves out of the VP, followed by movement of the remaining VP into SpecCP. At this point the VP's only phonetic exponent is its V-head, so that the end result is indistinguishable from the scenario where only the V-head had undergone movement. For our purposes, remnant movement is of interest because of the crucial role it plays in the proof that every instance of Move that spans arbitrary distances can be decomposed into a sequence of local Move steps (Thm. 1).

2.2 Minimalist Grammars, Derivation Trees, and Slices

As the focus of this paper is on the derivation trees of MGs rather than the phrase structure trees derived via Merge and Move, the details of both operations are of interest to us only in so far as they have ramifications for the shape of derivation trees or the string yield (which will be important for Thm 1). Nonetheless I give a full definition of the formalism here, staying close to the chain-based exposition of [19]. After that I formally define MDTLs and introduce the notion of slices.

Definition 2. *A* Minimalist grammar *is a 6-tuple* $G := \langle \Sigma, Feat, F, Types, Lex, Op \rangle$, *where*

- $\Sigma \neq \emptyset$ *is the alphabet,*
- *Feat is the union of a non-empty set* BASE *of* basic features *(also called* category features*) and its prefixed variants* $\{=f \mid f \in \text{BASE}\}$, $\{+f \mid f \in \text{BASE}\}$, $\{-f \mid f \in \text{BASE}\}$ *of* selector, licensor, *and* licensee *features, respectively,*
- $F \subseteq \text{BASE}$ *is a set of final categories,*

- *Types* $:= \{::, :\}$ *distinguishes* lexical *from* derived *expressions,*
- *the lexicon Lex is a finite subset of* $\Sigma^* \times \{::\} \times Feat^*$,
- *and Op is the set of generating functions to be defined below.*

A chain *is a triple in* $\Sigma^* \times Types \times Feat^*$, *and C denotes the set of all chains (whence* $Lex \subset C$*). Non-empty sequences of chains will be referred to as* expressions, *the set of which is called E.*

The set Op of generating functions consists of the operations merge and move, which are the respective unions of the following functions, with $s, t \in \Sigma^*$, $\cdot \in$ *Types*, $f \in$ BASE, $\gamma \in Feat^*$, $\delta \in Feat^+$, *and chains* $\alpha_1, \ldots, \alpha_k$, $\iota_1, \ldots, \iota_k$, $0 \leq k, l$:

$$\frac{s :: =f\gamma \qquad t \cdot f, \iota_1, \ldots, \iota_k}{st : \gamma, \iota_1, \ldots, \iota_k} \; merge1$$

$$\frac{s : =f\gamma, \alpha_1, \ldots, \alpha_k \qquad t \cdot f, \iota_1, \ldots, \iota_l}{ts : \gamma, \alpha_1, \ldots, \alpha_k, \iota_1, \ldots, \iota_l} \; merge2$$

$$\frac{s \cdot =f\gamma, \alpha_1, \ldots, \alpha_k \qquad t \cdot f\delta, \iota_1, \ldots, \iota_l}{s : \gamma, \alpha_1, \ldots, \alpha_k, t : \delta, \iota_1, \ldots, \iota_l} \; merge3$$

$$\frac{s : +f\gamma, \alpha_1, \ldots, \alpha_{i-1}, t : -f, \alpha_{i+1}, \ldots, \alpha_k}{ts : \gamma, \alpha_1, \ldots, \alpha_{i-1}, \alpha_{i+1}, \alpha_k} \; move1$$

$$\frac{s : +f\gamma, \alpha_1, \ldots, \alpha_{i-1}, t : -f\delta, \alpha_{i+1}, \ldots, \alpha_k}{s : \gamma, \alpha_1, \ldots, \alpha_{i-1}, t : \delta, \alpha_{i+1}, \ldots, \alpha_k} \; move2$$

Furthermore, all chains must satisfy the Shortest Move Constraint (SMC), according to which no two chains in the domain of move display the same licensee feature $-f$ *as their first feature. The string language generated by G is* $L(G) := \{\sigma \mid \langle \sigma \cdot c \rangle \in closure(Lex, Op), \cdot \in Types, c \in F\}$.

MGs and MCFGs [15] have the same weak generative capacity [6, 12]. In fact, for every MG there exists a strongly equivalent MCFG, but not the other way round. In a certain sense, then, one may view MGs as a narrowly restricted way of specifying MCFGs. A more peculiar fact about MGs is that in order for $l \in Lex$ to occur in a well-formed derivation, its feature string must be in $\{+f, =f \mid f \in \text{BASE}\}^* \times \text{BASE} \times \{-f \mid f \in \text{BASE}\}^*$. I will implicitly invoke this fact several times throughout the paper.

I now turn to the derivation trees of an MG, defining them in two steps.

Definition 3. *Given an MG* $G := \langle \Sigma, Feat, F, Types, Lex, Op \rangle$, *the largest subset of* T_E *satisfying the following conditions is called the* string-annotated derivation tree language sder(G) *of G:*

- *For every leaf node n,* $\ell(n) = \langle l \rangle$, $l \in Lex$.
- *For every subtree* $m(d_1, \ldots, d_n)$, $n \geq 1$, $op(d_1, \ldots, d_n)$ *is defined for exactly one* $op \in \{merge, move\}$ *and* $\ell(n) = op(d_1, \ldots, d_n)$.
- *For n the root node,* $\ell(n) = \langle \sigma : c \rangle$, *where* $\sigma \in \Sigma^*$ *and* $c \in F$.

Definition 4. *Given an MG G, its* Minimalist derivation tree language mder(G) *is the set of trees obtained from* sder(G) *by the map* μ:

- $\mu(\langle l \rangle) = l$, *where* $l \in Lex$
- $\mu(e(e_1, \ldots, e_n)) = op(\mu(e_1), \ldots, \mu(e_n))$, *where* $e, e_1, \ldots, e_n \in E$, $n \geq 1$, *and* op *is the unique operation in* Op *such that* $op(e_1, \ldots, e_n) = e$

As was first observed in [8], every MDTL is regular. The basic idea is to equip a bottom-up tree automaton with states corresponding to the feature components of the string-annotated derivation trees (there are only finitely many thanks to the SMC) and have its transition rules recast the conditions imposed on Merge and Move by the feature calculus. Interestingly, an MG's set of derived trees — which is not regular — can be obtained from its MDTL in an efficient way using a multi bottom-up tree transducer. This in turn means that MDTLs are the key to capturing MGs by finite-state means.

In [4], the concept of slices is introduced. Intuitively, slice(l) is the derivation tree equivalent of the phrase projected by l in the derived tree (using the standard linguistic notion of projection). That is to say, slices mark the subpart of the derivation that l has control over by virtue of its selector and licensor features (cf. Fig. 1 on page 212). From this perspective a derivation tree language is simply the result of combining slices in all possible ways such that none of conditions imposed on Merge and Move by the feature calculus are violated.

Definition 5. *Given a Minimalist derivation tree* $T := \langle D, \ell \rangle$ *and LI* l *occurring in* T, *the* slice of l *is the pair* slice$(l) := \langle S, \ell \rangle$, *where* $S \subseteq D$, $l \in S$, *and if node* $n \in D$ *immediately dominates a node* $s \in S$, *then* $n \in S$ *iff the operation denoted by* $\ell(n)$ *erased a selector or licensor feature on* l. *The unique* $n \in S$ *that is not dominated by any* $n' \in S$ *is called the* slice root *of* l.

The following properties hold of slices for every MG G [cf. 4]:

- Let $|\gamma|$ denote the length of the longest γ such that there is some $l \in Lex_G$ with feature string $\gamma c \delta$, $\gamma, \delta \in Feat^*$, $c \in$ BASE. Then for every $t \in$ mder(G) and slice $s := \langle S, \ell \rangle$ of t, $1 \leq |S| \leq |\gamma| + 1$.
- Every tree $t \in$ mder(G) is partitioned into slices.
- All slices are continuous (i.e. if node y is the child of x and the mother of z such that x and z belong to the same slice s, then y also belongs to s).

As the order of siblings is irrelevant in derivation trees, I stipulate for the sake of simplicity that all slices are strictly right-branching. Moreover, I assume that derivation trees are strictly binary branching, as this simplifies the math at various points in this paper — a small change which is easily accommodated by mapping $move(t)$ to $move(\natural, t)$, where $\natural$ is a new symbol not in $Lex \cup Op$.

Now given an LI l, node n is the kth node of slice(l) iff the shortest path from n to l is the sequence $\langle n_1, \ldots, n_k, l \rangle$ such that $n = n_1$ and no n_i is a left daughter, $1 \leq i \leq k$. Furthermore, n is associated to feature f iff n is the ith node of slice(l) and f is the ith feature of l. Two features g and h are said to *match* iff there is a feature $f \in$ BASE such that either $g = +f$ and $h = -f$ or $g = =f$ and $h = f$. Finally, a node n matches an LI l iff for some feature g of l, n is associated to a feature matching g.

2.3 Locality and Subclasses of Derivation Tree Languages

If one builds on the notion of slices, a natural way of imposing locality conditions on movement suggests itself. Take any MG G, $t \in \text{mder}(G)$, and suppose that l is an LI occurring in t with licensee features $-f_1, \ldots, -f_n$. Let $move[l] := \langle move_1, \ldots, move_n \rangle$ be the sequence of Move nodes in t such that the operation denoted by $move_i$ checked feature $-f_i$ on l, $1 \leq i \leq n$. An MG G is *k-local* or of *locality rank k*, $k \geq 1$, iff it holds for every $t \in \text{mder}(G)$ and LI l occurring in t with $move[l] := \langle move_1, \ldots, move_n \rangle$ that no more than $k-1$ slices intervene between m_i and m_{i+1} in $\langle m_0, m_1, \ldots, m_n \rangle := \langle l \rangle \cdot move[l]$, $0 \leq i < n$. Equivalently, the shortest path from $move_i$ to $move_{i+1}$ may contain at most k left branches. If G is k-local, we also say that movement in G is k-local or k-bounded. An MG G is *movement-free* iff no tree in mder(G) contains a node labeled *move*. It is *unrestricted* iff it is neither movement-free nor k-local for any $k \geq 1$. This terminology extends to MDTLs in the natural way. By MDTL[*merge*] and MDTL[*merge*, $move^{(k)}$] I denote, respectively, the class of all movement-free MDTLs and the class of all k-local MDTLs. The class of all MDTLs is simply denoted by MDTL[*merge*, *move*].

The notion of k-boundedness is motivated by the linguistic assumption that movement is successive-cyclic. In (3), for instance, the wh-phrase may not move from its base position to the beginning of the utterance in one fell swoop but rather has to land in every CP.

(3) Which professor$_i$ did John say [$_{\text{CP}}$ t_i that Bill told him [$_{\text{CP}}$ t_i that Mary has a crush on t_i]]

This assumption can be used to explain various syntactic, semantic and even morphological phenomena, for instance embedded inversion, stranding of quantifiers, binding ambiguities and complementizer agreement [cf. 3]. The following contrast provides a very simple example.

(4) a. [$_{\text{CP}_2}$ When$_i$ did John t_i tell you [$_{\text{CP}_1}$ who$_j$ Mary will meet t_j?]]
b. * [$_{\text{CP}_2}$ When$_i$ did John tell you [$_{\text{CP}_1}$ who$_j$ Mary will t_i meet t_j?]]

In (4a), *when* originates in the matrix clause and moves directly to the corresponding SpecCP position, while *who* does the same in the embedded clause. In (4b), *when* is supposed to originate in the embedded clause together with *who*, but now they cannot both move into CP-specifiers. In one case, *who* moves first, so that it fills SpecCP$_1$. As a consequence, *when* cannot move to SpecCP$_2$ due to successive cyclicity requiring it to go through the SpecCP$_1$ first, which is already filled by *who*. In the other case, *where* moves first but on its way to SpecCP$_2$ it leaves a trace in SpecCP$_1$ so that the latter is no longer available as a landing site to *who*. The details of the analysis have changed significantly over the years, and recently it has even been argued that CPs do not matter for cyclicity [3], but the basic idea that long-distance movement is in fact a sequence of local movement steps has remained unaltered.

If one ignores adjuncts, the kind of cyclicity involved in the examples above can be reinterpreted as wh-movement being 3-local: every clause consists of the slices CP, TP, VP, so movement from one CP to the next creates paths containing 3 left branches. The analogy between successive-cyclic movement and the restriction to k-locality is far from perfect, of course. Among other things, successive-cyclic movement is relativized to specific positions, whereas k-locality only cares about distance. But if the value of k is carefully chosen and combined with certain regular constraints on derivation trees [4, 7] that prevent movement from skipping, say, CP slices, a reasonably close approximation can be achieved.

Surprisingly, every MG can be translated into a weakly equivalent 1-local one. The idea is that instead of having subtree s move directly to its target position t, s can hitch a ride by being selected by another LI l. As long as the string component of l is empty, this will have no effect on the string yield. This reduces unbounded movement to k-local movement, which in turn can easily be reduced to 1-local movement.

Theorem 1. *For every unrestricted MG G, there is an MG G′ of locality rank* 1 *that defines the same string language.*

Proof. I sketch two linear (bottom-up) transducers τ and τ'. The former does most of the work by translating every $t \in \text{mder}(G)$ into its corresponding $\max(E)$-local t', where $\max(E)$ is the maximum length of expressions generated by G (i.e. the maximum number of chains per expression). The latter then rewrites t' as the 1-local t''. Since MDTLs are regular, and the image of a regular language under a linear transduction is itself regular, the output of $\tau \cdot \tau'$ is, too. This set is then intersected with $\text{mder}(G'')$, where $Lex_{G''} := \Omega_{\tau'} \setminus Op_{G''}$ and $F_{G''} := F_G$. The result of this intersection can then be automatically converted into a new MG, our G' [4, 7].

The idea underlying τ is that unbounded movement can be localized by creating intermediary landing sites which have no effect on the string language. The main task of the transducer is to insert those intermediate landing sites and transfer (a subset of) the features of each moving element to its landing site. By definition there are at most $n = \max(E) - 1$ moving elements at any given point in the derivation. Each state of τ consists of 1) n components q_i that memorize the feature string of the moving items and which of them have already been reinstantiated, 2) the expression the current node evaluates too (modulo the string component), 3) a boolean flag b that indicates whether some LI had the new licensee feature $-s_0$ added to its feature string.

Suppose we are at a node in the derivation tree t that belongs to the slice of LI $l := \sigma :: \gamma c \delta$ and that there are $0 \leq m \leq n$ moving LIs $l_i := \sigma :: \gamma_i c_i \delta_i$, where $1 \leq i \leq m$ and $c, c_i \in \textsc{Base}$ and $\sigma, \sigma_i \in \Sigma$ and δ_i is of the form $-f_{i_1}, \ldots, -f_{i_k}$, $k \geq 1$. For each i and $1 \leq u \leq v \leq k$, we define new LIs $l_i^c[u,v] := \varepsilon :: =c \; +f_{i_u} \; c \; -s_i \; -f_{i_u} \; \delta_i[u{+}1, v]$, where $-s_i$ is a new licensee feature and $\delta_i[u+1, v] := -f_{i_{u+1}}, \ldots, -f_{i_v}$. Given a feature string $\phi := f_1, \ldots, f_k$, we furthermore let $q(j, \phi) := f_1, \ldots, f_j \bullet f_{j+1} \ldots f_k$, $0 \leq j \leq k$.

The transducer τ now has to perform the following steps: First, if l is not itself among the moving elements, τ non-deterministically replaces it by $\hat{l} := \varepsilon :: \gamma c \; -s_0$ and switches the boolean flag b in its state from 0 to 1. If either $m = 0$ and $b = 1$ or $m \geq 1$ and $b = 0$, τ aborts at the slice root of $\hat{l}/l$.

Second, τ replaces each l_i carrying more than one licensee feature by $\hat{l}_i := \sigma_i :: \gamma c_i - f_{i_1}$ and stores $q(0, \delta_i[1,k])$ in q_i. Third, when τ reaches the slice root of $l/\hat{l}$, it inserts $C(l^c_m[u_m, v_m]) \cdot \ldots \cdot C(l^c_1[u_1, v_1])$, where $C(l^c_i[u_i, v_i]) := move(\natural, merge(\Box, l^c_i[u_i, v_i]))$, and u_i must be such that $q(u_i - 1, \delta_i[1,k])$ is the string stored in q_i. The value of v_i is chosen non-deterministically — in order for τ not to abort it must hold that every f_{i_j} can get checked later on without further intermediate movement sites for all $j \leq v_i$ but not for $j = v_i + 1$ (this is easily verified, as τ keeps track of licensor features in the second component of its states). With the insertion of $C(l^c_i[u_i, v_i])$, q_i is updated to $q(v_i, \delta_i[1,k])$. So far then, τ has introduced new slices slice($l^c_i[u_i, v_i]$) that function as the respective landing sites for each l_i, or rather, their impoverished counterpart $\hat{l}_i$. The fourth step requires τ to insert the context $C(s^c[m,b])$ above $C(l_m[u_m, v_m])$, where $s^c[j,b] := \varepsilon :: \ = c \ + s_{1-b} \ldots + s_j \ c$ for $1 \leq j \leq n$, and $C(s^c[j,b]) := \underbrace{move(\natural, \Box)\cdot}_{j+b \text{ times}} merge(\Box, s^c[j,b])$.

This enforces remnant movement of $l/\hat{l}$ and all $\hat{l}_i$, allowing each slice($l^c_i[k]$) to move freely later on without carrying along any other parts of the derived tree (which would induce a change in the string yield). The procedure as outlined above is iterated (with the value of c varying with l) until no more features need to be checked off. Since there are at most $n = \max(E) - 1$ moving elements in G, no LI l_i (including $l/\hat{l}$) has to cross more than n slices in order to check its $-f_{i_1}$ feature against $l^c_i[u,v]$ or its $-s_i$ feature against $s^c[j,b]$. Thus every instance of *move* is $\max(E)$-bounded.

All instances of $(k+1)$-bounded movement, $1 \leq k \leq \max(E)$, can be made 1-local as follows. Assume that slices slice(l_1), ... slice(l_k) intervene between $l := \sigma :: \gamma c \delta$ and its next occurrence. Then τ' has to prefix δ with k new movement licensee features $-l_1 \ldots - l_k$, and for each $l^c_i[u,v]$ $(1 \leq i \leq k)$ add $+l_i$ to the end of its feature string and $move(\natural, \Box)$ above its slice root. If several LIs move through a slice, the number of licensor features and Move nodes has to be adapted accordingly. Crucially, both the number of moving elements and the distance between LIs and individual occurrences is finitely bounded, so this strategy can easily be carried out by a non-deterministic linear transducer. □

3 (Un)Definability in Some Subregular Language Classes

3.1 Deterministic Top-Down Automata

Since MDTLs are regular, they can be recognized by non-deterministic top-down tree automata [cf. 5]. As we will see now, top-down non-determinism can be dispensed with only if it is compensated for by unbounded look-ahead. I consider two common variants of the standard deterministic top-down tree automaton (DTDA), both of which are more powerful than DTDAs but do not recognize all regular languages. One is the *sensing tree automaton* (STA) [11], which may also take the labels of a node's children into account in order to decide which states should be assigned to them, while the other is the l-r-deterministic DTDA (lrDTDA) [13], which allows for a limited kind of non-determinism. The classes of languages recognized by STAs and lrDTDAs are incomparable, but can easily be characterized in descriptive terms. For this reason, I focus on the languages themselves rather than the automata, and no further technical details of the latter will be discussed here (the interested reader is referred to [11] and references therein).

Definition 6. *Given a node v of some Σ-tree t, $\text{lsib}^t(u)$ is the string consisting of the label of u's left sister (if it exists) followed by the label of u, and $\text{rsib}^t(u)$ is the string consisting of the label of u and the label of its right sister (if it exists). Let $u_1, \ldots, u_n$ be the shortest path of nodes extending from the root to v such that u_1 is the root and $u_n = v$. Let ■ and ♣ be two new symbols not in Σ. By $\text{spine}^t(v)$ we denote the string recursively defined by*

$$\text{spine}^t(u_1) = \text{lsib}^t(u_1)\blacksquare\text{rsib}^t(u_1)$$

$$\text{spine}^t(u_1, \ldots, u_n) = \text{spine}^t(u_1, \ldots, u_{n-1}) \clubsuit \text{ lsib}^t(u_n) \blacksquare \text{ rsib}^t(u_n)$$

A regular tree language L is spine-closed *iff it holds for all trees $s, t \in L$ and nodes u and v belonging to s and t, respectively, that $\text{spine}^s(u) = \text{spine}^t(v)$ implies $s[u \leftarrow t|_v] \in L$.*

Definition 7. *A regular tree language L is* homogeneous *iff it holds that if $t[u \leftarrow a(t_1, t_2)] \in L$, $t[u \leftarrow a(s_1, t_2)] \in L$ and $t[u \leftarrow a(t_1, s_2)] \in L$, then also $t[u \leftarrow a(s_1, s_2)] \in L$.*

Proposition 1. *A regular tree language L is recognizable by*

- *an STA iff L is spine-closed [10].*
- *an lrDTDA iff L is homogeneous [13].*

Thanks to these characterizations, results for MDTLs are easily obtained.

Theorem 2. *MDTL[merge] and the class of tree languages recognized by STAs are incomparable.*

Proof. Let grammar G be defined by the following LIs (with names in square brackets for reference) and $F_G := \{a, b\}$:

$[a_0]$ a :: a	$[a_1]$ a :: =a a	$[a_2]$ a :: =a =a a
$[b_0]$ b :: b	$[b_1]$ b :: =b b	$[b_2]$ b :: =b =b b

Consider the derivation tree $t_a := merge_1(merge_2(a_0, a_1), merge_3(a_0, a_2))$ and its counterpart $t_b := merge_1(merge_2(b_0, b_1), merge_3(b_0, b_2))$ — the indices are for the reader's convenience. Even though $\text{spine}^{t_a}(merge_2) = \text{spine}^{t_b}(merge_2)$, it holds that $merge_1(merge_2(b_0, b_1), merge_3(a_0, a_2)) \notin \text{mder}(G)$, so $\text{mder}(G)$ is not spine-closed. □

Theorem 3. *Every $L \in$ MDTL[merge, move] is recognized by some lrDTDA.*

Proof. I show that every MDTL is homogeneous. For $a = move$, closure is trivially satisfied. So let $a = merge$. Merge depends only on the distribution of category and selector features, and there is no way to distribute these over t_1, t_2, s_1 and s_2 such that the fourth tree would be an illicit instance of Merge: In order for Merge to be licensed, one of t_1 or t_2 must have some category feature c as its first unchecked feature, and the other one the matching selector feature $=$c. Assume w.l.o.g. that t_2 carries the selector feature $=$c. Then s_1 must also have feature c, and s_2 feature $=$c. We also know that s_1 and t_1 on the one hand and s_2 and t_2 on the other agree on all features following these selector/licensor features, since the derivations differ only w.r.t. the subtree rooted by a. It follows that Merger of s_1 and s_2 is licit, so the required closure property obtains. □

Martens et al. [11] point out a peculiar property of languages recognized by lrTDAs but not by STAs: in order to determine which states should be assigned to the children of the root, one has to look arbitrarily deep into at least one of the subtrees dominated by the root. This is indeed typical of unrestricted MDTLs, where movement features at the very bottom of a derivation introduce dependencies that — given the impoverished nature of the interior node labels — cannot be predicted deterministically in a top-down fashion without unbounded look-ahead. The class of lrTDAs overshoots the mark, though, as it fails to draw a distinction even between MDTL[*merge*, *move*] and MDTL[*merge*].

3.2 Strictly Local and Locally Threshold Testable Languages

Let us now traverse the subregular hierarchy from the bottom instead, starting with LOC and subsequently moving on to SL and LTT. As lrTDAs before, local sets lack the granularity to distinguish any of the subclasses of MDTLs. But where the lrTDAs universally succeeded, local sets universally fail.

Theorem 4. *MDTL[merge] and the class of local sets are incomparable.*

Proof. Consider any movement-free MDTL L with a derivation containing a subtree of the form $t := merge(l, merge)$. As we require slices to be right-branching, l contains no selector or licensor features. Furthermore, the finiteness of the lexicon establishes an upper bound $|\gamma|+1$ on the size of slices. However, LOC $=$ SL$_2$, so if $L \in$ LOC, t could be composed with itself arbitrarily often, yielding slices of unbounded size. □

The shortcomings of LOC can be circumvented, though, by extending the size of the locality domain, i.e. by moving to SL$_k$ for some sufficiently large $k > 2$. Let $|\delta|$ be the maximum number of licensee features that may occur on a single LI, analogously to $|\gamma|$. Given a k-local MG, set $\kappa := (|\gamma|+1) * (|\delta| * k + 1) + 1$.

Theorem 5. *Every $L \in$ MDTL[merge, move$^{(k)}$] is strictly κ-local.*

Before we may proceed to the actual proof, the notion of *occurrences* must be introduced. Intuitively, the occurrences of an LI l are merely the Move nodes in the derivation tree that operated on one of l's licensee features. It does not take much insight to realize that the first occurrence of l has to be some Move node that dominates it (otherwise l's licensee features could not be operated on) and is not included in slice(l) (no LI may license its own movement). One can even require the first occurrence to be the very first Move node satisfying these properties, thanks to the SMC (the reader might want to reflect on this for a moment). The reasoning is similar for all other occurrences, with the sole exception that closeness is now relativized to the previous occurrence. In more formal terms: Given an LI $l := \sigma :: \gamma c \delta$ with $c \in \textsc{Base}$ and $\delta := -f_1, \ldots, -f_n$, its occurrences occ_i, $1 \leq i \leq n$, are such that

- occ_1 is the first node labeled *move* that matches $-f_1$ and properly dominates the slice root of l.
- occ_i is the first node labeled *move* that matches $-f_i$ and properly dominates occ_{i-1}.

Note that every well-formed MDTL obeys the following two conditions:

- *M1*: For every LI l with $1 \leq n \leq |\delta|$ licensee features, there exist nodes $m_1, \ldots, m_n$ labeled *move* such that m_i is the ith occurrence of l, $1 \leq i \leq n$.
- *M2*: For every node m labeled *move*, there is exactly one LI l such that m is an occurrence of l.

In fact, the implication holds in both directions.

Lemma 1. *For every MG G it holds that if $t \in T_{Lex_G \cup \{merge, move\}}$ is a combination of well-formed slices and respects all constraints on the distribution of Merge nodes, then it is well-formed iff M1 and M2 are satisfied.*

Proof. As just discussed the left-to-right direction poses little challenge. In the other direction, I show that μ^{-1} is well-defined on t and maps it to a well-formed $s \in \text{sder}(G)$. For LIs and Merge nodes, μ^{-1} is well-defined by assumption if it is well-defined for Move nodes. From the definition of *move* and the SMC it follows that $\mu^{-1}(move(\natural, t_2))$ (the expression returned by *move* when applied to $\mu^{-1}(t_2)$) is well-defined only if the root of t_2 is an expression consisting of at least two chains such that 1) its first chain has some feature $+f$ as its first feature and 2) the feature component of exactly one chain begins with $-f$. However, the former follows from the well-formedness of slices, while the latter is enforced by M2; in particular, if the SMC were violated, some Move node would be an occurrence for more than one LI. This establishes that μ^{-1} is well-defined for all nodes. Now $\mu^{-1}(t)$ can be ungrammatical only if the label of the root node contains some licensor or licensee features. The former is ruled out by M2 and the initial assumption that all slices are well-formed, whence every licensor feature is instantiated by a Move node. In the latter case, there must be some licensee feature without an occurrence in t, which is blocked by M1. □

Now we can finally move on to the proof of Thm. 5.

Proof. Given some k-local MG G with $L := \text{mder}(G) \in \text{MDTL}[merge, move^{(k)}]$, let κ-*factors*(L) be the set containing all κ-factors of L, and F the corresponding strictly κ-local language built from these κ-factors. It is obvious that $F \supseteq L$, so one only needs to show that $F \subseteq L$. Trivially, $t \in L$ iff $t \in F$ for all trees t of depth $d \leq \kappa$. For this reason, only trees of size greater than κ will be considered.

Assume towards a contradiction that $F \not\subseteq L$, i.e. there is a t such that $F \ni t \notin L$. Clearly $F \ni t \notin L$ iff some condition enforced by *merge* or *move* on the combination of slices is violated, as the general restrictions on tree geometry (distribution of labels, length and directionality of slices) are always satisfied by virtue of κ always exceeding $|\gamma| + 1$. I now consider all possible cases. In each case, I use the fact that the constraints imposed by *merge* and *move* operate over a domain of bounded size less than κ, so that if $t \in F$ violated one of them, one of its κ-factors would have to exhibit this violation, which is impossible as κ-*factors*$(F) = \kappa$-*factors*(L).

Case 1 [Merge]: *merge* is illicit only if there is an internal node n labeled *merge* in slice(l) such that the shortest path from n to l is of length $1 \leq i$, the ith feature of l is $+f$ for some $f \in$ BASE, and there is no LI l' such that the left daughter of n belongs to slice(l') and l' carries feature f. But the size of every slice is at most $|\gamma| + 1$, so the distance between n and l is at most $|\gamma|$, and that between n and l' at most $|\gamma| + 1$. Hence a factor of size $|\gamma| + 2$ is sufficient, which is less than κ. So if we found such a configuration, it would be part of some $\kappa_i \in \kappa$-*factors*$(F) = \kappa$-*factors*(L). Contradiction.

Case 2 [Move]: Conditions M1 and M2 can be split into three subcases.

Case 2.1 [Too few occurrences]: Assume that LI l has $j \leq |\delta|$ licensee features but only $i < j$ occurrences. Since $L \in$ MDTL[*merge*, *move*$^{(k)}$], the shortest path spanning from any LI to its last occurrence includes nodes from at most $|\delta| * k + 1$ distinct slices. Since the size of no slice minus its LI exceeds $|\gamma|$, some factor κ_i of size greater than $(|\gamma| * |\delta| * k) + (|\gamma| + 1) \leq \kappa$ must exhibit the illicit configuration, yet $\kappa_i \notin \kappa$-*factors*(L).

Case 2.2 [Too many Move nodes]: Assume that for some Move node m there is no LI l such that m is an occurrence of l. This is simply the reverse of Case 2.1, where we obtain a violation if it holds for no LI l in any κ_i that m is one of its occurrences. But then at least one of these κ_i cannot be in κ-*factors*(L).

Case 2.3 [SMC violation]: The SMC is violated whenever there are two distinct items l and l' for which Move node m is an occurrence. As 2.2, this is just a special case of 2.1. □

We now have a very good approximation of MDTL[*merge*, *move*$^{(k)}$] for any choice of $k > 0$. They are not local, or equivalently, strictly 2-locally testable, but they are strictly κ-locally testable, where κ depends on k and the maxima of licensor and licensee features, respectively. But what about MDTL[*merge*, *move*] in general?

To readers acquainted with MGs it will hardly be surprising that unrestricted MDTLs are not strictly locally testable. Nor is it particularly difficult to demonstrate that they even fail to be locally threshold testable. In [1], it was proved that closure under k-guarded swaps is a necessary condition for a language to be definable in $\text{FO}_{mod}[S_1, S_2]$ — that is to say, first-order logic with unary predicates for all labels, binary predicates for the left child and right child relations, respectively, and the ability to perform modulo counting. Evidently FO_{mod} $[S_1, S_2]$ is a proper extension of $\mathrm{FO}[S_1, S_2]$, and definability in the latter fully characterizes the locally threshold testable languages [20]. So no language that isn't closed under k-guarded swaps is locally threshold testable.

Definition 8. *Let $t := C \cdot \Delta_1 \cdot \Delta \cdot \Delta_2 \cdot T$ be the composition of trees $C := C_t[a, x)$, $\Delta_1 := C_t[x, y)$, $\Delta := C_t[y, x')$, $\Delta_2 := C_t[x', y')$ and $T := t|_{y'}$. The* vertical swap *of t between $[x, y)$ and $[x', y')$ is the tree $t' := C \cdot \Delta_2 \cdot \Delta \cdot \Delta_1 \cdot T$. If the subtrees rooted at x and x' are identical up to and including depth k, and the same holds for the subtrees rooted at y and y', then the vertical swap is* k-guarded.

Theorem 6. MDTL[*merge, move*] *and the class of tree languages definable in* $\mathrm{FO}_{mod}[S_1, S_2]$ *are incomparable.*

Proof. Consider a grammar containing (at least) the following four items:

$$\mathrm{a :: a} \quad \mathrm{a :: a - b} \quad \mathrm{a :: =a\ a} \quad \mathrm{a :: =a + b\ a}$$

I restrict my attention to those derivation trees in which movement occurs exactly once. Pick any $k \in \mathbb{N}$. Then there is some derivation tree that can be factored as above such that Δ_1 contains the movement node at some depth $m > k$, Δ_2 contains the corresponding LI a :: a $-$ b at some depth $n > k$, $C = \Delta = T$, and the depth of Δ and T exceeds k. Given this configuration, the vertical swap of Δ_1 and Δ_2 is k-guarded, yet $t' := C \cdot \Delta_2 \cdot \Delta \cdot \Delta_1 \cdot T$ is not a Minimalist derivation tree, as the movement node no longer dominates a :: a $-$ b, thereby negating closure under k-guarded swaps. □

The insufficiency of $\mathrm{FO}_{mod}[S_1, S_2]$ puts a strong lower bound on the complexity of MDTL[*merge, move*]. In the next section, I show that enriching $\mathrm{FO}[S_1, S_2]$ with proper dominance and equivalence is all it takes to make MDTL[*merge, move*] first-order definable.

4 Definability in First-Order Logic

I start with an $\mathrm{FO}[S_1, S_2]$ theory of MDTL[*merge*], which is then extended to $\mathrm{FO}[S_1, S_2, <, \approx]$ for MDTL[*merge, move*]. Given an MG G, $\mathrm{FO}[S_1, S_2]$ is defined over ordered binary branching trees in the standard way, with the signature containing a unary predicate p for each $p \in \Lambda := Lex_G \cup Op_G \cup \{\natural\}$ and binary predicates S_1 and S_2 for the left and right child relation, respectively. The equivalence relation is superfluous for MDTL[*merge*]. I write $x \triangleleft_1 y$ instead of $S_1(x, y)$, and similarly for S_2. Moreover, $x \triangleleft y$ iff $x \triangleleft_1 y \vee x \triangleleft_2 y$.

First a number of constraints are established to ensure that every node has exactly one label drawn from Λ, and that the arity of the labels is respected (it suffices only to restrict nullary symbols to leaves, as this entails that binary symbols can be assigned only to interior nodes). Furthermore, $\natural$ may be assigned to a node if and only if it is the left daughter of a Move node.

$$\forall x \left[\left(\bigvee_{u \in \Lambda} u(x) \right) \wedge \bigwedge_{u \in \Lambda} \left(u(x) \rightarrow \bigwedge_{v \in \Lambda \setminus \{u\}} \neg v(x) \right) \right]$$

$$\forall x\Big[\bigvee_{u \in \Lambda \setminus \{merge, move\}} u(x) \leftrightarrow \neg\exists y[x \triangleleft y]\Big]$$

$$\forall x \forall y[\natural(y) \leftrightarrow move(x) \wedge x \triangleleft_1 y]$$

As was pointed out in Sec. 2, MDTLs can be viewed as the result of combining the slices defined by LIs in all possible ways such that the constraints of the feature calculus are respected. Hence I first define the shape of slices before moving on to the feature conditions enforced by Merge. To simplify this task, I use $\searrow^n \phi(x)$ as a shorthand for "ϕ holds at the node reached from x by taking n steps down the right branch". The analogous $\swarrow^n \phi(x)$ moves us down the left branch instead, while $\nwarrow^n \phi(x)$ moves us upwards only along a right branch. Intuitively, $\searrow$, $\swarrow$ and $\nwarrow$ can be viewed as first-order implementations of modal diamond operators.

$$\searrow^0 \phi(x) \leftrightarrow \phi(x)$$

$$\searrow^n \phi(x) \leftrightarrow \exists y[x \triangleleft_2 y \wedge \searrow^{n-1} \phi(y)]$$

Recall that all slices are strictly right-branching and never exceed size $|\gamma| + 1$. This is equivalent to saying that there is no node that is at least $|\gamma| + 1$ S_2-steps away from a node satisfying a tautology $\top$.

$$\neg\exists x[\searrow^{|\gamma|+1} \top(x)]$$

Next, every interior node n must be licensed by a feature of the LI of the slice containing n. Again a special notational device proves useful: for any feature f, $f_i(x)$ holds iff for some $l \in Lex_G$ whose i^{th} feature is f, $l(x)$ is true (the index will be suppressed whenever the position of the feature is irrelevant). Now let $slr_i(x) \leftrightarrow \bigvee_{f \in \text{BASE}} {=}f_i(x)$ and $lcr_i(x) \leftrightarrow \bigvee_{f \in \text{BASE}} {+}f_i(x)$.

$$\forall x\Big[\Big(merge(x) \rightarrow \bigvee_{1 \le i \le |\gamma|} \searrow^i slr_i(x)\Big) \wedge \Big(move(x) \rightarrow \bigvee_{1 \le i \le |\gamma|} \searrow^i lcr_i(x)\Big)\Big]$$

Besides the evident restriction on the distribution of *merge* and *move*, the formula above also ensures that no $l \in \Lambda$ without selector or licensor features can ever be a right leaf.

We still have to establish a minimum size on slices, though, which is easily accomplished by requiring every selector/licensor feature to license a unique interior node.

$$\forall x\Big[\bigwedge_{1 \le i \le |\gamma|} \Big(\big(slr_i(x) \rightarrow \nwarrow^i merge(x)\big) \wedge \big(lcr_i(x) \rightarrow \nwarrow^i move(x)\big)\Big)\Big]$$

Note that this also prevents every LI with selector or licensor features from occurring on a left branch. The topmost slice in the derivation is also subject to the condition that the category of its LI must be final.

$$\forall x\Big[\neg\exists y[y \triangleleft x] \rightarrow \bigvee_{\substack{c \in F \\ 0 \le i \le |\gamma|}} \searrow^i c(x)\Big]$$

So far, then, our first-order theory enforces the correct minimum/maximum size of slices for every $l \in Lex_G$ and fixes their branching direction and node labels. For MDTL[*merge*], it only remains to capture the feature dependencies imposed by Merge: the category feature of the LI of the slice on the left branch has to match the selector feature of the LI found along the right branch.

$$\forall x \left[merge(x) \rightarrow \bigwedge_{c \in \text{BASE}} \left(\swarrow^1 \bigvee_{0 \leq i \leq |\gamma|} \searrow^i c(x) \leftrightarrow \bigvee_{1 \leq j \leq |\gamma|} \searrow^j = c_j(x) \right) \right]$$

Extending this basis to unrestricted MDTLs is surprisingly easy using the notion of occurrences we encountered earlier on. First, proper dominance and equivalence are added to the signature of FO[S_1, S_2], yielding FO[$S_1, S_2, <, \approx$]. As before, I use infix notation for all binary relations, so instead of $< (x, y)$ I write $x \triangleleft^+ y$. For every $i \leq |\delta|$, $match_i(x, y)$ denotes that x is associated to a feature that matches the ith licensee feature of y.

$$match_i(x,y) \leftrightarrow$$
$$\bigvee_{f \in \text{BASE}} \left(\bigwedge_{\substack{c \in \text{BASE} \\ 1 \leq j \leq |\gamma|+1}} (c_j(y) \rightarrow -f_{j+i}(y)) \wedge move(x) \wedge \bigvee_{1 \leq g \leq |\gamma|} \searrow^g +f_g(x) \right)$$

Furthermore, the predicate $x \blacktriangleleft y \leftrightarrow \exists z, \exists z' [(x \triangleleft^+ z \vee x \approx z) \wedge z \triangleleft_1 z' \wedge (z \triangleleft^+ y \vee z \approx y)]$ holds of x and y iff x properly dominates y and they belong to different slices. Building on these two notions, it is a straightforward task to recast the definition of occurrences in first-order terms.

$$occ_1(x,l) \leftrightarrow match_1(x,l) \wedge x \blacktriangleleft l \wedge \neg \exists y \left[x \triangleleft^+ y \wedge match_1(y,l) \wedge y \blacktriangleleft l \right]$$

$$occ_i(x,l) \leftrightarrow x \triangleleft^+ l \wedge match_i(x,l) \wedge \exists y \Big[x \triangleleft^+ y \wedge occ_{i-1}(y,l) \wedge$$
$$\neg \exists z \left[x \triangleleft^+ z \wedge z \triangleleft^+ y \wedge match_i(z,l) \right] \Big]$$

In line with Lem. 1, constraining the distribution of *move* requires but three formulas that demand, respectively, that every licensee feature has a matching move node, that every mode node has a matching licensee feature, and that no movement node can be matched against more than one licensee feature (SMC). It is only this very last condition that depends on the equivalence predicate as there is no other first-order definable way of distinguishing nodes (the use of equivalence in the definition of $\blacktriangleleft$ is merely a matter of convenience and can easily be avoided).

$$\forall x \left[\bigwedge_{\substack{c \in \text{BASE} \\ 1 \leq i \leq |\gamma|+1}} \left(c_i(x) \rightarrow \bigwedge_{\substack{f \in \text{BASE} \\ 0 \leq j \leq |\delta|}} \left(-f_{i+j}(x) \rightarrow \exists y \left[occ_j(y,x) \right] \right) \right) \right]$$

$$\forall x\Big[move(x) \rightarrow \exists l\big[\bigvee_{1\leq i\leq|\delta|} occ_i(x,l)\big]\Big]$$

$$\forall x \forall l\Big[\bigwedge_{1\leq i\leq|\delta|}\Big(occ_i(x,l) \rightarrow \forall l'\big[\bigwedge_{j\in[|\delta|]\setminus\{0,i\}} \neg occ_j(x,l') \wedge \big(occ_i(x,l') \rightarrow l \approx l'\big)\big]\Big)\Big]$$

5 Conclusion

The results reported herein highlight the rather indirect relation between MDTLs and the string languages they derive. MGs without movement yield context-free string languages, whereas even bounded movement is sufficient to generate all multiple context-free languages. At the level of tree languages, however, both movement-free and k-local MGs are strictly locally testable, whereas unrestricted movement leads to an increase in complexity that pushes MDTLs out of the realm of local threshold testability (see Fig. 2 on the next page).[1]

As my results posit a split between k-local and unrestricted MGs on the level of derivation trees, they seem to vindicate the assumption commonly made by syntacticians that locality restrictions on movement are a fundamental property of natural language that keeps computational complexity in check. On the other hand, weak generative capacity remains unaffected, and the locality rank is immaterial, as all local grammars can be made 1-local. Further work is needed before a full understanding can be reached as to how derivational complexity may interact with string language complexity, what measure of complexity should be used, and how this relates to syntactic proposals.

It must also be pointed out that alternative representations of Minimalist derivation trees could conceivably paint a different picture. Eventually, one would like to have a better understanding as to which aspects of a derivation tree language genuinely reflect the complexity of the derivational machinery underlying the MG formalism and which are just notational quirks. By probing different formats for Minimalist derivation trees we might also unearth new connections between MGs and Tree Adjoining Grammar, an area that has recently enjoyed increased interest.

[1] The strictly local nature of movement-free and k-local MDTLs also implies that they can be recognized by deterministic tree-walking automata. I conjecture that this does not carry over to unrestricted MDTLs unless the automata are enriched with two weak pebbles. In particular, non-deterministic tree-walking automata cannot recognize unrestricted MDTLs: The fundamental problem one faces while sifting through a derivation tree with unrestricted movement in a sequential manner is that either 1) the automaton has to keep track of an unbounded number of features when performing a brute-force search for an LI matching a given movement node, or 2) it gets lost in the derivation tree and cannot make its way back to the movement node in question. This makes it impossible to ensure that every movement node is an occurrence for exactly one LI, and non-determinism offers no remedy. The addition of two pebbles, on the other hand, allows the automaton to mark the movement node and the LI that was inspected last, so that the automaton can always find its way back and can infer from the position of the second pebble which LIs have already been looked at.

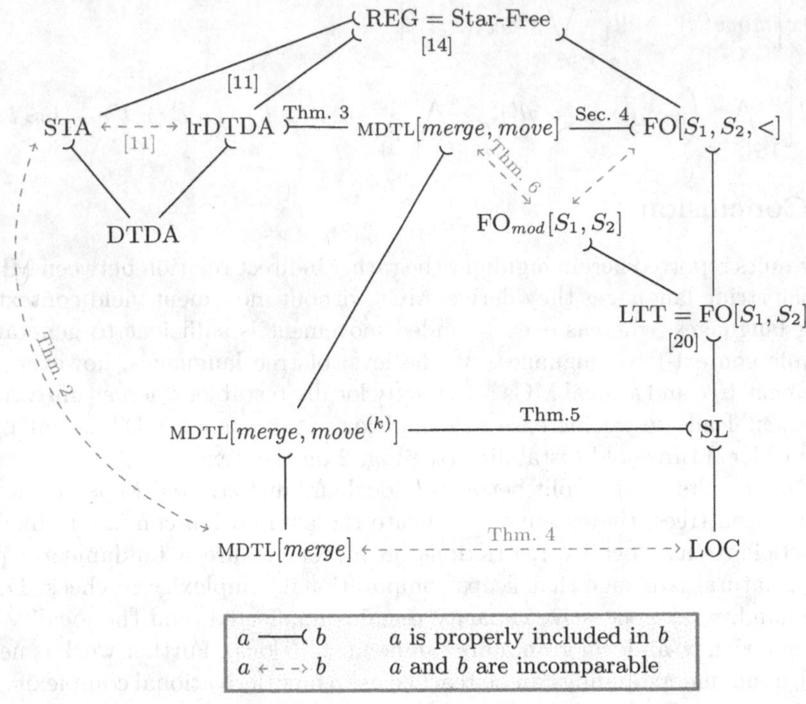

Fig. 2. MDTLs in the subregular space of strictly binary branching tree languages (references omitted for obvious relations)

Acknowledgments. My thanks go to Ed Stabler and the three anonymous reviewers for their helpful criticism. The research reported herein was supported by a DOC-fellowship of the Austrian Academy of Sciences.

References

[1] Benedikt, M., Segoufin, L.: Regular tree languages definable in FO and in FOmod. ACM Transactions in Computational Logic 11, 1–32 (2009)

[2] Chomsky, N.: The Minimalist Program. MIT Press, Cambridge (1995)

[3] den Dikken, M.: Arguments for successive-cyclic movement through SpecCP. A critical review. Linguistic Variation Yearbook 9, 89–126 (2009)

[4] Graf, T.: Closure properties of minimalist derivation tree languages. In: Pogodalla, S., Prost, J.-P. (eds.) LACL 2011. Lecture Notes in Computer Science (LNAI), vol. 6736, pp. 96–111. Springer, Heidelberg (2011)

[5] Gécseg, F., Steinby, M.: Tree Automata. Academei Kaido, Budapest (1984)

[6] Harkema, H.: A Characterization of Minimalist Languages. In: de Groote, P., Morrill, G., Retoré, C. (eds.) LACL 2001. LNCS (LNAI), vol. 2099, pp. 193–211. Springer, Heidelberg (2001)

[7] Kobele, G.M.: Minimalist Tree Languages Are Closed Under Intersection with Recognizable Tree Languages. In: Pogodalla, S., Prost, J.-P. (eds.) LACL 2011. LNCS, vol. 6736, pp. 129–144. Springer, Heidelberg (2011)

[8] Kobele, G.M., Retoré, C., Salvati, S.: An automata-theoretic approach to minimalism. In: Rogers, J., Kepser, S. (eds.) Model Theoretic Syntax at 10, pp. 71–80 (2007)

[9] Mainguy, T.: A probabilistic top-down parser for Minimalist grammars (2010), arXiv:1010.1826v1

[10] Martens, W.: Static Analysis of XML Transformation- and Schema Languages. Ph.D. thesis, Hasselt University (2006)

[11] Martens, W., Neven, F., Schwentick, T.: Deterministic top-down tree automata: Past, present, and future. In: Proceedings of Logic and Automata, pp. 505–530 (2008)

[12] Michaelis, J.: Transforming linear context-free rewriting systems into minimalist grammars. In: de Groote, P., Morrill, G., Retoré, C. (eds.) LACL 2001. LNCS (LNAI), vol. 2099, pp. 228–244. Springer, Heidelberg (2001)

[13] Nivat, M., Podelski, A.: Minimal ascending and descending tree automata. SIAM Journal on Computing 26, 39–58 (1997)

[14] Potthoff, A., Thomas, W.: Regular tree languages without unary symbols are star-free. In: Proceedings of the 9th International Symposium on Fundamentals of Computation Theory, pp. 396–405 (1993)

[15] Seki, H., Matsumura, T., Fujii, M., Kasami, T.: On multiple context-free grammars. Theoretical Computer Science 88, 191–229 (1991)

[16] Stabler, E.P.: Derivational minimalism. In: Retoré, C. (ed.) LACL 1996. LNCS (LNAI), vol. 1328, pp. 68–95. Springer, Heidelberg (1997)

[17] Stabler, E.P.: Computational perspectives on minimalism. In: Boeckx, C. (ed.) Oxford Handbook of Linguistic Minimalism, pp. 617–643. Oxford University Press, Oxford (2011)

[18] Stabler, E.P.: Top-down recognizers for MCFGs and MGs. In: Workshop on Cognitive Modeling and Computational Linguistics, pp. 39–48. ACL, Portland (2011)

[19] Stabler, E.P., Keenan, E.: Structural similarity. Theoretical Computer Science 293, 345–363 (2003)

[20] Thomas, W.: Languages, automata and logic. In: Rozenberg, G., Salomaa, A. (eds.) Handbook of Formal Languages, vol. 3, pp. 389–455. Springer, New York (1997)

[21] Verdú-Mas, J.L., Carrasco, R.C., Calera-Rubio, J.: Parsing with probabilistic strictly locally testable tree languages. IEEE Transactions on Pattern Analysis and Machine Intelligence 27, 1040–1050 (2005)

Building a Formal Grammar for a Polysynthetic Language

Petr Homola*

Codesign, s.r.o.
phomola@codesign.cz

Abstract. We present the results of a project of building a lexical-functional grammar of Aymara, an Amerindian language. There was almost no research on Aymara in computational linguistics to date. The goal of the project is twofold: First, we want to provide a formal description of the language. Second, NLP resources (lexicon and grammar) are being developed that could be used in machine translation and other NLP tasks. The paper presents formal description of selected properties of Aymara which are uncommon in well-researched Western languages. Furthermore, we present an abstract linguistic representation in the LFG framework which is less language specific than f-structures.

1 Introduction

Aymara is an Amerindian language spoken in Bolivia, Chile and Peru by approx. two million people. It is a polysynthetic language that has many lexical and structural similarities with Quechua but the often suggested genetic relationship between these languages is still disputed.

The only research on Aymara in the field of computational linguistics we know about is the project described in [2].[1] The presented project uses Lexical-Functional Grammar (LFG) [15, 3] to formally describe the lexicon, morphology and syntax of Aymara in a manner suitable for natural language processing (NLP). The grammar we have implemented is capable of parsing complex sentences with embedded clauses.

Aymara is a polysynthetic language with a very complicated system of polypersonal agreement (see Section 3.5 for a brief description). A rare property of words in Aymara is the so-called vowel elision (sometimes called 'subtractive morphology') which is quite hard to describe formally. We show how vowel elision can be dealt with in the lexicon.

The paper is organized as follows: Section 2 is a brief introduction to LFG. Section 3 presents selected properties of Aymara, many of them absent from well-researched languages such as English, and their formal analysis in LFG. Section 4 introduces a dependency-based abstraction of f-structures which brings formal grammars closer cross-linguistically. Finally, we conclude in Section 5 and give an outlook for further research.

* I am very indebted to the anonymous reviewers for their valuable comments.

[1] There is also the system Atamiri [8, 9, 10] which uses Aymara as internal representation of translated sentences.

P. de Groote and M.-J. Nederhof (Eds.): Formal Grammar 2010/2011, LNCS 7395, pp. 228–242, 2012.

2 Lexical Functional Grammar

LFG is a linguistic formalism suitable for theoretical linguistics as well as NLP. An LFG grammar consists of a lexicon and a set of context-free phrase structure rules that are annotated with functional constraints. The lexicon deals with the morphology of languages (which is particularly important for languages with rich inflection) whereas phrase structure rules deal with syntax.

For example, the English sentence *The dog chases a cat*, given the rules and lexical entries in (1), would yield the c(constituent)-structure in (2).

(1)
S → NP VP
VP → V NP
NP → D N
D → *the* | *a*
N → *dog* | *cat*
V → *chases*

(2)
```
              S
         /         \
       NP           VP
      /  \        /    \
     D    N      V      NP
     |    |      |     /  \
    the  dog  chases  D    N
                      |    |
                      a   cat
```

After having added functional annotations to the rules (illustrated in (3)[2]), we get the f(unctional)-structure in (4).

(3)
S → NP (↑ SUBJ) = ↓ VP ↑=↓
VP → V ↑=↓ NP (↑ OBJ) = ↓
NP → D (↑ SPEC) = ↓ N ↑=↓

(4)
$$\begin{bmatrix} \text{PRED} & \text{`chase}\langle(\uparrow\text{SUBJ})(\uparrow\text{OBJ})\rangle\text{'} \\ \text{TENSE} & \textsc{pres} \\ \text{SUBJ} & \begin{bmatrix} \text{PRED} & \text{`dog'} \\ \text{SPEC} & \begin{bmatrix} \text{DEF} & + \end{bmatrix} \end{bmatrix} \\ \text{OBJ} & \begin{bmatrix} \text{PRED} & \text{`cat'} \\ \text{SPEC} & \begin{bmatrix} \text{DEF} & - \end{bmatrix} \end{bmatrix} \end{bmatrix}$$

[2] The symbol ↑ designates the f-structures associated with the mother node in the c-structure and ↓ designates the f-structure associated with the current node (on the right-hand side of the rule).

It should be noted that while f-structures are somewhat universal across languages, c-structures are language specific (since they encode synsemantic words, word order and inner structure of phrases). In Aymara, which does not have a VP, (5)[3] would have the c-structure given in (6) but its f-structure would be the same as the English one in (4).

(5) *Anux* *phis-w* *kat-u*
dog-TOP cat-ELI,FOC chase-NFUT$3\rightarrow3$
"The dog chases a cat."

(6)

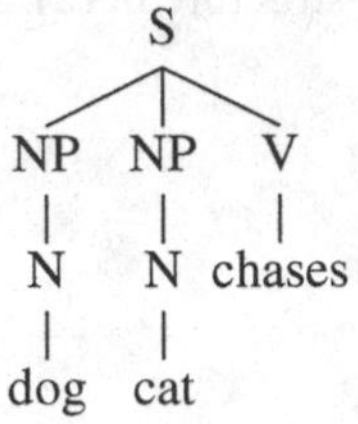

```
           S
      /    |    \
    NP    NP     V
    |      |     |
    N      N   chases
    |      |
   dog    cat
```

As can be seen, the c-structure in (6) is flat. Moreover, Aymara does not have articles, but there are discourse markers (the suffixes *-x* for topic and *-w* for focus in (5)) that are part of the word, hence their function is encoded in the lexicon (see Section 3.10 for examples).

We agree with [20] that c-structures represent the process of syntactic derivation whereas f-structures (which roughly correspond to dependency trees in depedency-based grammars, see Section 4) are the result of this derivation. As has been suggested by [11], at least for some languages, phrase structures encode only word order (at the clause level).

A very good description of various practical problems associated with writing a formal grammar is [5].

3 Some Properties of Aymara

In this section, we focus on some properties of Aymara at the level of morphology and syntax which are mostly absent from Western languages such as English, and sketch their analysis in LFG. A detailed description of the language can be found in [12, 1, 6, 4].

3.1 Agglutinative Morphology

Aymara has a very rich inflection. Suffixes of various categories can be chained to build up long words that would be expressed by a sentence in languages like English. For example, *alanxarusksmawa* (*ala-ni-xaru-si-ka-sma-wa*) means "I am preparing myself to go and buy it for you".

In concordance with the principle of lexical integrity [3], we deal with morphology in the lexicon. [14] has suggested to use word-internal (sublexical) rules to analyze structurally complex words in agglutinative languages. We have adopted this analysis.

[3] In the glosses, NFUT$3\rightarrow3$ means non-future tense. The numbers express the person of the subject and an additional argument, mostly object.

3.2 Vowel Elision

Aymara uses vowel elision as morphosyntacic marking. As illustrated in (7) and (8), there are minimal pairs that make phrases differ syntactically and semantically.

(7) *aycha manq'a-ni*
meat eater
"who eats much meat"

(8) *aych manq'a-ni*
meat-ELI eat-FUT$3{\rightarrow}3$
"(s)he will eat meat"

There are three types of vowel elision that interact with each other. *Object elision* marks a noun or pronoun as direct object, such as in (9) (as opposed to (10)).

(9) *khit-s uñj-i*
whom-ELI,FOC see-NFUT$3{\rightarrow}3$
"Whom does he/she see?"

(10) *khiti-s uñj-i*
who see-NFUT$3{\rightarrow}3$
"Who does see him/her?"

Noun compound elision occurrs in NPs. The final vowel of noun attributes gets elided if they have three or more syllables, as illustrated in (11) and (12).

(11) *aymar aru* (vs. **aymara aru)*
Aymara-ELI language
"the Aymara language"

(12) *qala uta* (vs. **qal uta)*
stone house
"stone house"

Complement elision is applied to all words that are arguments or adjuncts of a verb except for the final word of a clause, as in (13).[4]

(13) *ut sara-sk-ta* (vs. **uta saraskta)*
house-ELI go-PRG-NFUT$3{\rightarrow}3$
"stone house"

Whereas object elision concerns the nucleus of a word (the stem with an optional possessive and/or plural suffix), noun compound and complement elisions concern the final vowel of a word (the vowel of the last suffix or the stem if there are no suffixes). Vowel elision is dealt with in the lexicon. As for noun compound elision, all nouns with more than two syllables get (↑ COMPEL) = + if the final vowel of the word nucleus is elided and (↑ COMPEL) = − if it is not. Nouns with two vowels do not define this attribute, i.e., it can be unified with both values. The corresponding rule for compound nouns is given in (14).

[4] Object and noun compound elision has the gloss ELI in examples.

$$\text{(14)} \quad \begin{array}{lll} N' \rightarrow & (N') & N \\ & (\uparrow \text{MOD}) = \downarrow & \uparrow = \downarrow \\ & (\downarrow \text{COMPEL}) = + & \end{array}$$

3.3 Differential Object Marking

In Aymara, animate and inanimate direct objects are marked differently. Animate objects get the allative suffix (in other cases, the allative has the function of the dative which is a common case of grammaticalization, cf. [13]).[5] For example, the object in (15) is marked whereas the object in (16) is unmarked:

(15) *jila-ma-r* *uñj-ta*
brother-POSS2-ALL see-NFUT$_{1\rightarrow 3}$
"I see/saw your brother."

(16) *uta-m* *uñj-ta*
house-POSS2,ELI see-NFUT$_{1\rightarrow 3}$
"I see/saw your house."

The corresponding rules for direct objects are presented in (17).[6] Note that the first rule can apply to both animate and inamimate objects since differential object marking is optional.

$$\text{(17)} \quad \begin{array}{lll} VP \rightarrow & V \;, & NP \\ & \uparrow = \downarrow & (\uparrow \text{OBJ}) = \downarrow \\ & & (\downarrow \text{OBJEL}) = + \\ & & (\downarrow \text{CASE}) = - \\ VP \rightarrow & V \;, & NP \\ & \uparrow = \downarrow & (\uparrow \text{OBJ}) = \downarrow \\ & & (\downarrow \text{ANIM}) = + \\ & & (\downarrow \text{CASE}) = \text{ALL} \end{array}$$

3.4 Case Stacking

Case stacking occurrs with coordinated nouns, as in (18), where both nouns have two case suffixes: comitative and allative.

(18) *jila-ma-mpi-r* *kullaka-ma-mpi-ru*
brother-POSS2-COM-ALL sister-POSS2-COM-ALL
"to your brother and sister"

In our grammar, we deal with case stacking in the lexicon (such a wordform is assigned the allative case but it is also marked as a member of NP coordination). As with differential object marking, the use of the comitative suffix *-mpi* is not obligatory.

[5] Differential object marking in Aymara is widespread but not obligatory.

[6] The comma means that the order of V and NP is not significant.

3.5 Polypersonal Agreement

Being a polysynthetic language, Aymara has polypersonal conjugation, i.e., the finite verb agrees with the subject and with another argument which may be the object (direct or indirect) or an oblique argument. An example is given in (19).

(19) *Uñj-sma*
see-NFUT$_{1\rightarrow 2}$
"I see/saw you."

The morpholexical entry for *uñjsma* is given in (20).[7] Note that the PRED value for both subject and object is optional.[8]

(20)
uñjsma V (↑PRED) ='uñjaña⟨(↑SUBJ)(↑OBJ)⟩'
(↑TAM TENSE) = NON-FUT
(↑TAM MOOD) = INDIC
((↑SUBJ PRED) = 'PRO')
(↑SUBJ PERS) = 1
((↑OBJ PRED) = 'PRO')
(↑OBJ PERS) = 2

The verb agrees with the subject and with the most animate argument which may be a patient, addressee or source, e.g., *um chur-äma*-FUT$_{1\rightarrow 2}$ "I will give you water" (addresse), *aych al-äma*-FUT$_{1\rightarrow 2}$ "I will buy meat from you" (source) etc. However, there are verbal suffixes which can make the verb agree with other arguments, such as the beneficiary, e.g., *jupa-r aych chura-rap-itäta*-BEN-FUT$_{2\rightarrow 1}$ "You will give him bread for me" (the verb agrees with the beneficiary instead of the addressee *jupa-r*-ALL "him"). All these agreement rules are encoded in the lexicon.

3.6 Causatives

Causative constructions are analyzed as biclausal in our grammar because the causative suffix *-ya* can be used recursively, as illustrated in (21).

(21) *yat-ta,* *yati-y-ta,*
know-NFUT$_{2\rightarrow 3}$ inform-CAUS-NFUT$_{2\rightarrow 3}$
yati-ya-y-ta
make-to-inform-CAUS-CAUS-NFUT$_{2\rightarrow 3}$
"you know/knew (it), you inform(ed) (someone about something), you make/made (someone$_i$) inform (somebody$_j$ about something)"

The causee has the comitative case if the verb is transitive, as illustrated in (22).

(22) *Naya-x Mariya-mp Juwanti-r lich chura-y-ä.*
I-TOP Maria-COM Juan-ALL milk-ELI give-CAUS-NFUT$_{1\rightarrow 3}$
"I will make Maria give milk to Juan."

[7] TAM means Tense-Aspect-Mood.
[8] Both arguments can be dropped.

Causative verbs are analyzed by a sublexical grammar. Hence the sentence in (22) has the f-structure given in (23).

(23)
$$\begin{bmatrix} \text{PRED} & \text{`CAUS}\langle(\uparrow\text{SUBJ})(\uparrow\text{OBJ})(\uparrow\text{XCOMP})\rangle\text{'} \\ \text{TENSE} & \text{NFUT} \\ \text{SUBJ} & \begin{bmatrix}\text{"nayax"}\end{bmatrix} \\ \text{OBJ} & \boxed{1}\begin{bmatrix}\text{"Mariya"}\end{bmatrix} \\ \text{XCOMP} & \begin{bmatrix} \text{PRED} & \text{`churaña}\langle(\uparrow\text{SUBJ})(\uparrow\text{OBJ})(\uparrow\text{OBL})\rangle\text{'} \\ \text{SUBJ} & \boxed{1} \\ \text{OBJ} & \begin{bmatrix}\text{"lichi"}\end{bmatrix} \\ \text{OBL} & \begin{bmatrix}\text{"Juwanti"}\end{bmatrix} \end{bmatrix} \end{bmatrix}$$

3.7 Converbs

Converbs are non-finite verb forms which express a secondary process as an adjunct of the process expressed by the main (finite) verb. In Aymara, converbs usually have the suffix *-sina* or *-sa*, as illustrated in (24) and (25).[9]

(24) *Uta-r* *juta-sin* *phay-i.*
house-ALL come-CONV cook-NFUT$3{\rightarrow}3$
"After coming home, (s)he cooked."

(25) *Jacha-sa-x* *sarx-i.*
cry-CONV-TOP leave-NFUT$3{\rightarrow}3$
"While crying, (s)he left."

The subject of the converb is usually the subject of the main verb, so the f-structure of the converb is an open adjunct (XADJ) in the f-structure of the main verb and the corresponding rule contains the constraint (↑ SUBJ) = (↑ XADJ SUBJ).[10]

3.8 Free Word Order

At the clause level, the word order in Aymara is not restricted although SOV is preferred. There is also no evidence for a VP, thus we assume a flat phrase structure. The rule for matrix clauses is given in (26).[11]

---

[9] There are other converb suffixes such as *-ipana*, but we ignore them in our grammar for now as they are rarely used in the dialect of La Paz which we are focusing on.

[10] Their can be more than one converb in an f-structure. In LFG, (X)ADJs are sets.

[11] In the functional annotation, κ is either '–' (no case) or a semantic case and GF is the corresponding grammatical function. Note that there may be a complementizer at the beginning or at the end of the clause. At most one complementizer can be present due to the LFG uniqueness condition (otherwise there would be conflicting PRED values).

$$\text{(26)}\quad \text{S} \rightarrow \underset{\uparrow=\downarrow}{(\text{C})}\ \mathcal{C}^{+}\ \underset{\uparrow=\downarrow}{(\text{C})}$$

$$\text{where } \mathcal{C} \text{ is } \underset{\uparrow=\downarrow}{\text{V}} \vee \underset{(\downarrow \text{CASE}) = \kappa \Rightarrow (\uparrow \text{GF}) = \downarrow}{\text{NP} \mid \text{S}}$$

As can be seen, word order in a clause is free with the exception of an optional complementizer (see (27) and (28)) which can be placed at the beginning of the clause or at its end.

(27) *Ukat jut-i*
then come-NFUT$_{3\rightarrow3}$
"Then (s)he came."

(28) *Jut-ät ukaxa...*
come-FUT$_{2\rightarrow3}$ if
"If you will come..."

There are no discontinuous constituents and complement clauses can be embedded in the matrix sentence. Since Aymara is not discourse-configurational (see the next subsection), the word order, despite of being free, is usually unmarked (SOV) and if it is different then mostly for stylistic reasons.

3.9 Relative and Complement Clauses

Relative and complement clauses are formally almost identical so we describe them together. There are two types of them: Finite and non-finite (with a nominalized verb). Non-finite verbs agree only with the subject (unlike finite verbs, see Section 3.5 above). A relative clause is illustrated in (29).[12]

(29) *Qillqa-ña-j liwr chur-äma*
write-NREL-1POSS book-ELI give-NFUT$_{1\rightarrow2}$
"I will give you the book which I will write."

Relative and complement clauses begin usually with the subject (which is optional) in locative and end with the verb which is marked for tense (realized vs. non-realized event) and person of the subject, as desrcribed by the rule in (30).[13]

$$\text{(30)}\quad \text{S} \rightarrow \underset{\substack{(\uparrow \text{SUBJ}) = \downarrow \\ (\downarrow \text{CASE}) = \text{LOC}}}{(\text{NP})}\ \ \mathcal{C}^{*}\ \ \underset{\substack{\uparrow=\downarrow \\ (\uparrow \text{PERS}) = (\uparrow \text{SUBJ PERS})}}{\text{V}}$$

$$\text{where } \mathcal{C} \text{ is } \underset{\uparrow=\downarrow}{\text{V}} \vee \underset{(\downarrow \text{CASE}) = \kappa \Rightarrow (\uparrow \text{GF}) = \downarrow}{\text{NP} \mid \text{S}}$$

[12] NREL designates a nominalized verb which expresses a non-realized (future) event (marked with the suffix *-ña*).

[13] The category S is used because a standalone clause with a nominalized verb expresses the obligative mood.

Thus (31) has the c-structure shown in (32). Note that the complement clause is a COMP (i.e., it has its own SUBJ) in the f-structure of the main verb *yattwa* "I know".

(31) *Naya-x* *Mariya-n* *Chukiawu-r* *kuti-ta-p* *yat-t-wa*
I-TOP Maria-LOC La Paz-ALL return-REL-3POSS know-NFUT$_{1\to3}$-FOC
"I know that Maria came back to La Paz."

(32)

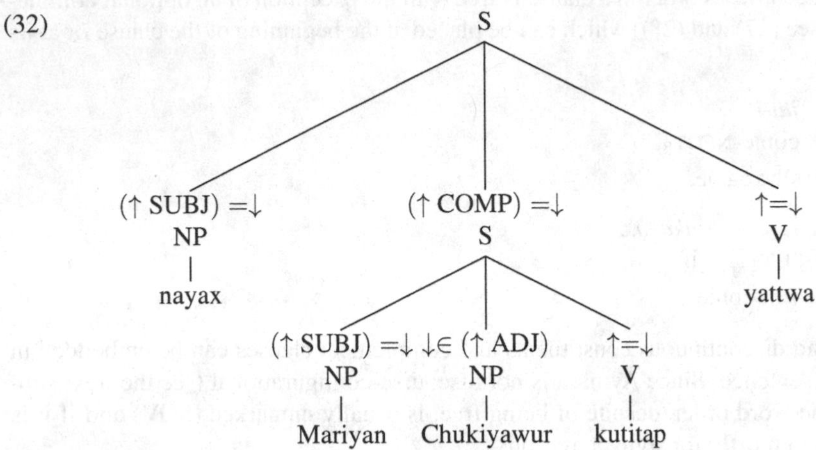

Finite relative and complement clauses end respectively with the complementizer *uka* or *uk*,[14] as in (33).

(33) *Mariya-x* *utj-k-i* *uka* *uta-r* *sara-ni*
Maria-TOP live-SUB-NFUT$_{3\to3}$ COMPL house-ALL go-NFUT$_{3\to3}$,FOC
"(S)he will go to the house where Maria lives."

3.10 Topic-Focus Articulation

We have adopted the approach proposed by [18]. Thus we use an i(nformation)-structure to approximate topic-focus articulation (TFA).[15]

A simple example of two sentences which differ only in TFA is given in (34) (the word *qullqirï* is a verbalized noun).

(34) *Juma-x* *qillq-irï-ta-wa*
you-SG,TOP be-a-writer-AG-NFUT$_{2\to3}$-FOC
"You are a writer."

Juma-w *qillq-irï-ta-xa*
you-SG,FOC be-a-writer-AG-NFUT$_{2\to3}$-TOP
"It is you who is the writer."

[14] Additionally, the verb is marked as subordinate (SUB).

[15] The difference is that we use only two discourse functions, TOP and FOC, with the possibility for words being discourse-unspecified (the term 'discourse-neutral' is used sometimes). This is exactly how morphological marking of TFA works in Aymara.

The morpholexical entries for *jumax* and *jupaw* and corresponding i-structures for the sentences in (34) are given in (35) and (36), respectively.[16]

(35) *jumax* PRON $(\uparrow\text{PRED}) = \text{`PRO'}$
$(\uparrow\text{PERS}) = 2$
$(\uparrow\text{PRED FN}) \in (\uparrow_i\text{TOP})$

$$\begin{bmatrix} \text{TOP} & \left\{\text{`jumax'}\right\} \\ \text{FOC} & \left\{\text{`qillqiri'}\right\} \end{bmatrix}$$

(36) *jumaw* PRON $(\uparrow\text{PRED}) = \text{`PRO'}$
$(\uparrow\text{PERS}) = 2$
$(\uparrow\text{PRED FN}) \in (\uparrow_i\text{FOC})$

$$\begin{bmatrix} \text{TOP} & \left\{\text{`qillqiri'}\right\} \\ \text{FOC} & \left\{\text{`jumax'}\right\} \end{bmatrix}$$

The i-structure is very important for correct translation. For example, the sentence *Chachax liwrw liyi* would be translated as "The man read(s) a book" whereas *Chachaw liwrx liyi* would be better translated as "The book is/was read by a man".[17]

4 Lexical Mapping Theory and D-Structures

Although f-structures abstract to some extent from language specific features (such as differential object marking, see (37) where the Spanish dative phrase and the Polish genitive phrase would be in accusative in German), there are still many differences even between relatively closely related languages.[18]

(37) *Ayer* *visité* *a* *Juan*
yesterday visit-PAST,1SG to Juan
"I visited Juan yesterday."

Nie *mam* *samochodu*
NEG have-PRES,1SG car-SG,GEN
"I don't have a car."

[16] According to a LFG convention, FN represents the lemma of the PRED value (the subcategorization information is omitted).

[17] Unlike some other languages with morphological topic and/or focus markers, such as Japanese (cf. examples from [19]: *Taroo-wa*-TOP *sono hon-o*-ACC *yondeiru* "Taroo is reading that book." vs. *Sono hon-wa*-TOP *Taroo-ga*-NOM *yondeiru* "That book, Taroo is reading"), Aymara allows their co-occurrence with case suffixes without limitation.

[18] For example, the East Baltic language Latvian has only agent-less passives (i.e., in LFG, it completely lacks OBL_{ag}, cf. [7]), whereas its closest and partially mutually intelligible relative Lithuanian has and frequently uses agents in passives.

[16] have suggested a method of translating f-structures between languages.[19] However, their approach has been heavily criticized [21, 22]. [24] examine the use of a(rgument)-structures in machine translation (MT). In LFG, a-structures are another level of linguistic representation which provides the lexico-syntactic interface. The mapping between a-structures and f-structures is defined by the so-called Lexical Mapping Theory (LMT; see [3]). We will give a brief overview of LMT here.

LFG assumes that there is a prominence hierarchy of semantic roles. We use the hierarchy shown in (38) (proposed by [3]):

(38) agent $\succ$ beneficiary $\succ$ experiencer/goal $\succ$ instrument $\succ$ patient/theme $\succ$ locative

Argument grammatical functions (GF) are assigned features *objective* and *restricted* as in (39). The markedness hierarchy of GFs is given in (40).

(39)

	-r	+r
-o	SUBJ	OBL_θ
+o	OBJ	OBJ_θ

(40) SUBJ $\succ$ OBJ, $OBL_\theta \succ OBJ_\theta$

Verbs in LFG have an a-structure that expresses their valence. The arguments of each verb are ordered according to the hierarchy in (38) and annotated with *-o, -r, +o, +r*. General LMT principles determine how the arguments are mapped onto GFs. The initial role is mapped onto SUBJ if classified with $[-o]$. Otherwise, the leftmost role classified $[-r]$ is mapped onto SUBJ. Other roles are mapped onto the lowest compatible GF according to the hierarchy in (40). There are two other constraints: Every verb must have a SUBJ and each role must be associated with a unique function, and conversely.

For example, the verb *pound* would have the a-structure and mapping shown in (41).

(41)

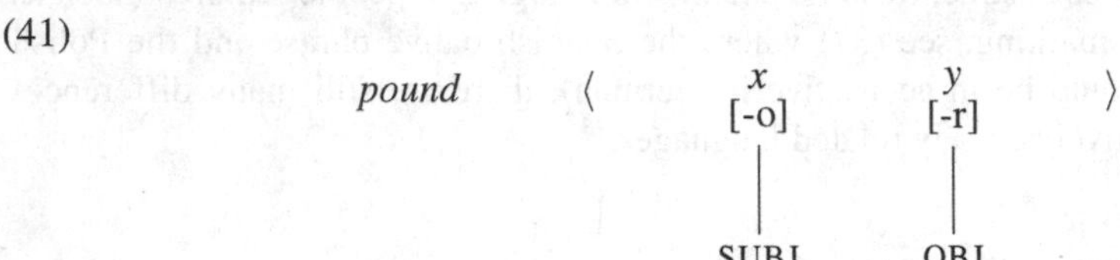

[3] argues that LMT allows for natural treatment of passives, ditransitives and other constructions which have been handled by lexical rules in earlier version of LFG.

We use the information provided by f-structures, i-structures, c-structures and a-structures to create a dependency-based representation of parsed sentences (a tectogrammatical tree in the terminology of [23]). The main reason is that we already have a module that generates English and Spanish sentences from (tectogrammatical) syntax trees. Furthermore, [25] present promising results of MT using tectogrammatics.

In the following, we will use the term d(ependency)-structure to refer to dependency trees. Table 1 gives a brief overview of which information at different levels of linguistic representation in LFG is used in d-structures.

[19] The idea is to parse the source sentence, adapt the f-structure to the target language and generate the target sentence. [17] have shown that LFG generation produces context-free languages.

Table 1. Information provided by LFG layers to d-structures

LFG layer	**information reflected in d-structures**
c-structure	original word order
f-structure	dependencies and coreferences between phrases
i-structure	topic-focus articulation
a-structure	valence (semantic roles and their mapping to GFs)

D-structures do not carry any additional information except for the data already present at the four levels given in Table 1 but they are less language specific. The skeleton of a d-structure is provided by the f-structure. According to a generally accepted principle of deep syntax (tectogrammatics) only autosemantic (content) words are represented by nodes in d-structures. In LFG, autosemantic words are associated with projections of lexical categories, i.e., f-structures with the PRED attribute (see [3] for a detailed discussion of lexical and functional categories and the so-called 'coheads'). Thus a d-structure derived from (4) (repeated here as (42) for convenience) would have three nodes for the words *dog, chases* and *cat*.

(42)
$$\left[\begin{array}{ll} \text{PRED} & \text{`chase}\langle(\uparrow\text{SUBJ})(\uparrow\text{OBJ})\rangle\text{'} \\ \text{TENSE} & \text{PRES} \\ \text{SUBJ} & \left[\begin{array}{ll} \text{PRED} & \text{`dog'} \\ \text{SPEC} & \left[\text{DEF} \;\; +\right] \end{array}\right] \\ \text{OBJ} & \left[\begin{array}{ll} \text{PRED} & \text{`cat'} \\ \text{SPEC} & \left[\text{DEF} \;\; -\right] \end{array}\right] \end{array}\right]$$

The edges are labelled with semantic roles. This is possible due to the bi-uniqueness of the mapping between roles and GFs (see above). However, there is one exception: The initial role is assigned a special label which we call 'actor' (ACT, which is equvalent to what [3] calls 'logical subject'). This partially reflects the shifting of actants in tectogrammatics as defined by [23].

So far, we have an unordered tree (f-structures are unordered by definition).[20] We define an ordering based on information structure, as proposed for deep syntax by [23]. Thus we use the i-structure to define a partial ordering on the nodes of the d-structure (TOP $\prec$ 'discourse-unspecified' $\prec$ FOC). The nodes in each of the three topic-focus do-

[20] Generally, the skeleton rendered by f-structures may contain a cycle, i.e., a node with more the one mother node. This is how LFG handles coreferences, such as in the sentence *I want to go home* where the complement clause is an open complement (XCOMP) in the f-structure of 'want' and $(\uparrow \text{SUBJ}) = (\uparrow \text{XCOMP SUBJ})$. To obtain a well-formed tree, we reflect the path of length 1 in the f-structure as an edge and the remaining (conflicting) functional paths as co-references.

mains are ordered according to their original ordering in the sentence (which is captured by c-structures).[21]

The resulting d-structure is given in (43).[22]

(43)

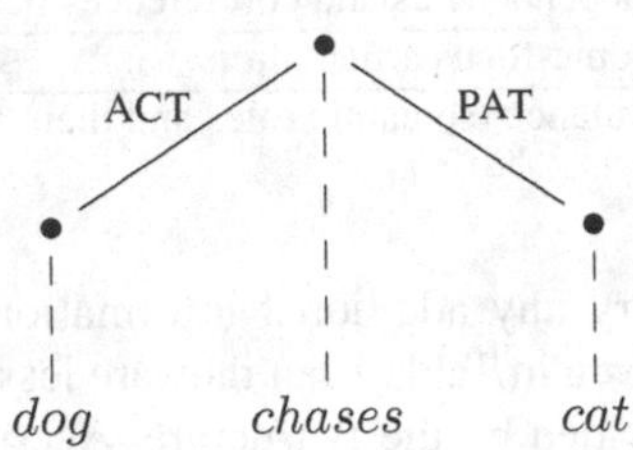

Let us briefly point out some properties of d-structures as defined above. Most of them directly correspond to properties of deep syntax (tectogrammatical) trees.

1. There is a bi-unique mapping between d-structure nodes and autosemantic (content) words. Synsemantic (auxiliary/function) words are represented as attributes of nodes. This is a direct consequence of LFG 'coheads'.
2. 'Dropped' words (e.g., subject and/or object pronouns in so-called pro-drop languages) are re-established in d-structures as a consequence of the LFG Principle of Completeness since PRED attributes are instantiated in the lexicon if needed (cf. [3]).
3. Edge labels in d-structures reflect semantic relations rather the GFs which are more language specific.
4. The ordering of d-structure nodes is partially determined by topic-focus articulation.

However, there are several differences. Note that d-structures can be non-projective (tectogrammatical trees are projective by definition [23]) which is a direct consequence of how long-distance dependencies are represented in f-structures. Furthermore, one word can be represented by more than one d-structure node (for example, in languages with incorporation).

[5] give a detailed description of the process of parallel grammar development. In our approach, the correspondence between original LFG structures and d-structures poses some (mostly technical) limitations on grammar writers. For example, f-structures of synsemantic words (functional categories) must be 'coheads' of their functional categories (however, this is a general requirement in modern LFG according to [3]). Also, GFs must conform to the strict constraints imposed by LMT.

Table 2 shows how many c-structures, f-structures and d-structures are identical (two d-structures are identical if they have the same structure and edge labels) in a parallel

[21] In free word-order languages, NPs and PPs usually have more rigid word order than clause arguments and adjuncts, thus in an MT system, the module for syntactic synthesis of the target language would reorder the d-structure according to language specific word-order rules.

[22] For the sake of simplicity, we present only the tree structure. Attributes in the original f-structure other then PRED are attached directly to the corresponding nodes.

Table 2. Identical c-, f- and d-structures in a parallel corpus

level	identical representation
c-structure	7.8%
f-structure	38.3%
d-structure	69.5%

Aymara-Spanish corpus of 1,000 sentences. The Spanish grammar has been developed for this experiment.

5 Conclusions and Further Research

We have presented a formal grammar for Aymara and pointed out some interesting properties of the language and how they can be dealt with in the LFG framework.

As can be seen, the LFG framework can be easily used to develop formal grammars of polysynthetic languages such as Aymara. While the rules we have developed cover a large part of the Aymara syntax, the lexicon we have now needs to be expaned. Currently, we are focusing on refining sublexical rules, i.e., rules that handle word-internal morphosyntax.

We have chosen LFG for our grammar because it has a solid formal linguistic foundation while providing grammars that can be directly used in NLP. However, we are developing the grammar for use in MT and LFG's f-structures are still relatively language-specific. To overcome this limitation, we have developed a fully automatic procedure which induces d(ependency)-structures (deep syntax trees) that represent a higher level of abstraction. Our d-structures are not only more suitable for cross-lingual NLP tasks such as MT but they also disclose that LFG is, in its core, a dependency-based formalism.

References

[1] Adelaar, W.: The Languages of the Andes. Cambridge University Press (2007)

[2] Beesley, K.R.: Finite-state Morphological Analysis and Generation for Aymara. In: Proceedings of the Global Symposium on Promoting the Multilingual Internet (2006)

[3] Bresnan, J.: Lexical-Functional Syntax. Blackwell Textbooks in Linguistics, New York (2001)

[4] Briggs, L.T.: Dialectal Variation in the Aymara Language of Bolivia and Peru. Ph.D. thesis. University of Florida (1976)

[5] Butt, M., King, T.H., Niño, M.E., Segond, F.: A Grammar Writer's Cookbook. CSLI Publications (1999)

[6] Cerrón-Palomino, R., Carvajal, J.C.: Aimara. In: Crevels, M., Muysken, P. (eds.) Lenguas de Bolivia. Plural Editores, La Paz, Bolivia (2009)

[7] Forssman, B.: Lettische Grammatik. Verlag J.H. Roell, Dettelbach (2001)

[8] Guzmán de Rojas, I.: ATAMIRI — interlingual MT using the Aymara language. In: New Directions in Machine Translation (1988)

[9] Guzmán de Rojas, I.: El Software de Traducción Multilingüe ATAMIRI. In: Proceedings of the VII Simposio Ibero-Americano de Terminologia e Indústrias da Língua (2000)

[10] Guzmán de Rojas, I.: Experience with language implementations in ATAMIRI. In: Proceedings of the Workshop on Bolivian & Rhondonian Languages (2006)
[11] Hale, K.L.: Warlpiri and the grammar of non-configurational languages. Natural Language & Linguistic Theory 1, 5–47 (1983)
[12] Hardman, M., Vásquez, J., de Dios, J.Y.: Aymara. Compendio de estructura fonológica y gramatical. Instituto de Lengua y Cultura Aymara (2001)
[13] Heine, B., Kuteva, T.: World Lexicon of Grammaticalization. Cambridge University Press (2002)
[14] Ishikawa, A.: Complex Predicates and Lexical Operations in Japanese. Ph.D. thesis. Stanford University (1985)
[15] Kaplan, R.M., Bresnan, J.: Lexical-Functional Grammar: A formal system for grammatical representation. In: Bresnan, J. (ed.) Mental Representation of Grammatical Relations. MIT Press, Cambridge (1982)
[16] Kaplan, R.M., Netter, K., Wedekind, J., Zaenen, A.: Translation By Structural Correspondences. In: Proceedings of 4th EACL, pp. 272–281 (1989)
[17] Kaplan, R.M., Wedekind, J.: LFG Generation Produces Context-free Languages. In: Proceedings of COLING 2000, Saarbrücken (2000)
[18] King, T.H.: Focus Domains and Information Structure. In: Butt, M., King, T.H. (eds.) Proceedings of the LFG Conference (1997)
[19] Kroeger, P.R.: Analyzing Syntax. Cambridge University Press (2004)
[20] Kruijff, G.K.: A Dependency-based Grammar. Tech. rep. Charles University, Prague, Czech Republic (2000)
[21] Sadler, L., Crookston, I., Arnold, D., Way, A.: LFG and Translation. University of Texas at Austin, pp. 11–13. LRC (1990)
[22] Sadler, L., Thompson, H.S.: Structural Non-Correspondence In Translation. In: Proceedings of the 5th Conference of the European Chapter of the Association for Computational Linguistics, pp. 293–298 (1991)
[23] Sgall, P., Hajičová, E., Panevová, J.: The Meaning of the Sentence in Its Semantic and Pragmatic Aspects. D. Reider Publishing Company (1986)
[24] Wong, S.H.S., Hancox, P.: An Investigation into the Use of Argument Structure and Lexical Mapping Theory for Machine Translation. In: Proceedings of the 12th Pacific Asia Conference on Linguistics, Information and Computation, Singapore (1998)
[25] Žabokrtský, Z., Ptáček, J., Pajas, P.: TectoMT: Highly Modular MT System with Tectogrammatics Used as Transfer Layer. In: ACL 2008 WMT: Proceedings of the Third Workshop on Statistical Machine Translation, pp. 167–170. Association for Computational Linguistics, Columbus (2008)

Eliminating Ditransitives

András Kornai

Harvard University Institute for Quantitative Social Science
and Computer and Automation Research Institute, Hungarian Academy of Sciences
andras@kornai.com

Abstract. We discuss how higher arity verbs such as *give* or *promise* can be treated in an algebraic framework that admits only unary and binary relations and does not rely on event variables.

Introduction

Until the groundbreaking work of Russell (1900), ideas of semantic representation centered on the Aristotelian notion that the predicate inheres in (is attributed to) the subject. In modern terminology, this amounts to admitting only unary relations such as `dog(x)` or `barks(y)` and treating binary relations such as MARRY(X,Y) as the conjunction of unaries `marry(x) & marry(y)`. (For greater clarity, unaries will be given in `typewriter` and binaries in SMALL CAPS font.) As Russell pointed out, such an analysis will of necessity treat all binary relations as symmetrical, with intolerable consequences for those relations that are asymmetrical such as GREATER_THAN(X,Y) or FATHER_OF(X,Y). A Davidsonian analysis trivially eliminates ditransitives and higher arity verbs, but only at the price of introducing an event variable, a step of dubious utility for statives like *has.* We follow Russell in admitting at least one asymmetric relation, which we will denote '<', and perhaps a handful of others such as HAS(x,y) 'x possesses y' AT(x,y) 'x is at location y', CAUSE(x,y), etc.

While we are obviously not disputing Russell's key observation, we believe the remedy he proposed was far too radical, throwing out all the linguistic insight that comes with the subject/predicate analysis. In this paper we propose to retrench, both in terms of drastically reducing the number of available binary relations and in terms of eliminating ternary and higher order relations entirely. To illustrate the main ideas in Section 1 we begin with a typical higher arity verb, *promise,* which is generally treated as involving at least three, but possibly as many as five, open slots: an agent, the promissor; a recipient, to whom the promise is made; the object of the promise; and perhaps an issue date and a term date as in *Alice promised Carol on Monday that she will get her twenty bucks back before Friday.* In Section 2 we present the tectogrammar, which has its roots in the decomposition technique long familiar from generative semantics (Lakoff 1968), whereby *kill* is analyzed as 'cause to die' and *give* as 'cause to have' – we discuss what makes the current model immune to Fodor's (1970) critique. In Section 3 we present the formal model using a classic generalization

P. de Groote and M.-J. Nederhof (Eds.): Formal Grammar 2010/2011, LNCS 7395, pp. 243–261, 2012.

of finite state automata (FSA) and finite state transducers (FST), *machines* (Eilenberg 1974). In the concluding Section 4 we discuss how this approach differs both from the standard model-theoretic approach and the less standard, but widely used systems of knowledge representation by semantic networks such as presented in Quillian (1969), Brachman (1979), or Sowa (2000), which retain a fundamentally Aristotelian character. We argue that the elimination of ditransitives makes possible a fundamental simplification in the network mechanism in that we no longer need to deal with hypergraphs where 'edges' could be node sets of arbitrary size – ordinary graphs will suffice.

1 The Semantics of *Promise*

What does it mean to make, and keep promises? As Rawls (1955:16) puts it, "The point of the practice is to abdicate one's title to act in accordance with utilitarian and prudential considerations in order that the future may be tied down and plans coordinated in advance". Our goal is not to dispute what Rawls says, indeed we take this to be a perfectly reasonable explanation of why the social practice of promise keeping is useful, our goal here is simply to explicate all the hidden implicational background assumed by Rawls and by users of English in general.

A *promise* is a commitment to some future action or some state of affairs that can be brought about by such action. It is assumed that the promissor is someone who can either perform the action in an agentive fashion, or that the promise pertains to the actions of someone or something under the control of the promissor. Thus *I will have the car ready by 8AM tomorrow* or *No, he won't make a mess* are well-formed promises, while *Water boils at 100 degrees centigrade I promise* is dubious usage, and *You will win the lottery/I will cure your cancer* are suspect on their face. To make an explicit promise encompasses an implicit statement by the promissor that they be capable of either performing the action themselves, or be capable of inducing someone/something to perform it for them. We will not have much to say about those cases, such as promising the boiling point of water, that can be paraphrased as 'I'm informing you', beyond the simple observation that this pertains to the knowledge state of the promissor, and in fact the promissor would be the first to admit this. But we are crucially interested in cases such as *I will cure your cancer* or *I promise eternal life* where the ability of the promissor to deliver is in grave doubt.

Let X be a predicate of some sort, and let $P(A, X, T_0, T)$ be the statement 'at time T_0 A promises X will hold at time T'. We need at least a concept of linear order of time (since *I promise you won't have to wear a scarf tomorrow* is meaningful in a way that *I promise you didn't have to wear a scarf yesterday* is not) and the condition $T_0 < T$. Further, we need a notion of agency that restricts the overall set of promises to keepable ones, thus distinguishing *I promise I will bring the book tomorrow* from *I promise I will win the lottery tomorrow*. Broadly speaking, there are actions (or states of affairs – from here on we will just speak of 'matters') that are within our power, and there are matters that are not: fetching a physical object generally falls in the first category, suddenly becoming wealthy falls in the second. We need a predicate $C(A, X, T)$ which means 'agent A can control matter X at time T'. Such control can be physical, as in the case of bringing the book, or purely notional, as in the case of a judge declaring some contract null and void. It is here that the emptiness of the promise about the boiling point of water becomes evident: clearly, whatever this boiling point is (actually, it is 99.97 °C at normal atmospheric pressure), there is no person who can change it.

So far, we have $P(A, X, T_0, T) \Rightarrow C(A, X, T)$ where the $\Rightarrow$ is some sort of normative implication: $U \Rightarrow V$ means that if U is reasonable we can reasonably expect V or, what is the same, if V does not hold U cannot be reasonably expected. Thus, a reasonable person A will not promise that someone will win the lottery because we (any reasonable person, A inclusive) don't expect that A can control the outcome of the drawing. If A is an employee of the sweepstakes company the expectations are different, and a (criminal) promise can possibly be made, but we'd still want to know a great deal more about the causal chain whereby this control over the drawing (or perhaps over the recording or the announcement of the results) is exerted. Notice that the test of reasonableness is not any different for those cases where our default assumption is the presence, rather than the absence, of control: we assume owners control their dogs and parents control their babies, yet we remain slightly dubious in regards to promises such as *He won't make a mess* precisely because we don't necessarily see the promissor as having the requisite degree of control over the matter.

For control, at minimum we need a matter that can be both ways: unless we have $M(X, T)$ and $M(\neg X, T)$ (where M is some possibility operator 'might') there cannot be any controller of X. What does $M(X, T)$ mean? Certainly $X(T)$, the fact that X holds at T, is sufficient to guarantee that X *might* hold at T, but it is either the case that $X(T)$ or it is the case that $\neg X(T)$ so knowing the state of the matter X at T is insufficient – this is well-traveled ground in modal logic. If the only possible worlds are the states of the actual world at different time instances, $M(X, T)$ implies $\exists T_1 X(T_1) \wedge \exists T_2 \neg X(T_2)$. If there are different alternatives with different timelines this becomes more complicated, but for our purposes we can get by with the simple view and our simple notion of natural or default implication $\Rightarrow$. Fortunately, we already have a different time instance at hand, namely the time T_0 when the promise is made. The thesis we will defend here includes the somewhat radical abductive inference that this is all that is required: the whole modal apparatus can be dispensed with in favor of the view that a promise is actually a promise to change, $P(A, X, T_0, T) \Rightarrow (\neg X(T_0) \wedge X(T))$.

At first blush, such a view seems to disallow all promises aimed at keeping some state of affairs intact. Since our goal is to offer a theory of ordinary language use, ignoring canonical cases of promises, such as marital vows, which are rather clearly aimed at preserving a certain state of affairs, is not an option, and we need to discuss how these fit in our model. The key issue, as we shall see, is the semantics of the modal operator *might,* which, as we will argue, already carries this implication of change. Before turning to this, let us simplify the example a bit. Marital vows are rather complex in that they require the presence of two agents and have an aspect of mutuality, so to simplify matters we use a promise of (continued) non-smoking as our example. We claim that the difference between a promise to quit, *This was my last cigarette*, where $\texttt{smokes}(T_0)$ is to be followed by $\neg$ $\texttt{smokes}(T)$ for $T > T_0$, and a promise to stay the course, where the expectation is the exact same $\neg$ $\texttt{smokes}(T)$ for $T > T_0$, is a matter of accommodation: what is hearer assumes in such cases is that the non-smoking behavior at T_0 (the time of making the promise) was accidental. We make this argument indirectly: suppose that $\neg$ $\texttt{smokes}(T_0)$ was not accidental, it was already the result of a promise. But renewal of a promise would be an empty gesture, for either the original promise was valid, in which case it remains binding for all future times, or it was not, in which case we cannot reasonably expect the promissor to upheld the renewed promise in light of non-performance on the earlier one. Therefore, by the usual quality implicature, we assume that any promissor is a non-accidental non-smoker for the first time. A general consequence of this line of argument is that it is pragmatically impossible to re-promise something.

Turning to the modal M we see that $\exists T_1 X(T_1) \wedge \exists T_2 \neg X(T_2)$ does not exhaust the meaning of $M(X,T)$. First of all, if this were sufficient, from $\exists T_1 X(T_1) \wedge \exists T_2 \neg X(T_2)$ we could conclude $M(X,T')$ with any time T', whereas when we say *John might come Tuesday* this is certainly not implicationally equivalent with *John might come Wednesday.* Rather, *might* implies both agency and causal control, so that when John might come this means both that it is within his power to come and that unless he sets his mind on this it won't happen. This logic, being embedded in the lexical definition of the word *might*, is so strong that it extends even to cases where our contemporary thinking fails to see causal control, let alone agency and free will, to be at play. Consider the weather. When we say *The sun might shine* what this means is that the Sun, as an agent, can decide to come out from hiding behind the tree. The reference to the traditional children's song "Oh Mister Sun, Sun, Mister Golden Sun" may imply to some readers that the primitive animistic viewpoint whereby the Sun has the power to change its behavior is a vestigial remnant of a mode of thought restricted to kindergarten, yet the Wall Street Journal will use the exact same language about how stocks may rise or how the market can wipe out the gains it made in the past two weeks.

So far, we have a unary modal operator $M(X)$ that simply abbreviates the fact that some matter X *might* come about, a binary modal operator $M(X,T)$ that says it might come about at time T, and a ternary operator $M(A,X,T)$ that says that it might come about at time T by the agency of A. For the sake of completeness we could also add a binary operator $M(A,X)$ that says X might happen because of the agency of A but leaves the time unspecified. The standard approach would be to take the operator with the maximum arity as basic and define the others as special cases with some of the argument slots of the basic operator filled by some default value or quantified over. Here we take the opposite tack, and argue that the basic operator has just one slot, for the matter X, and that the other slots are inherited from this simply because 'matters' in our sense can have agents, times, etc. But before getting into the details of this mechanism in Section 2, let us summarize what we have so far: *promise* is an ordinary verb whose agent X is also assumed, by default, to be the causal agent who brings about the promised matter X. The object X of the promise is typically expressed by an infinitival (as in *She promised to come*), a future tensed *that*-clause (as in *She promised that she will come*) or simply as some noun phrase or combination of noun phrases (as in *She promised complete immunity in return for a full confession*). The time of making the promise, T_0 is in the past relative to the time T that is relevant for the object of the promise, and from $P(A, X, T_0, T)$ we can conclude ($\Rightarrow$) both $C(A, X, T)$ and $\neg X(T_0)$.

Under the assumptions made here predicate arguments are handled quite differently from the way one would naively assign the participant roles. In the case of immunity, we assume the promissor p is in a position to cause some suspect s to have immunity against prosecution q for some misdeed d, and that it is s who needs to confess to d. Yet the sentence is perfectly compatible with a more loose assignment of roles, namely that the actual misdeed was committed by some kingpin k, and s is merely a witness to this, his greatest supposed crime being the withholding of evidence. This d', being an accessory after the fact, is of course also a misdeed, but the only full-force implication from the lexical content of *immunity* is that there is some misdeed m that could trigger prosecution against which s needs immunity, not that $m = d$ or $m = d'$. The hypothesis $m = d$ is merely the most economical one on the part of the hearer (requiring a minimum amount of matters to keep track of) but one that can be defeased as soon as new evidence comes to light.

2 The Tectogrammar of *Promise*

Our method of analysis relies on unary (intransitive) predicates such as `promise(X)`, `prosecute(Y)`, `commit(Z)`, `misdeed(W)`, `immune(V)` and so forth, and on some lexical implications, expressed in terms of binary (transitive) predicates of what it means to DO or HAVE these things. (For now, we retain function/argument notation with variables to present these, but the formal system defined in Section 3 will not make use of variables.) Since to the mathematical logician the temptation to look at these as instances of Currying is almost irresistible, we want to make clear at the outset that in what follows the operation $A(B)$ 'apply A to B' does not imply in any way that some intermediate function which takes functions as arguments was created. In fact, there is no implication that A or B are functions, and as we argue in Section 3, it is better to think of them as algebraic structures of a particular kind, *machines* (Eilenberg 1974). Yet somehow, with or without variables, the function-argument structure needs to be specified, which is precisely the task of tectogrammar (Curry 1961).

In order to deal with the external (subject) argument, we introduce an operator MAKE for which the external argument is obligatory. Taking the nominal meaning of *promise* as basic, this means that *to promise* is derived from this nominal by application of the (morphologically implicit) MAKE: the expression *s promises* X will be analyzed as MAKE(s,promise(X)). The use of implicit operators has a long tradition, going back at least to generative semantics where the standard analysis of *kill* was 'cause to die'. The use of unary operators is less widespread, and implies a significant departure from the standard mode of analysis whereby *She promised immunity for a confession* would be analyzed as *immunity* being the object of the promise, and *confession* as a free adverbial, outside the subcategorization frame of *promise*. The unary mode of analysis forces us assume that there is a single element, *immunity for a confession*, that is the object of the promise. What this means is that we must recognize another silent element, one that we will call `deal`, 'something for something', as an integral part of the analysis. This is confirmed by the communicative ease of introducing a definite description in a following sentence *The deal was rejected.* Further analysis of *deal* as 'trade presented by the offeror as advantageous to the other party' would be possible, but we do not pursue this here, since the main idea, that a promise has a single matter as its object, is already clear.

The same analysis is offered in regards to the time parameters, which are also standardly viewed as free adverbials. It is clear that the making of the promise has a temporal parameter. All finite verbs have an inflectional slot for this purpose, so this much is clear irrespective of one's stance on using an implicit MAKE operator. This is the parameter we denoted by T_0 above. A consequence of our analysis is that if the object X has a time parameter T this is part of the promise, rather than being a free adverbial: if $\neg X(T)$ the promise is considered unfulfilled.

Again, the same analytic method can be applied to the causation predicate $C(A, X, T)$: instead of three direct arguments, we assume that the agent A is the subject of a head operator MAKE and the object X may, but need not, carry a temporal parameter of its own. There are many subtle issues concerning temporal causation, e.g. when by placing a bomb in Bob's car on Monday Alice causes Bob to die on Tuesday, but we can largely skirt these as the central issue here is the promise, rather than the causal control required to keep it. It is worth keeping in mind that the typical failure mode of promises is not by failure to exert causal control but rather bad faith or forgetting: in most cases of broken promises the promissor could have done the right thing but didn't, out of forgetfulness, or simply because the promise was not in earnest to begin with.

Finally, the same method works for M: there is a single argument, some matter X that might come about, but there is no time parameter other than the one that X may bring in, and for agentless cases there is no agent either. Thus *It might rain* is formulated M(rain) and *It might not rain* is formulated $M(\neg$rain$)$. Based on the analysis offered so far, these two mean the same. However, if we consider the agentful cases, such as *John might insist on a vegetarian meal*, which is M(John insists) and *John might not insist on a vegetarian meal*, which is M(John $\neg$ insists) the implications are very different: in the first case *we better tell the caterers* is reasonable, in the second *maybe we don't have to bother the caterers* is. Notice, however, that these implications concern our future plans not those of the agent: for the *might rain* case *we better set up a tent* is reasonable, for the *might not rain* case *maybe we don't have to set up a tent* is. What is really at stake are the plans of the hearer (irrespective of whether the act is by God or by John) to which we turn now.

In Rawls' words, promises are means to tie down the future. Simply put, $P(A, X, T_0, T)$ is *kept* by $X \Rightarrow X(T)$ or, by contraposition, it is reasonable to infer that the promise was not kept (or no promise was made) if we observe $\neg X(T)$. By the analysis presented above, both time and agent parameters can be eliminated from the argument structure: a promise X is kept if X, broken if $\neg X$. If Alice promises Carol twenty dollars, and Bob, a mutual friend, gives it to Carol the next day saying that it came from Alice, Carol will consider Alice's promise kept. If Bob just leaves the money on Carol's desk, Carol will not particularly know (or care) whether it came directly from Alice or not, she will likely assume that it did. However, if Carol finds the twenty dollar bill on the pavement she will not assume that Alice kept her promise. What this little example shows that the assumption of causation is still very much part of the meaning of promise. But if $P(A, X, T_0, T) \Rightarrow C(A, X, T)$ is now replaced by $P(X) \Rightarrow C(X)$, what means do we have to guarantee the identity of the promissor and the causer?

To answer this question we must invoke the external argument (Roeper 1987, Sichel 2009). Recall that the object of the promising, the matter X, is a promise because the promissor *A made* this promise. How did *A* make the promise? Obviously, she was doing things with words, she said *I promise*. It is evident that the agent of a performative is the performer, and the way to create a performative is by saying it. Rather than analyzing *s promises X* as MAKE(s, `promise(X)`) we will take into account the specific manner of making and analyze it as *say(s, promise(X))* or better yet, *say(s, P)* where the object of the saying happens to be a promise *P*. Notice that the exact same analysis is available for other performatives such as *deny* or *name* (as in *I name this ship Marie Celeste*): all that is required is to have a denial, or a name, as the object of saying.

Saying requires a recipient the same way causation requires an agent. It is possible that the default recipient is everyone, as in *proclaim*, or some higher power, as in *swear*, and in fact swearing (an oath) is meaningless without the assumption of such a higher power. But in the cases of central interest, communication between individuals for the purpose of making plans, promises are made to the hearer by the speaker, and the implication $P(X) \Rightarrow C(X)$ can be kept: the maker of the promise, the sayer, is the person held responsible for causing X to come about. Given our larger commitment to eliminate higher arity predicates, introducing a ditransitive *say(A,O,R)* is a step of dubious utility. To simplify the analysis, we therefore take *say* to be analogous to *give* and analyze it as 'give words'. By giving a physical object X to R we create a situation where HAS(R, X) will be true. By giving our word, we create a promise.

Adding the recipient to the picture, the analysis becomes *s promises X to R* meaning *s causes R to have s's word that X* or simply CAUSE(R, HAS(s,word(X)).

It is not necessary for the promise to be addressed to the recipient, in fact a strong promise may explicitly invoke some higher recipient such as God. The real issue is how this giving of words, especially to beings whose very existence is doubtful, can nevertheless facilitate 'tying down the future'. As Rawls argues, a promise is a promise to refrain from reevaluating later i.e. to go with the valuation at the time of the promising. When Alice says on Monday *Carol you will get your twenty bucks back before Friday* what this means that on Monday Alice values highly Carol's having the money by Friday, and will do things to make this happen, such as going to the ATM and withdraw cash on Tuesday, or begging Bob to loan her a twenty on Wednesday so that she can pay Carol back.

Generative semanticists were largely content to use natural language paraphrases, saying *kill* means 'cause to die'. Here we sketched a theory that is only slightly more formal, saying x *kill* y means 'x CAUSE `die(`y`)`. By introducing explicit role variables, and typographically encoding the distinction between unary and binary predicates, the notation is more capable of exposing the tectogrammar than reliance on the infinitival *to*. This actually neutralizes a central point of Fodor's (1970) critique of the generative semantics analysis, because arguments concerning the placement of pronouns are no longer applicable. (As a matter of fact, subsequent developments in binding theory also rendered this kind of criticism irrelevant.)

The key reason for using *to* in the paraphrase was the commitment that generative semantics had to utilizing phrase-markers (context-free trees) as underlying structures, and the assumption that deep structure is the appropriate place to fix the lexical category of the words (Lakoff 1968). It is clear from the foregoing that we are quite content treating *promise* as entirely neutral between nominal and verbal, and forming the verbal version by zero affixation of MAKE. This is one point where the work presented here departs quite strikingly from the generative semantics tradition, reaching back straight to Pāṇini, who also was a generative semanticist in the sense of deriving surface form from underlying meaning, but was also more of a morphologist, deriving both nominal and verbal forms from the same root.

Fodor's final argument is based on on the perceived arbitrariness of the decomposition: why stop at 'cause to die', why not go to 'cause not to live' or 'cause not to have life functions' and so on? This criticism is pertinent not just to generative semantics, but in fact to any system where the meaning of one entity is described in terms of other entities. There are two known ways out: first, designating a fixed set of primitives where decomposition stops. This is the approach taken both by the Longman Dictionary of Contemporary English, where a set of about two thousand primitives is used (Boguraev and Briscoe 1989), and by the NSM school (Wierzbicka 1985). The second way out is to use an algebraic, rather than logic-based, theory of decomposition (Kornai 2010a), which is immune to the charge of arbitrariness of primitives the same way linear spaces are independent of the choice of basis we use to present them: the choice is arbitrary, but one choice is just as good as the other.

3 The Formal Model

For Russell, whose chief interest was with providing logical foundations for mathematics and the sciences, the Aristotelian maxim of Leibniz that predicates are inherent in their subject was completely untenable, since such an assumption would make it impossible to handle asymmetric cases like the predicate *father*. The differences between *Mick fathered Mixon* and *Mixon fathered Mick* are easily seen in the implications (defaults) associated with the superordinate (parent) and subordinate (child) slots: the former is assumed to be independent of the latter (already existed before the act of fathering took place), the latter is assumed to be dependent on the former, the former controls the latter (in the same everyday sense of control that we used so far, not in the grammatical sense), and not the other way around, etc.

In our treatment of verbs, it will indeed be necessary to admit at least one asymmetric relation, which we will denote '<', and perhaps a handful of others such as HAS(x,y) 'x possesses y' or AT(x,y) 'x is at location y'. At the same time, we are more parsimonious with relations than Russell, for whom the existence of a single asymmetrical relation was sufficient reason to open the floodgates and admit all kinds of relations, and presented a theory in which no ternary relations are used in the definiens. We illustrated our method of analysis on a hard case, *promise*, that is standardly thought to require at least three, and possibly as many as five, arguments, and argued that at the tectogrammatical level it has only *one* argument, the thing that is being promised. All other arguments are linked in either externally (the promissor, by the matrix verb MAKE) or recursively, by invoking the frame of the act of promise-making (which we analyzed as an act of giving words), or the frame of the matter being promised.

To round out this picture what we need is a theory of the representational objects, one that describes how semantic representations are formed, maintained, and destroyed (see 3.1) and a theory of bookkeeping that tells us how such objects can act as slot-fillers in the tectogrammar (see 3.2). (Ideally, we would also want an account of the phenogrammar, how all these steps are realized on the surface, but this is clearly beyond the scope of this paper.)

3.1 Representation by Machines

Fortunately, a good theory of representational objects is already at hand: these are the *machines* of Eilenberg (1974). In brief, *a machine is a mapping between the alphabet of some FSA and the relation monoid of some set X*. Eilenberg intended machines to be an algebraic formulation of the flowcharts widely used at the time for describing the structure of computer programs – we will use them to represent the meaning of morphemes, words, phrases, sentences, and texts alike. The FSA is used as the *control* of the device just as in Turing Machines, and the relations are best thought of as transformations of the base set X that the machine is about.

Definition 1. A *machine* with an alphabet Σ over a base set X is given by an *input set* Y; an *output set* Z; a relation $\alpha : Y \to X$ called the *input code*; a relation

$\omega : X \to Z$ called the *output code*; a finite state automaton $\langle S, T, I, F\rangle$ over Σ called the *control* FSA; and a mapping M of each $\sigma \in \Sigma$ to some $\phi \in \Phi \leq 2^{X \times X}$.

Since our objects are semantic representations for natural language expressions rather than flowcharts, we need to tweak this definition a bit. As we are not dealing with the phenogrammar, we can safely ignore the input and output mappings, which are primarily formal tools for transducing input to, and output from, the machine. This will simplify the definition, but we also need to complicate it a bit: we need to be more specific about the base set X, whose elements will be called *partitions,* and we will need to designate one of these partitions as the *head.* One partition (conventionally numbered as the 0th member of the set X) will contain the phonological form (printname) of the machine, the other(s) will store information relating to the argument(s).

We will call the machines so defined *lexemes*, and informally it is best to think of these as monolingual dictionary entries (see Kornai 2010). One characteristic difference between the model-theoretic and the more cognitively inspired theories of lexical semantics is the type structure: Montague Grammar relies on a strict set of intensional and extensional types, with n-ary predicates and relations, while lexical semantics is generally conceived of in network terms, with only two main types, graph *nodes* corresponding to lexemes, and graph *edges* corresponding to various links, directed or undirected. From the perspective of strict typing, it is natural to ask how property bundles are composed: for example, if properties correspond to qualia, is it simply the case that adjectives are qualia and nouns are bundles of qualia? From the perspective of the essentially type-free network theory, the main question is to sort out the kinds of links permitted by the model (Woods 1975). Here we will try to sketch an answer to both kinds of questions.

Primitive lexemes come in two subvarieties, *unary* and *binary*: the classes will be denoted by U and B and the instances written in `typewriter font` and SMALL CAPS respectively. Most lexical entries, not just nouns, adjectives, and intransitive verbs, but also verbs of higher arity (transitives, ditransitives, etc.), both in predicative and in substantive forms, are viewed as unary, and the binary category is reserved primarily for adpositions (both pre- and postpositions) and case markers. With adpositions, it is very hard to see how expressions signifying pure spatial relations such as *under* or *near* could be given a satisfying model without reference to the pairs of objects standing in the named relation, and from a grammatical perspective it is quite clear that case markers behave very similarly (for a modern summary, see Anderson 2006). There are a few stray examples elsewhere in the system of grammatical formatives, such as the possessive relation, generally not regarded a true case, and the comparative morpheme *-er,* but it is clear that on the whole binary lexemes are restricted to a small, closed subset of function words, while the large, productive classes of content words are all unary under the analysis offered here.

Definition 2. The surface syntax of lexemes can be summarized in a Context-Free Grammar (V, Σ, R, S) as follows. The nonterminals V are the start symbol S; the binary relation symbols B which can include '<', CAUSE, HAS, ... etc.

taken from some small fixed inventory of deep cases, thematic roles, grammatical functions, or similarly conceived linkers; and the unary relation symbols collected in U. Variables ranging over V will be taken from the end of the Latin alphabet, v, w, x, y, z. The terminals are the grouping brackets '[' and ']', the derivation history parentheses '(' and ')', and we introduce a special terminating operator ';' to form a terminal *v;* from any nonterminal v. The rule $S \rightarrow U|B|\lambda$ handles the decision to use unary or binary predicates, or perhaps none at all. The operation of *attribution* is captured in the rule schema $w \rightarrow w; [S^*]$ which produces the list defining w. (This requires the CFG to be *extended* in the usual sense that regular expressions are permitted on the right hand side, so the rule really means $w \rightarrow w; []|w; [S]|w; [SS]|...$) Finally, the operation of *predication* is handled by $u \rightarrow u; (S)$ for unary, and $v \rightarrow Sv; S$ for binary nonterminals.

Our interest is both with the terminal yield of the grammar (V, Σ, R, S) and the sentential forms that still contain nonterminals. The meaning postulates are specific instances of the attributive rule schema $w \rightarrow w; [S^*]$ which produces the list defining w and the predicative schemas $u \rightarrow u; (S)$ and $v \rightarrow Sv; S$. Whenever such a postulate is used, the definiendum x is terminated (replaced by the terminal x; and thus no longer available for further rewriting), but the substantive terms that occur in the definiens are still in nonterminal form. Before drawing many conclusions from the fact that the syntax is defined as context-free it is worth emphasizing that this is *pure* syntax. Thus, `dog EQ four-legged, animal, hairy, barks, bites, faithful, inferior` is a well-formed equational formula defining the dog, but so is `cat EQ barks` – the syntax is entirely neutral as to whether this is true or what sense it makes. The standard method of trying to make sense of such formulas would be to interpret them in model structures, and failure to do so is generally seen as failure of connecting language to reality (Lewis 1970, Andrews 2003). Yet, as we have argued elsewhere (Kornai 2010b), such an effort is bound to misfire wherever we encounter language that is not about reality.

Consider *Pappus tried to square the circle/trisect the angle/swallow a melon.* In one case, we see Pappus intently studying the works of Hippocrates, in the other we see him studying Apollonius, and in the third case we see him in the vegetable patch desperately looking for an undersized melon in preparation for the task – clearly the truth conditions are quite different. We may very well imagine a possible world where throats are wider or melons are smaller, but we know it for a fact that squaring the circle and trisecting the angle are logically impossible tasks. Yet to search for a proof, be it positive or negative, is quite feasible, and the two searches lead us into different directions early on: squaring the circle begins with the Hippocratic lunes, and culminates in Lindemann's 1882 proof, while trisecting the angle begins with the *Conics* of Apollonius and does not terminate until Wantzel's 1832 proof. The problem is not with nonexistent objects such as superwide throats, for which the intensional treatment of opacity works fine, but also *necessarily* nonexistent objects whose extension is empty at every index. (To make matters worse, we rarely know in advance whether something fails to exist by accident or of necessity.)

In truth, it is not just the existence of hard hyperintensionals that stands in the way of ever completing the program of model-theoretic semantics – the failure of this approach is more evident from ordinary sentences than from subtle technical notions concerning hyperintensionals, which may yet get resolved by work such as Pollard (2008). Consider, for example, the following statement, (Jonathan Raban, NYRB 04/12/07): *There is in Sullivan's makeup [] an Oxford debater's ready access to the rhetoric of condescending scorn.* Clearly, this is a completely meaningful, non-paradoxical sentence, which conveys good information about Sullivan to the readers of the New York Review, yet attempts to analyze it in terms of satisfaction in model structures are fruitless. It is quite unclear who is, and who isn't, an Oxford debater, or how we could go about distinguishing an Oxford from a Harvard debater in terms of the set of people involved (especially as most debaters are perfectly capable of switching between the various styles of debate). The same can be asked about every constituent of the sentence: where is, in a model structure, *someone's makeup*, and what kind of objects r are we sifting through to determine whether r is or is not part of Sullivan's makeup? What is *scorn*, and are Lewis' (1970) remarks on Markerese really exemplars of the *condescending* variety, or are they, perhaps, well reasoned and not at all scornful?

The semantics that attaches to the lexeme-based representations defined above by purely syntactic means is of a different kind. We may not have a full understanding of the relation x *has ready access to* y, but we do know that having ready access to something means that the possessor can deploy it swiftly and with little effort. What the sentence means is simply that Raban has studied the writings of Sullivan and found him capable of doing so, in fact as capable as those highly skilled in the style of debate practiced at the Oxford Union where condescension and scorn are approved, even appreciated, rhetorical tools. It is basically left to the reader to supply their own understanding of *condescension* and *scorn*, and there is no reason to believe that this understanding is framed in terms of specifying at every index whether something is condescending or scornful. Rather, these terms are either primitives, or again defined by meaning postulates.

A defining characteristic of this network of definitions is that little semantic distinction can be made between verbs like *to promise, to prosecute, to commit, to (be/make) immune, to *misdo*, their substantive forms *promising, prosecuting/prosecution, commitment, immunity, *misdoing*, and their cognate objects *the promise, the prosecution, the commitment, the misdeed.* In this respect, the underlying type system proposed here is considerably less strict than that of Lakoff (1968), where deep structure was assumed to be the appropriate place for fixing the lexical categories of the words. But this kind of loose typing, the necessity of which is a central claim in Turner (1983, 1985), is quite suitable for a purely lexical theory, like that of Pāṇini, which can capture the essential grammatical parallelism between active, passive, and stative constructions (see Kiparsky 2002:2.2). We also stay close to the Pāṇinian model in assuming that the argument structure, such as it is, is created by the linkers. To illustrate the mechanism, consider *give(x,y,z)*, which is standardly analyzed as as 'transferring possession of y from x to z'. From our perspective, such an analysis is assuming too much, because when we say *The idea gave him the shivers* one cannot reasonably conclude that the shivers were originally in the idea's possession, and when we say *Mary gave him typhoid*, we cannot conclude that Mary ceased to have typhoid just by giving it to him. Thus we have a simpler analysis, CAUSE(x,HAS(z,y)) 'cause to have' where CAUSE is used to denote the agentive linker.

It is worth noting that the formalism offered above does not rely on function/argument notation and variables at all. To do away with these entirely, we already fixed the notation: since the binary operators can be written infix, while unary operators are written prefix, parens are sufficient to fix the location (though not the identity) of the variables: a formula such as x CAUSE(z HAS y), can be reduced to CAUSE(HAS). The example is only illustrative of the formal mechanism – this is not the place to recapitulate the subtleties of causation discussed in Talmy (1988), Jackendoff (1990:72) and elsewhere in the linguistic literature. By assuming right association most parens can be omitted, only those signaling left association need be retained to disambiguate application order if necessary (so far we have not found actual examples). For grouping, braces will be used, so that the conjunctive feature bundles defining nouns can be kept together. Such a tight notation does not leave a great deal of room for scope ambiguities, but as we have argued in some detail elsewhere (Kornai 2010a), this entails little loss in that universally quantified expressions, outside the technical language of mathematics, are read generically rather than episodically.

Eliminating variables is a significant step toward bringing the formalism closer to the network diagram notation familiar from many works in lexical semantics and Knowledge Representation (for a good selection, see Findler 1979, Brachman and Levesque 1985). We cannot discuss the network aspect of the theory here in sufficient detail, but we note that in the machine formalism the proliferation of links, characteristic of many network theories, is kept under strict control. This is achieved by two means: first, IS_A links are derived rather than primitive (see Kornai 2010), and second, by the elimination of ditransitives.

Were we to permit ditransitives and higher arity predicates as primitives, we would need as many kinds of links as the maximum arity predicate has arguments, and to the extent this number is treated as an unlimited resource (as in some analyses of serial verbs) we would need to countenance an infinite number of link types. As it is, we are restricting the theory to only two kinds of links: those corresponding to substitution of the first argument, and those corresponding to the substitution of the second (as a matter of fact, ergative/absolutive classification of links would be just as feasible, but we do not pursue this alternative here).

3.2 Slot-Filling

The only fundamental aspect of the theory not discussed so far is the bookkeeping, how to specify which empty slot in a machine corresponds to which verbal argument, how to guarantee that no slot gets filled twice, and in case of obligatory arguments, how to guarantee that the slot does get filled. Recall that Definition 1 contains two moving parts, an FSA and a base set X, as well as a mapping from the alphabet of the automaton to the set of relations over X. This, we claim, is already sufficient for the purposes of tectogrammar. Unaries, by their very nature, have only one slot to be filled, so linking something there requires no traffic signals: wherever X is an unary and Y is an arbitrary machine $X(Y)$ is obtained by placing an instance of Y on the one and only non-phonological partition of X.

For the binary case, consider *Mick fathered Mixon* and assume that *father* is a relational noun or that *to father* is a transitive verb. What we wish to obtain (using infix notation) is `Mick` FATHER `Mixon` rather than `Mixon` FATHER `Mick` or `Mixon, Mick` FATHER or something else. We will ignore the tense marking, and we will assume a rather sophisticated phenogrammar that has already succeeded in turning the surface expression into `Mick`-NOM, `Mixon`-ACC, FATHER. In English, the nominative and accusative linking is provided by word order, in other languages it may very well be provided by overt case marking. (In fact, it is slightly wrong to use the terms nominative and accusative in that the two slots may as well be linked by ergative and absolutive case, but this affects only the phenogrammar of the language in question, not the mechanism proposed here.)

It is sufficient for the alphabet of the control automaton of the FATHER machine to distinguish three elements, those NPs that are nominatively marked, for which we use the letter n, those accusatively marked, for which we use the letter a, and all others, denoted by o (see Fig 1). Since *to father* is transitive, the control FSA will be a square, with a start state we denote by ◎, an accepting state •, and two other states serving as counters for unfilled valences. The language accepted by the automaton is the shuffle product of exactly one a, exactly one n, and an arbitrary number of os.

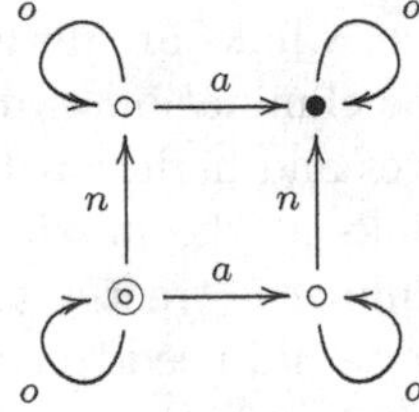

Fig. 1. FSA for transitive verbs

The control is used to define a mini-language that checks the tectogrammatic conditions: for example for verbs that alternate in transitivity such as *eat* the top left state could also be defined accepting, so that *Mick ate*, unlike **Mick fathered*, would come out as grammatical.

The mapping M is also part of the bookkeeping mechanism. Continuing with the example of FATHER, let us denote the two partitions 1 and 2. The relations possible over these include $F = \{(1,1),(1,2),(2,1),(2,2)\}$; $I = \{(1,1),(2,2)\}$;

$P = \{(2,2)\}$ and $Q = \{(1,1),\}$ (there are a total of 16 relations over two elements, but the others need not concern us here). Here we map by M the letter o on the identity relation I, the letter a on the projection P and the letter n on the projection Q. As we build up a string, we are also building up a product of relations, so from starting the full relation F, by the time we multiplied with exactly one P, one Q, and any number of Is, we arrive at the empty relation. The mechanism is flexible enough to handle complex relation-changing verbal affixation rules such as passivization or causativization.

Finally, let us consider how the 'cause to have' analysis of *give* is formalized using machines. The square FSA of Fig. 1 is replaced by a cube, whose edges are now labeled n(ominative), d(ative), a(ccusative), and o(ther), though the loops labeled o that appear over each vertex are omitted from the figure for clarity.

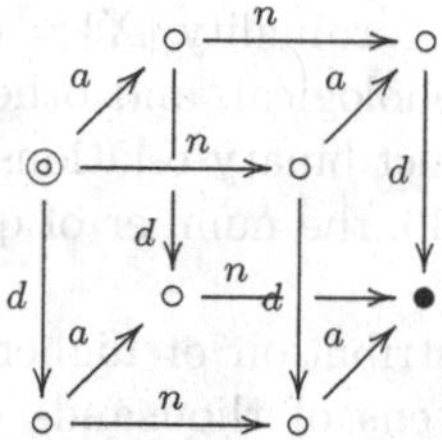

Fig. 2. FSA for *give*

Assuming all three arguments are obligatory, there is only one accepting state, the bottom back right corner of the cube. The base set X has three members (not counting the phonological partition), which are obtained by substituting the HAS machine in the second (subordinate) partition of the CAUSE machine. In a network diagram, this is depicted as Fig 3 below, with nodes both for binary and unary machines, and different coloring (straight vs. dotted) of the edges to make clear which edge originates in the first, and which in the second partition of the binaries.

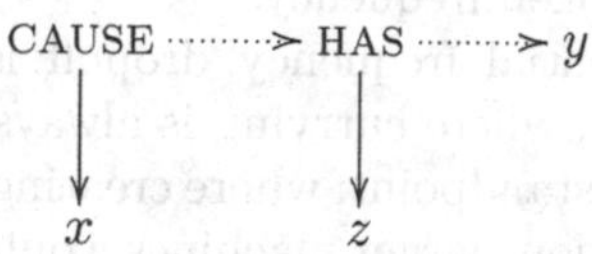

Fig. 3. Base set for *give(x,y,z)*

4 Conclusions

Since grammars need to capture tectogrammatical generalizations, some form of slot-filling mechanism, such as the f-structure of LFG or the SUBCAT mechanism of HPSG, is clearly needed for dealing with predicate-argument structure. Indeed, the need is felt so strongly that a variety of linguistic theories such as case grammar (Anderson 2006), valency theory (Somers 1987) and tagmemics (Pike

1960) posited slot filling as the basic (and in some cases, the only) mechanism for describing syntactic phenomena.

¿From a formal standpoint the most immediate mechanism for slot-filling is to use some kind of variable binding term operators, typically lambdas, as in $\lambda x \lambda y \lambda z$ *give*(x, y, z). Once we take this step, the elimination of ditransitives, and indeed the elimination of transitives, becomes a trivial matter of currying, and attention is shifted to other aspects of the system: as is well known (Marsh and Partee 1984), variable binding itself is a formally complex operation, with attendant difficulties for creating effective parsing/generation/acquisition algorithms.

In the machine formalism propounded here it would actually be possible to have ditransitives or even higher arity predicates, but only at a computational cost that increases superexponentially. For technical reasons n-ary predicates require machines with base set cardinality $|X| = n + 1$ (the 0th slot is used for storing the phonological, morphological, and other position-independent information) so the number of distinct binary relations ϕ is $2^9 = 512$, the number of ternaries would be $2^{16} = 65,536$, the number of quaternaries $2^{25} = 33,554,432$ and so on.

Note that the empirical distribution of higher arity verbs drops off rather sharply: in English we have tens of thousands of intransitive and transitive verbs, but only a few hundred ditransitives, and only a handful of candidates for tritrasitive or higher arity. Following Schank (1973), the single most frequent class is physical transfer (PTRANS) verbs such as *give, get, bring* and negative PTRANS such us *bar, block, keep* – altogether less than thirty examples including portmanteau manner-qualified forms such as *throw, toss* and *mail* where the indirect object is arguably optional. The next most frequent class is mental transfer (MTRANS) verbs like *signal, promise, inform, show* followed by transfer of possession (ATRANS) verbs such as *award, bequeath, remit* and their negatives such as *begrudge, deny.* or *refuse.* The M and A classes already show signs of morphological complexity, and in languages that have overt causative or benefactive morphology the higher arity classes are somewhat larger, but still a small fraction in terms of token frequency.

This faster than exponential frequency dropoff is hard to grasp from the variable-binding standpoint, where currying is always available, but makes perfect sense from the machine standpoint, where creating (acquiring) and operating (during parsing and generation) larger machines would require disproportionally larger resources. In this regard, the current work fits far better with variable-free (Szabolcsi 1987, Jacobson 1999, Steedman 2001) than with mainstream semantics. However, the fit is far from perfect, in that machines are best thought of as a means of capturing the structure of meaning postulates, rather than as a calculus for compositional meaning. Of the two, we actually consider lexical (non-compositional) structure the higher priority task, given that the primary information source in a sentence, responsible for over 85% of the information conveyed, is the choice of the words, rather than the grammatical structure, which accounts of less than 15% (see Kornai 2010 for how these numbers are obtained).

Altogether, the theory presented here fits better with the 'cognitive' approach pursued by Jackendoff, Talmy, Langacker, Fauconnier, Lakoff, Wierzbicka, and many others, and with the whole network tradition of Knowledge Representation originating with Quillian (1967) and Schank (1973). One issue that has put the cognitive work on a less than equal footing with the Montague Grammar tradition was the naive formalism (famously dubbed 'markerese' by Lewis 1970), and part of our goal is to provide a formal apparatus that is capable of restating the linguistic insights of the cognitive work in a theory that is sufficiently formal for computer implementation.

Readers familiar with the history of network theories will know that one of the key implementational issues is the variety of links permitted in the system (see in particular Woods 1975), and in this regard the elimination of ditransitives is a key step. In a network graph, every edge from a node x to some node y and bearing the label l is of necessity an ordered triple (l, a, b) i.e. an information structure with three slots. A theory that makes the claim that these are not unanalyzed primitives but can be built from simpler, binary structures enables reduction of complexity across the whole system. Specifically, we claim that there are only *two* kinds of links (depicted by full vs. dotted lines in Fig.3), corresponding to the superordinate (first) and the subordinate (second) slot of binary relations. There is no claim that first always means '1' or subject, and second means '2' or object, the formal theory presented here is quite capable of handling mismatches such as experiencer subjects. The claim is simply that there is never a '3' or indirect object on a par with the first two arguments.

To summarize, we have repurposed Eilenberg's machines as a simple, variable-free mechanism for decomposing the meaning of higher arity relations and keeping track of the tectogrammar (function-argument structure). This is the hard case: extending the system to adjectival and adverbial modifiers is trivial and requires no further machinery (see Kornai 2010). The result is a formalism conducive to the style of grammatical analysis familiar from Pāṇini and from generative semantics, and capable of encoding the semantic insights developed from Aristotle to contemporary knowledge representation and cognitive semantics.

Acknowledgments. We thank Donca Steriade (MIT) for comments on an earlier draft. Work supported by OTKA grants #77476 (Algebra and algorithms) and #82333 (Semantic language technologies).

References

References

Anderson, J.: Modern grammars of case: a retrospective. Oxford University Press (2006)

Andrews, A.: Model-theoretic semantics as structural semantics. ms, ANU (2003)

Boguraev, B.K., Briscoe, E.J.: Computational Lexicography for Natural Language Processing. Longman (1989)
Brachman, R.: On the epistemological status of semantic networks (1979)
Brachman, R., Levesque, H.: Readings in knowledge representation. Kaufman Publishers Inc., Los Altos (1985)
Curry, H.B.: Some logical aspects of grammatical structure. In: Jakobson, R. (ed.) Structure of Language and its Mathematical Aspects, pp. 56–68. American Mathematical Society, Providence (1961)
Eilenberg, S.: Automata, Languages, and Machines, vol. A. Academic Press (1974)
Findler, N.: Associative Networks: Representation and Use of Knowledge by Computers. Academic Press (1979)
Fodor, J.: Three reasons for not deriving "kill" from "cause to die". Linguistic Inquiry 1(4), 429–438 (1970)
Graham, A.C.: Two Chinese Philosophers, London (1958)
Jackendoff, R.S.: Semantic Structures. MIT Press (1990)
Jacobson, P.: Towards a variable-free semantics. Linguistics and Philosophy 22, 117–184 (1999)
Kiparsky, P.: On the Architecture of {P}\={a}\d{n}ini's grammar. ms, Stanford University (2002)
Kornai, A.: Mathematical Linguistics. Springer (2008)
Kornai, A.: The algebra of lexical semantics. In: Jäger, G., Michaelis, J. (eds.) Proceedings of the 11th Mathematics of Language Workshop. FoLLI Lecture Notes in Artificial Intelligence. Springer (2010a)
Kornai, A.: The treatment of ordinary quantification in English proper. Hungarian Review of Philosophy 54(4), 150–162 (2010b)
Lakoff, G.: Pronouns and reference (1968)
Lewis, D.: General semantics. Synthese 22(1), 18–67 (1970)
Marsh, W., Partee, B.: How non-context-free is variable binding? In: Cobler, M., MacKaye, S., Wescoat, M. (eds.) Proceedings of the West Coast Conference on Formal Linguistics III, pp. 179–190 (1984)
Pike, K.: Language in Relation to a Unified Theory of the Structure of Human Behavior. Mouton, The Hague (1960)
Pollard, C.: Hyperintensions. Journal of Logic and Computation 18(2), 257–282 (2008)
Quillian, M.R.: The teachable language comprehender. Communications of the ACM 12, 459–476 (1969)
Rawls, J.: Two concepts of rules. The Philosophical Review 64(1), 3–32 (1955)
Roeper, T.: Implict arguments and the head-complement relation. Linguistic Inquiry 18, 267–310 (1987)
Russell, B.: The Philosophy of Leibniz. Allen und Andwin (1900)
Schank, R.: The Fourteen Primitive Actions and Their Inferences. Stanford AI Lab Memo 183 (1973)
Sichel, I.: New evidence for the structural realization of the implicit external argument in nominalizations. Linguistic Inguiry 40(4), 712–723 (2009)
Somers, H.L.: Valency and case in computational linguistics. Edinburgh University Press (1987)
Sowa, J.: Knowledge representation: logical, philosophical, and computational foundations, vol. 594. MIT Press (2000)
Steedman, M.: The Syntactic Process. MIT Press (2001)
Szabolcsi, A.: Bound variables in syntax – are there any? In: Gronendijk, J., Stokhof, M., Veltman, F. (eds.) Proceedings of the 6th Amsterdam Colloquium. Institute for Language, Logic, and Information, Amsterdam, pp. 331–351 (1987)

Talmy, L.: Force dynamics in language and cognition. Cognitive Science 12(1), 49–100 (1988)

Turner, R.: Montague semantics, nominalisations and Scott's domains. Linguistics and Philosophy 6, 259–288 (1983)

Turner, R.: Three theories of nominalized predicates. Studia Logica 44(2), 165–186 (1985)

Wierzbicka, A.: Lexicography and conceptual analysis. Karoma, Ann Arbor (1985)

Woods, W.A.: What's in a link: Foundations for semantic networks. Representation and Understanding: Studies in Cognitive Science, 35–82 (1975)

Lambek Grammars with the Unit

Stepan Kuznetsov

Department of Mathematical Logic and Theory of Algorithms,
Faculty of Mechanics and Mathematics, Moscow State University
skuzn@inbox.ru

Abstract. Pentus' theorem states that any language generated by a Lambek grammar is context-free. We present a substitution that reduces the Lambek calculus enriched with the unit constant to the variant of the Lambek calculus that does not contain the unit (but still allows empty premises), and use this substitution to prove that any language generated by a categorial grammar based on the Lambek calculus with the unit is context-free.

1 L*-Grammars

We consider the calculus L* (the Lambek calculus that allows empty premises) — a variant of the calculus L introduced in [4]. The set $\mathrm{Pr} = \{p_1, p_2, p_3, \dots\}$ is called the set of *primitive types. Types* of L* are built from primitive types using three binary connectives: $\backslash$ *(left division)*, $/$ *(right division)*, and $\cdot$ *(multiplication)*; we shall denote the set of all types by Tp. Capital letters $(A, B, \dots)$ range over types. Capital Greek letters range over finite (possibly empty) sequences of types; Λ stands for the empty sequence. Expressions of the form $\Gamma \to C$ are called *sequents* of L*.

Axioms: $A \to A$.

Rules:

$$\frac{A\Pi \to B}{\Pi \to A \setminus B}\ (\to \setminus) \qquad \frac{\Pi \to A \quad \Gamma B \Delta \to C}{\Gamma \Pi (A \setminus B) \Delta \to C}\ (\setminus \to)$$

$$\frac{\Pi A \to B}{\Pi \to B \,/\, A}\ (\to /) \qquad \frac{\Pi \to A \quad \Gamma B \Delta \to C}{\Gamma (B \,/\, A) \Pi \Delta \to C}\ (/ \to)$$

$$\frac{\Pi \to A \quad \Delta \to B}{\Pi \Delta \to A \cdot B}\ (\to \cdot) \qquad \frac{\Gamma A B \Delta \to C}{\Gamma (A \cdot B) \Delta \to C}\ (\cdot \to)$$

Now let us define the notion of an L*-grammar. We call an *alphabet* an arbitrary finite non-empty set. The set of all words over the alphabet Σ (i. e., finite sequences of elements of Σ) is denoted by Σ^*. Any subset of Σ^* is called a *formal language* over Σ.

Definition 1. *An* L*-grammar *is a triple* $\mathcal{G} = \langle \Sigma, H, \rhd \rangle$, *where* Σ *is an alphabet,* $H \in \mathrm{Tp}$, *and* $\rhd$ *is a finite correspondence between* Tp *and* Σ *(i. e.,* $\rhd \subset \mathrm{Tp} \times \Sigma$*). The* language generated by $\mathcal{G}$ *is the set of all words* $a_1 \dots a_n$ *over* Σ *for which there exist types* $B_1, \dots, B_n$ *such that* $\mathrm{L}^* \vdash B_1 \dots B_n \to H$ *and* $B_i \rhd a_i$ *for all* $i \le n$. *We shall denote this language by* $\mathfrak{L}(\mathcal{G})$.

P. de Groote and M.-J. Nederhof (Eds.): Formal Grammar 2010/2011, LNCS 7395, pp. 262–266, 2012.

We also consider context-free grammars:

Definition 2. *A* context-free grammar *is a quadruple $G = \langle N, \Sigma, P, S\rangle$, where N and Σ are two disjoint alphabets, $P \subset N \times (N \cup \Sigma)^*$, P is finite, and $S \in N$. We define a binary relation $\Rightarrow_G$ as follows: for all $\omega, \psi \in (N \cup \Sigma)^*$ we have $\omega \Rightarrow_G \psi$ if and only if $\omega = \eta A \theta$, $\psi = \eta \beta \theta$, and $\langle A, \beta\rangle \in P$ for some $A \in N$, $\beta, \eta, \theta \in (N \cup \Sigma)^*$. The binary relation $\Rightarrow_G^*$ is the reflexive transitive closure of $\Rightarrow_G$. The language $\mathfrak{L}(G) = \{w \in \Sigma^* \mid S \Rightarrow_G^* w\}$ is the* language generated by *G. Such languages are called context-free.*

These two notions of formal grammar are equivalent in the following sense (this is proved in [3] using methods from [1], [2], and [7]):

Theorem 1. *A formal language is context-free if and only if it is generated by some* L*-*grammar.*

The original Lambek calculus L is obtained from L* by adding the restriction $\Pi \neq \Lambda$ on the rules $(\to \backslash)$ and $(\to /)$. The notion of L-grammar is defined similarly to the notion of L*-grammar.

Theorem 2. *A formal language without the empty word is context-free if and only if it is generated by some* L*-grammar.*

The "if" part in Theorem 2 is proved in [2] using the construction from [1] and the "only if" part is proved in [7].

The "only if" part in Theorem 1 is proved by the same argument as the "only if" part in Theorem 2. Now we sketch the proof of the "if" part. If a language M is context-free and does not contain the empty word, then, by Theorem 2, there exists an L-grammar $\mathcal{G}$, such that $M = \mathfrak{L}(\mathcal{G})$. Moreover, this grammar has the property that the rules $(\to \backslash)$ and $(\to /)$ do not appear in derivations used to check whether particular words belong to $\mathfrak{L}(\mathcal{G})$. Hence, $\mathfrak{L}(\mathcal{G})$ will not change if we consider $\mathcal{G}$ as an L*-grammar. Some extra work is needed to handle the case where M contains the empty word (it is done in the unpublished paper [3]).

2 L_1-Grammars

In this section we consider the calculus L_1 (the Lambek calculus with the unit). By Tp_1 we denote the set of types generated from primitive types and the constant **1** (unit) using the connectives $\backslash$, $/$, and $\cdot$. We shall now use Tp_1 instead of Tp. The calculus L_1 is obtained from L* by adding an extra axiom $\to \mathbf{1}$ (denoted by $(\to \mathbf{1})$) and an extra rule

$$\frac{\Gamma\Delta \to C}{\Gamma \mathbf{1} \Delta \to C}\ (\mathbf{1} \to).$$

The calculus L_1 was introduced by Lambek in [5].

It is easy to see that L_1 is a conservative extension of L*. Therefore, due to Theorem 2, the class of context-free languages is contained in the class of L_1-languages. We shall prove the converse statement.

Theorem 3. *Every* L_1*-language is context-free.*

Due to Theorem 2 it is sufficient to prove that for any L_1-language there exists an L^*-grammar. To do this, we shall build a substitution that reduces derivability in L_1 to derivability in L^*.

Let $\mathscr{A}$ be a syntactic object (a type, a sequence of types, a sequent, or a grammar). By $\mathscr{A}[z := A]$ we denote $\mathscr{A}$ with type A substituted for z (here $z \in \mathrm{Pr} \cup \{\mathbf{1}\}$). The notation $\mathscr{A}[z_1 := A_1, z_2 := A_2, \ldots]$ means that all substitutions are performed simultaneously.

Theorem 4. *For any sequent $\Pi \to C$ built from types that belong to* Tp_1 *and for any primitive type q not occurring in $\Pi \to C$, the following equivalence holds:*

$$\mathrm{L}_1 \vdash \Pi \to C \iff \mathrm{L}^* \vdash (\Pi \to C)[p_i := (\mathbf{1} \cdot p_i) \cdot \mathbf{1}][\mathbf{1} := q \setminus q].$$

(Here and further the shorthand "$p_i := (\mathbf{1} \cdot p_i) \cdot p_i$" means that the substitution is performed for *every* i.)

Before proving Theorems 3 and 4, we introduce some notions and establish several lemmas.

Two types $A, B \in \mathrm{Tp}_1$ are *equivalent* (denotation: $A \leftrightarrow B$), iff $\mathrm{L}_1 \vdash A \to B$ and $\mathrm{L}_1 \vdash B \to A$. In L_1 there holds the rule of equivalent substitution: if $\mathrm{L}_1 \vdash (\Pi \to C)[z := A]$ and $A \leftrightarrow B$, then $\mathrm{L}_1 \vdash (\Pi \to C)[z := B]$. In particular, if $z \leftrightarrow B$ and $\mathrm{L}_1 \vdash \Pi \to C$, then $\mathrm{L}_1 \vdash (\Pi \to C)[z := B]$.

The rule $(\mathbf{1} \to)$ can be considered a special case of the weakening rule. Any L_1-derivation can be rebuilt in such a way that all applications of this rule will immediately follow the axioms. In other words, the L_1 calculus can be equivalently formulated without the $(\mathbf{1} \to)$ rule and with two extra series of axioms: $\mathbf{1}^k \to \mathbf{1}$ $(k \geq 0)$ and $\mathbf{1}^k p_i \mathbf{1}^m \to p_i$ $(k, m \geq 0,\ i \geq 1)$. We denote them $(\to \mathbf{1})_{\mathrm{w}}$ and $(\mathrm{ax})_{\mathrm{w}}$, respectively. Further we shall use this new calculus for L_1.

Consider an auxiliary calculus L_1^- which is obtained from L^* by adding axioms $(\to \mathbf{1})_{\mathrm{w}}$. It is clear that L_1^- is a fragment of L_1.

Lemma 1. *For every sequent $\Pi \to C$ built from types that belong to* Tp_1 *the following equivalences hold:*

$$\begin{aligned} \mathrm{L}_1 \vdash \Pi \to C \iff \mathrm{L}_1 \vdash (\Pi \to C)[p_i := (\mathbf{1} \cdot p_i) \cdot \mathbf{1}] \\ \iff \mathrm{L}_1^- \vdash (\Pi \to C)[p_i := (\mathbf{1} \cdot p_i) \cdot \mathbf{1}]. \end{aligned}$$

Proof. The first equivalence follows from the fact that $p_i \leftrightarrow (\mathbf{1} \cdot p_i) \cdot \mathbf{1}$.

In the second equivalence the right-to-left implication is obvious. Let us prove the other one: we shall deduce the third statement from the first one (which is equivalent to the second one). We substitute $(\mathbf{1} \cdot p_i) \cdot \mathbf{1}$ in the L_1-derivation of $\Pi \to C$. It is easy to see that this substitution conserves the $(\to \mathbf{1})_{\mathrm{w}}$ axioms and all rules. Now it is sufficient to check that the result of such a substitution in $(\mathrm{ax})_{\mathrm{w}}$ is derivable in L_1^-:

$$\dfrac{\dfrac{\dfrac{\dfrac{\mathbf{1}^{k+1} \to \mathbf{1} \quad p_i \to p_i}{\mathbf{1}^{k+1}\, p_i \to \mathbf{1} \cdot p_i}(\to\cdot) \quad \mathbf{1}^{m+1} \to \mathbf{1}}{\mathbf{1}^{k}\,\mathbf{1}\, p_i\, \mathbf{1}^{m+1} \to (\mathbf{1} \cdot p_i) \cdot \mathbf{1}}(\to\cdot)}{\mathbf{1}^{k}\,(\mathbf{1} \cdot p_i)\,\mathbf{1}\,\mathbf{1}^{m} \to (\mathbf{1} \cdot p_i) \cdot \mathbf{1}}(\cdot\to)}{\mathbf{1}^{k}\,((\mathbf{1} \cdot p_i) \cdot \mathbf{1})\,\mathbf{1}^{m} \to (\mathbf{1} \cdot p_i) \cdot \mathbf{1}}(\cdot\to)$$

The next two lemmas essentially repeat the argument from [6] about the closed (without variables but with constants) fragment of multiplicative cyclic linear logic.

Lemma 2. *If* $\mathrm{L}_{\mathbf{1}}^{-} \vdash \Pi \to C$ *and* $q \in \mathrm{Pr}$, *then* $\mathrm{L}^{*} \vdash (\Pi \to C)[\mathbf{1} := q \setminus q]$.

Proof. Perform the substitution in the $\mathrm{L}_{\mathbf{1}}^{-}$-derivation of $\Pi \to C$. Axioms $(\mathrm{ax})_{\mathrm{w}}$ and rules of inference will remain untouched. Axioms $(\to \mathbf{1})_{\mathrm{w}}$ will transform into sequents $(q \setminus q)^k \to q \setminus q$, which are derivable in L^{*}:

$$\dfrac{\dfrac{\dfrac{q \to q \quad \dfrac{\vdots}{q\,(q \setminus q)\,\ldots\,(q \setminus q) \to q}(\setminus\to)}{q\,(q \setminus q)\,(q \setminus q)\,\ldots\,(q \setminus q) \to q}(\setminus\to)}{q\,(q \setminus q)\,(q \setminus q)\,(q \setminus q)\,\ldots\,(q \setminus q) \to q}(\setminus\to)}{(q \setminus q)\,(q \setminus q)\,(q \setminus q)\,\ldots\,(q \setminus q) \to q \setminus q}(\to\setminus)$$

where the derivation starts with

$$\dfrac{q \to q \quad \dfrac{q \to q \quad q \to q}{q\,(q \setminus q) \to q}(\setminus\to)}{\ldots}(\setminus\to)$$

Lemma 3. *If* $\mathrm{L}^{*} \vdash (\Pi \to C)[\mathbf{1} := q \setminus q]$ *and q is a primitive type that does not occur in* $\Pi \to C$, *then* $\mathrm{L}_{\mathbf{1}} \vdash \Pi \to C$.

Proof. Let $\mathrm{L}^{*} \vdash (\Pi \to C)[\mathbf{1} := q \setminus q]$. Consider the sequent $(\Pi \to C)[\mathbf{1} := q \setminus q][q := \mathbf{1}]$. On the one hand, it is derivable in $\mathrm{L}_{\mathbf{1}}$, since $(\Pi \to C)[\mathbf{1} := q \setminus q]$ is derivable in $\mathrm{L}_{\mathbf{1}}$ (due to the conservativity of $\mathrm{L}_{\mathbf{1}}$ over L^{*}) and the substitution rule is valid in $\mathrm{L}_{\mathbf{1}}$. On the other hand, the sequent involved is actually $(\Pi \to C)[\mathbf{1} := \mathbf{1} \setminus \mathbf{1}]$, because occurrences of q could appear only inside the types $q \setminus q$ that are substituted for $\mathbf{1}$. Therefore the derivability of this sequent in $\mathrm{L}_{\mathbf{1}}$ is equivalent to the derivability of $\Pi \to C$ (since $\mathbf{1} \leftrightarrow \mathbf{1} \setminus \mathbf{1}$).

Proof (of Theorem 4).

$$\begin{aligned} \mathrm{L}_{\mathbf{1}} \vdash \Pi \to C &\Longrightarrow \mathrm{L}_{\mathbf{1}}^{-} \vdash \big(\Pi \to C\big)[p_i := (\mathbf{1} \cdot p_i) \cdot \mathbf{1}] \\ &\Longrightarrow \mathrm{L}^{*} \vdash \big(\Pi \to C\big)[p_i := (\mathbf{1} \cdot p_i) \cdot \mathbf{1}][\mathbf{1} := q \setminus q] \\ &\Longrightarrow \mathrm{L}_{\mathbf{1}} \vdash \big(\Pi \to C\big)[p_1 := (\mathbf{1} \cdot p_i) \cdot \mathbf{1}] \Longrightarrow \mathrm{L}_{\mathbf{1}} \vdash \Pi \to C. \end{aligned}$$

Here the first and the fourth implications hold due to Lemma 1, the second one holds due to Lemma 2, and the third one holds due to Lemma 3.

Proof (of Theorem 3). Let $M = \mathfrak{L}(\mathcal{G})$ for some $\mathrm{L}_{\mathbf{1}}$-grammar $\mathcal{G}$. The language M is generated by the L^{*}-grammar $\mathcal{G}[p_i := (\mathbf{1} \cdot p_i) \cdot \mathbf{1}][\mathbf{1} := q \setminus q]$ where q is a primitive type that is not used in $\mathcal{G}$.

Acknowledgments. I am most grateful to Prof. M. Pentus for guiding me into the subject and constant attention to my studies.

This work was supported by the Russian Foundation for Basic Research [08-01-00399], by the Presidential Council for Support of Leading Scientific Schools [NSh-65648.2010.1], and by the Scientific and Technological Cooperation Programme Switzerland–Russia [STCP-CH-RU].

References

1. Bar-Hillel, Y., Gaifman, C., Shamir, E.: On categorial and phrase-structure grammars. Bull. Res. Council Israel Sect. F 9F, 1–16 (1960)
2. Buszkowski, W.: The equivalence of unidirectional Lambek categorial grammars and context-free grammars. Zeitschr. für math. Logik und Grundl. der Math. 31, 369–384 (1985)
3. Kuznetsov, S.: Lambek grammars with one division and one primitive type. Unpublished manuscript (2010)
4. Lambek, J.: The mathematics of sentence structure. American Math. Monthly 65(3), 154–170 (1958)
5. Lambek, J.: Deductive systems and categories II: Standard constructions and closed categories. In: Hilton, P. (ed.) Category Theory, Homology Theory and Their Applications I. Lect. Notes Math., vol. 86, pp. 76–122. Springer, Berlin (1969)
6. Métayer, F.: Polynomial equivalence among systems LLNC, $LLNC_a$ and $LLNC_0$. Theor. Comput. Sci. 227(1), 221–229 (1999)
7. Pentus, M.: Lambek grammars are context free. In: Proc. of the 8th Annual IEEE Symposium on Logic in Computer Science, pp. 429–433. IEEE Computer Society Press, Los Alamitos (1993)

Resolving Plural Ambiguities by Type Reconstruction

Hans Leiß

Centrum für Informations- und Sprachverarbeitung
Universität München
Oettingenstr. 67, D-80538 München, Germany

Abstract. We describe a type reconstruction algorithm for a fragment of natural language. It is based on Hindley's algorithm for simple types, but extends it with subtyping and overloading. We extend one of Montague's fragments of English by plural noun phrases which may have several types and by overloaded verbs to allow for distributed and non-distributed readings of noun phrases and verb arguments. We demonstrate how type reconstruction can select suitable meanings of subject noun phrases depending on the meaning of verb phrases. Thus, type reconstruction enables us to handle some violations of Frege's compositionality principle.

1 Introduction

Some plural noun phrases have different readings:

- distributive: *John and Mary dream a nightmare (each)*
- reciprocal: *John and Mary like each other*
- collective: *John and Mary play chess (together)*

Montague[12] only treated the distributive reading: the noun phrase meaning is obtained by abstraction from the predicate, and the predicate must be a property of individuals. However, this abstraction cannot be applied to a symmetric relation between individuals or to a group of individuals (assuming groups are not individuals of the same kind), so we need different meanings for the reciprocal and collective reading. This poses the question how to represent the meaning of the noun phrase: do we need different meanings for different contexts, and if so, how many differences have to be made? Frege's compositionality principle appears to be violated here, since the context has an impact on the meaning of the noun phrase.

We approach this problem in the following way. Based on a distinction between types of verb arguments –individual, pair and group arguments–, we infer types of verb phrases. Assuming there are only finitely many base types (subtypes of individuals), there are only finitely many types of verb phrases or types of contexts for a noun phrase. Abstracting from those, we arrive at finitely many λ-terms that represent the meanings of a noun phrase in these types of contexts.

P. de Groote and M.-J. Nederhof (Eds.): Formal Grammar 2010/2011, LNCS 7395, pp. 267–286, 2012.

The exact number and form of these terms depend on the structure of the noun phrase, of course, and each has a different type. Since we have to assume that the same verb can have different types –for example, because it is applicable to groups as well as to individuals, or to individuals of different type–, verb phrases will also have a number of different meanings. To analyse a sentence, we compute the possible meanings of noun phrase and context and compare their types. Only those meanings whose types match are suitable meanings in the given situation; here, matching means that the type of an argument of a function is a subtype of the functions argument-type.

We develop this idea to some extent in an extension of Montague's PTQ-grammar, focussing on the distinction between distributive and reciprocal readings of plural noun phrases. Some actual programming examples are presented at the end.

2 The Extensional Version of Montague's PTQ

We first recall part of (the extensional fragment of) Montague's PTQ-fragment of English[12]. The grammar has two basic syntactic categories, individual names e and sentences t. From these, complex syntactic categories can be built using two binary constructors, $/$ and $/\!/$. Among the complex categories are common nouns $CN := t/\!/e$, intransitive verbs and verb phrases $IV := t/e$, noun phrases or terms $T := t/IV$, transitive verb phrases $TV := IV/T$, and determiners $DET := T/CN$.

There are atomic expressions of some of these categories, like

$$\text{he}_n, \text{John} : e, \quad \text{walk} : IV, \quad \text{love} : TV, \quad \text{woman} : CN, \quad \text{every} : DET.$$

Complex expressions of a given category can be built according to a number of syntactic construction rules of the form

$$(\text{S Nr.})\ \frac{\alpha_1 : A_1, \ldots, \alpha_k : A_k}{\alpha : A},$$

where $\alpha_1, \ldots, \alpha_k$ are expressions (strings) of categories $A_1, \ldots, A_k$, respectively, and from these the complex expression α of category A is formed by some functions to modify and concatenate strings.

To give meanings to the expressions, each syntactic category A is translated to a simple type A' over the basic types e and t, using

$$e' := e, \qquad t' := t, \qquad (A/B)' := (B' \to A'), \qquad (A/\!/B)' := (B' \to A').$$

Each expression e of category A is translated to a λ-term e' of type A'. For example, pronouns are translated to variables, $\text{he}_n' := x_n : e$, individual names, common nouns and intransitive verbs to constants

$$\text{John}' := \text{john'} : e, \quad \text{woman}' := \text{woman'} : e \to t, \quad \text{walk}' := \text{walk'} : e \to t.$$

Constants of categories DET and TV with complex argument category are translated to appropriate λ-terms of higher-order types:

$$\text{every}' := \lambda N^{e\to t}\lambda P^{e\to t}\forall x^e(Nx \to Px) : DET' = (e \to t) \to ((e \to t) \to t).$$

Since we are only dealing with the extensional fragment, we assume that for each constant $v : TV$ there is a first-order extensional predicate $v_* : e \to (e \to t)$ which determines the translation of v to a higher-order object $v' : T' \to (e \to t)$ by $v' := \lambda P^{(e\to t)\to t}\lambda x^e.P(\lambda y^e.v_*(y)(x))$, for example

$$\text{love}' := \lambda P^{(e\to t)\to t}\lambda x^e.P(\lambda y^e.\text{love}_*(y)(x)).$$

Writing, as usual, $v(x_1, x_2)$ for $v(x_2)(x_1)$ when $v : (C/C_1)/C_2$, this amounts to a "meaning postulate"

$$\exists v_*^{e\to(e\to t)}\forall P^{(e\to t)\to t}\forall x^e(v'(x, P) \leftrightarrow P(\lambda y^e.v_*(x, y))).$$

The translation of a complex expression is defined by translation rules

$$(\text{T Nr.})\ \frac{\alpha'_1 : A'_1,\ \ldots, \alpha'_k : A'_k}{\alpha' : A'}$$

which build the translation of α from the translation of its constituent expressions $\alpha_1, \ldots, \alpha_k$ in the corresponding syntactic construction (S Nr.). We use the following subset of Montague's rules (with rule numbers as in [12]):

$$(\text{S 1})\ \frac{\alpha : e}{\alpha : T} \qquad (\text{T 1})\ \frac{\alpha' : e}{\lambda P^{e\to t}.P(\alpha') : (e \to t) \to t}$$

$$(\text{S 2})\ \frac{\delta : DET,\quad \xi : CN}{\delta\ \xi : T} \qquad (\text{T 2})\ \frac{\delta' : (CN' \to T'),\quad \xi' : CN'}{\delta'(\xi') : T'}$$

$$(\text{S 3})_n\ \frac{\xi : CN,\quad \varphi : t}{\xi \text{ such that } \varphi[he_n/he] : CN} \qquad (\text{T 3})_n\ \frac{\xi' : CN',\quad \varphi' : t}{\lambda x_n^e(\xi'(x_n) \wedge \varphi') : CN'}$$

$$(\text{S 4})\ \frac{\alpha : T,\quad \delta : IV}{\alpha\ \delta^{3.sg} : t} \qquad (\text{T 4})\ \frac{\alpha' : T',\ \delta' : IV'}{\alpha'(\delta') : t}$$

$$(\text{S 5})\ \frac{\delta : TV,\quad \beta : T}{\delta\ \beta^{acc} : IV} \qquad (\text{T 5})\ \frac{\delta' : (T' \to IV'),\quad \beta' : T'}{\delta'(\beta') : IV'}$$

$$(\text{S 11})_a\ \frac{\varphi : t,\quad \psi : t}{\varphi \text{ and } \psi : t} \qquad (\text{T 11})_a\ \frac{\varphi' : t,\quad \psi' : t}{(\varphi' \wedge \psi') : t}$$

$$(\text{S 11})_b\ \frac{\varphi : t,\quad \psi : t}{\varphi \text{ or } \psi : t} \qquad (\text{T 11})_b\ \frac{\varphi' : t,\quad \psi' : t}{(\varphi' \vee \psi') : t}$$

$$(\text{S 12})_a\ \frac{\delta : IV,\quad \gamma : IV}{\delta \text{ and } \gamma : IV} \qquad (\text{T 12})_a\ \frac{\delta' : e \to t,\quad \gamma' : e \to t}{\lambda x^e(\delta'(x) \wedge \gamma'(x)) : e \to t}$$

$$(\text{S } 12)_b \ \frac{\delta : IV, \quad \gamma : IV}{\delta \text{ or } \gamma : IV} \qquad (\text{T } 12)_b \ \frac{\delta' : e \to t, \quad \gamma' : e \to t}{\lambda x^e(\delta'(x) \vee \gamma'(x)) : e \to t}$$

$$(\text{S } 13) \ \frac{\alpha : T, \quad \beta : T}{\alpha \text{ or } \beta : T} \qquad (\text{T } 13) \ \frac{\alpha' : T', \qquad \beta' : T'}{\lambda P^{e \to t}(\alpha'(P) \vee \beta'(P)) : T'}$$

$$(\text{S } 14)_n \ \frac{\alpha : T, \quad \varphi : t, \quad he_n \in \mathit{free}(\varphi)}{\varphi[he_n/\alpha] : t} \quad (\text{T } 14)_n \ \frac{\alpha' : T', \quad \varphi' : t, \quad x_n \in \mathit{free}(\varphi')}{\alpha'(\lambda x^e_n . \varphi') : t}$$

The two substitution operations $[he_n/he]$ in $(\text{S } 3)_n$ and $[he_n/\alpha]$ in $(\text{S } 14)_n$ are not spelled out here and differ slightly; depending on the gender of the common noun ξ in $(\text{S } 3)_n$, $\varphi[he_n/he]$ is to replace the leftmost occurrence of he_n in φ by *he* or *she* and adjust the remaining ones by suitable (personal or reflexive) pronouns, while $\varphi[he_n/\alpha]$ is to replace the leftmost occurrence of he_n in φ by α and adjust the remaining ones by suitable (personal or reflexive) pronouns.

Where needed, we assume a category of ditransitive verb phrases $DTV := TV/T$, with the construction and translation rules as for transitive verbs:

$$(\text{S } 5') \ \frac{\delta : DTV, \quad \beta : T}{\delta \text{ to } \beta^{dat} : TV} \qquad (\text{T } 5') \ \frac{\delta' : (T' \to TV'), \quad \beta' : T'}{\delta'(\beta') : TV'}$$

For constants $v : DTV$, an extensional ternary relation $v_* : e \to (e \to (e \to t))$ is assumed that determines the translation –using $v_*(x, y, z)$ for $v_*(z)(y)(x)$– via

$$v' = \lambda Q^{T'} \lambda P^{T'} \lambda x^e . Q(\lambda z^e . P(\lambda y^e . v_*(x, y, z))) : T' \to (T' \to (e \to t)) = DTV'.$$

This gives the dative object of a ditransitive verb scope over the accusative object, and both are in the scope of the subject. However, since we are not dealing with intensional verbs here, where it is important that quantified object noun phrases are "in the scope of the verb", we can restrict $\alpha : T$ in (S 4) and $\beta : T$ in (S 5) and (S 5') to be pronouns, and use $(\text{S } 14)_n$ to add noun phrases with the scope at will.

2.1 Different Meanings of Plural Noun Phrases

Montague's rules deal with singular noun phrases only. For example, noun phrases can be combined by *or* with (S 13), but there is no rule to combine noun phrases by *and.* By adding conjunction and further constructions of plural noun phrases, expressions with several meanings arise. In general, a plural noun phrase translates to *several* λ-terms of different types. We distinguish three different readings of plural noun phrases:

(i) In the distributive reading, the plural noun phrase acts as a function applied to properties of individuals, hence $NP' = (e \to t) \to t$. For example, for a common noun N and a predicate P, ((most N) P) means $most(x : e, N(x), P(x))$, i.e. the application of the predicate is distributed to the individuals in the restriction set of the noun phrase.

(ii) In the group reading, the plural noun phrase acts as a function applied to properties of groups of individuals, hence $NP' = (g(e) \to t) \to t$, where $g(e)$ is the type of groups of individuals. Treating groups as finite sets, ((the N) P) means $P(\{x : e \mid N(x)\})$, and ((a, b and c) P) means $P(\{a, b, c\})$.

(iii) In the reciprocal or pair reading, the plural noun phrase acts as a function applied to properties of pairs of individuals, hence $NP' = (p(e) \to t) \to t$, where $p(e)$ is the type of unordered pairs of objects of type e. The pair reading is often triggered by a reciprocal pronoun, as in She introduced the guests to each other, but need not be, as in John and Bill are neighbours.

Which of the three readings a plural noun phrase $np : NP_{pl}$ can have depends on its form; some forms, like the conjunction of proper names, admit all three readings. Thus, the meaning of a noun phrase, or at least the selection among its possible meanings, depends on the context of use, and does not strictly obey Frege's compositionality principle.

For simplicity, below we use ordered pairs instead of unordered ones, and lists instead of finite sets. That is, we use $\langle x, y \rangle : e \times e$ instead of $\{x, y\} : p(e)$ and quantifiers over ordered pairs instead of quantifiers over unordered pairs. Likewise, we use the type e^* of lists of individuals as $g(e)$, and assume a predicate $member : e \to (e^* \to t)$ and list comprehension $[x \mid N(x)]$ to go from a predicate $N : e \to t$ to a group. However, the only essential point for our purpose is that we can distinguish individuals, pairs and groups by their type.

2.2 Verb Types and Overloaded Verbs

An important syntactic difference between formal and natural languages is that in logic, quantifiers construct *formulas*, while in natural languages, they construct noun phrases, and these, like *terms* in logic, occur as arguments of predicates resp. complements of verbs. Montague[12] interpreted intransitive verbs and verb phrases as properties of individuals, $IV', VP' := e \to t$, and noun phrases as properties of those, $T' := (e \to t) \to t$, so that the predication construction $np \cdot vp : t =: S$ becomes the application of a function $np : T'$ to an argument $vp : VP'$. Montague's translation rules for simple sentences with compound subject, for example

$$\begin{aligned}((\text{every man}) \text{ walks})' &= (\lambda P^{e\to t} \forall x^e (\text{man}'(x) \to P(x)))(\lambda x^e.\text{walk}'(x)),\\ ((\text{John or Mary}) \text{ talks})' &= \lambda P^{e\to t}(P(\text{john}') \vee P(\text{mary}'))(\lambda x.\text{talk}'(x)),\end{aligned}$$

always distribute the predicate to the variable or constant individual parts of the noun phrase, to which the predicate can be applied in accordance with the types. Thus, Montague's PTQ has noun phrases in singular only, which are given the distributive reading.

For plural noun phrases, however, as mentioned in the previous section, we can distinguish at least three readings. Which of these is the correct one for a particular occurrence? Since the noun phrase occupies an argument position of a verb (or adjective or noun), it is often possible to choose among the meanings according to the argument type of this verb (or adjective or noun, respectively).

We distinguish between individual type e, pair type $p(e)$, or group type $g(e)$, respectively. For example, we may have love : $e \to (e \to t)$, but be alike : $p(e) \to t$, and meet : $g(e) \to t$. Moreover, we admit *overloaded* verbs, i.e. verbs with several types. For example, we may have the transitive marry : $e \to (e \to t)$ as well as the intransitive reciprocal marry : $p(e) \to t$ (i.e. marry each other), and both welcome : $e \to (e \to t)$ with individual and welcome : $e \to (g(e) \to t)$ with group object type.[1]

Predicates of type $p(e) \to t$ may either be basic verbs, nouns, or adjectives, as in marry, resemble, neighbour, sibling, or parallel, similar, or constructed from transitive or ditransitive verbs and a reciprocal pronoun. Often[4,13], the reciprocal pronoun is interpreted as a generalized quantifier of type $\langle 1, 2\rangle$, i.e. $reci(A, R)$ expresses a relation between sets A and binary relations R of individuals, such as

$$|A| \geq 2 \wedge A^2 \setminus Id \subseteq R,$$
$$|A| \geq 2 \wedge A^2 \setminus Id \subseteq R^+, \qquad \text{or}$$
$$|A| \geq 2 \wedge \forall x \in A \exists y \in A\, (y \neq x \wedge \langle x, y\rangle \in R \cap \breve{R} \setminus Id),$$

where R^+ is the transitive closure and $\breve{R}$ the converse of R. We deviate from this view and consider the reciprocal pronoun as a predicate transformer

$$reci : (e \to (e \to t)) \to (p(e) \to t)$$

which takes a binary relation to a unary predicate on unordered pairs. Below, we interprete *reci* as mapping a binary relation R to its symmetric irreflexive kernel $R \cap \breve{R} \setminus Id$, so that $reci(R) : p(e) \to t$ is the predicate P obtained from R by

$$P\{x, y\} :\iff R(x, y) \wedge R(y, x) \wedge x \neq y.$$

Other interpretations, like the "ordering reciprocal" of spatial prepositions, viz. on top of each other, might be chosen depending on semantic properties of the input relation R, cf. [13]. How the predicate $reci(R)$ interacts with the subject noun phrase is left to the predication construction and may depend on the structure of the noun phrase. Basically, we distribute $reci(R)$ to *pairs* of individuals of the restriction set of the plural noun phrase: for example, `((most N) (R (each other)))` means[2]

$$most(\{x, y\} : p(e), N(x) \wedge N(y) \wedge x \neq y, R(x, y) \wedge R(y, x) \wedge x \neq y).$$

Of course, nouns also can be overloaded, for example child : $e \to (e \to t)$ vs. child : $e \to (p(e) \to t)$. Hence the plural John and Mary can have different readings in the object position of child, resp. the noun child is distributed or not:

[1] See Franconi[5] for an approach where different readings of a verb are coded by applying operators to the verb.

[2] Note that we need binary determiners with symmetric restriction relations anyway, viz. most siblings are alike.

the children of John and Mary versus the common children of John and Mary. In the grammar below, we only deal with overloaded verbs.

Besides distinguishing between individuals e, pairs $p(e)$ and groups $g(e)$, it is often useful to structure the type of individuals by subtypes. Subtypes give another source of overloading. For example, with a subtype $m \leq e$ of humans and a subtype $s \leq e$ of unanimated objects, a verb like to rise ought to be overloaded as rise : $s \to t$, rise : $m \to t$ for The sun rises vs. John rose from the table. For each subtype $\sigma \leq e$, subtypes $p(\sigma) \leq p(e)$ and $g(\sigma) \leq g(e)$ arise. We postpone subtypes to section 3.1.

2.3 Choosing among Noun Phrase Meaning

For each occurrence of a plural noun phrase $np : NP_{pl}$ a particular one of its possible meanings, distributive, pair or group reading, has to be chosen. The proper choice depends on the noun phrase's types as well as the type of the verb argument place at which it occurs. In the predication construction $np \cdot vp : S$, say, we inspect the type of the subject position: if the $vp : VP$ is a property of individuals, we read the np distributively, if the $vp : VP$ is a symmetric binary relation, the np is read as (quantification over) a set of pairs; and if the vp is a group property, then the np is read as a group or quantification over groups. So we need three readings of $np : NP_{pl}$, of types $(p(e) \to t) \to t$, $(e \to t) \to t$, and $(g(e) \to t) \to t$.

While the pairwise reading largely depends on overt reciprocals, distinctions between distributive and group reading depend on the type of verb argument at which the plural noun phrase occurs; if the verb is overloaded, several readings may still be possible for the noun phrase.

2.4 Extension of PTQ by Plural Noun Phrases and Reciprocal Pronouns

In order to extend the PTQ-fragment to some plural noun phrases, we need to refine syntactic categories by features, add constants and syntax rules to construct expressions of these categories, and add translations into typed λ-calculus. We only add the number feature, which is relevant for the semantics; morphological features like case are handled implicitly through the string combination functions.

(i) We need to refine syntactic categories by features and translate them to suitable types.

First, we split the category IV_{pl} into IV_e, $IV_{e\times e}$ and IV_{e^*} according to the argument type: e for distributive, $e \times e$ for reciprocal, and e^* for group reading, and translate them to

$$IV'_e := e \to t, \quad IV'_{e\times e} := e \times e \to t, \quad IV'_{e^*} := e^* \to t.$$

This then gives rise to a corresponding splitting of complex categories, like

$$T_e := t/IV_e, \quad T_{e\times e} := t/IV_{e\times e}, \quad T_{e^*} := t/IV_{e^*},$$
$$TV_{e,e} := IV_e/T_e, \quad TV_{e,e^*} := IV_e/T_{e^*}, \quad TV_{e^*,e} := IV_{e^*}/T_e$$

etc., so that

$$T'_e := T' = (e \to t) \to t, \quad T'_{e\times e} := (e \times e \to t) \to t, \quad T'_{e^*} := (e^* \to t) \to t$$

and, for example,

$$TV'_{e,e^*} = (IV_e/T_{e^*})' = (T'_{e^*} \to IV'_e) = ((e^* \to t) \to t) \to (e \to t).$$

If we split noun phrases in singular noun phrases T_{sg} and plural noun prases T_{pl}, we can identify Montague's category T with T_{sg} and further split T_{pl} into T_e[3] for those with distributive reading, $T_{e\times e}$ for those with reciprocal reading, and T_{e^*} for those with group reading.

(ii) We need to add constants of these new categories and give their translation to typed λ-terms.

a) constants for plural determiners:[4]

$$\begin{aligned}
(the : DET_e)' &:= \lambda N^{e\to t}\lambda P^{e\to t}.all(x^e, N(x), P(x))\\
(the : DET_{e\times e})' &:= \lambda N^{e\to t}\lambda P^{e\times e\to t}.all(\langle x,y\rangle^{e\times e}, N\{x,y\}, P(x,y))\\
(the : DET_{e^*})' &:= \lambda N^{e\to t}\lambda P^{e^*\to t}.P[x^e \mid N(x)]\\
(all : DET_e)' &:= \lambda N^{e\to t}\lambda P^{e\to t}.all(x^e, N(x), P(x))\\
(all : DET_{e\times e})' &:= \lambda N^{e\to t}\lambda P^{e\times e\to t}.all(\langle x,y\rangle^{e\times e}, N\{x,y\}, P(x,y))\\
(some : DET_e)' &:= \lambda N^{e\to t}\lambda P^{e\to t}.ex(\langle x,y\rangle^{e\times e}, N\{x,y\}, P\{x,y\}),\\
(some : DET_{e\times e})' &:= \lambda N^{e\to t}\lambda P^{e\times e\to t}.ex(\langle x,y\rangle^{e\times e}, N\{x,y\}, P(x,y))\\
(most : DET_e)' &:= \lambda N^{e\to t}\lambda P^{e\to t}most(x^e, N(x), P(x))\\
(most : DET_{e\times e})' &:= \lambda N^{e\to t}\lambda P^{e\times e\to t}most(\langle x,y\rangle^{e\times e}, N\{x,y\}, P(x,y))
\end{aligned}$$

Here, $N^{e\to t}\{x,y\}$ is an abbreviation for the formula $N(x)\wedge N(y)\wedge x \neq y$, and $P^{e^*\to t}[x^e \mid N(x)]$ means application of P to the group formed from the extension of $N^{e\to t}$.

b) constants for verbs with reciprocal or collective argument types:

$$marry : IV_{e\times e}, \quad be\ similar : IV_{e\times e}, \quad collect : TV_{e,e^*}, \quad agree\ on : TV_{e^*,e}.$$

[3] Sometimes we use T_e as if it included the singular noun phrases, so that $v : DTV_{e,e,e}$ can be used with singular as well as plural complements; in subject position, agreement with the verb must be achieved by the string combination functions.

[4] We don't add determiners of arbitrary type to suppress searching through higher-order domains. But for subtypes $\sigma \leq e$, it may be useful to have special determiners DET_σ, for example to distinguish interrogatives who and what.

Note that some of the verbs with reciprocal subject have transitive analogs:

$$marry : TV, \quad be\ similar\ to : TV.$$

From the transitive analogs the reciprocal version arise by adding an implicit "each other".

c) constants for common nouns in plural that denote symmetric binary relations:

$$neighbours : CN_{e\times e}, \quad friends : CN_{e\times e}, \quad siblings : CN_{e\times e}.$$

They can be used in combination with an optional reciprocal pronoun complement. To build definite noun phrases from these relational nouns, variants like

$$(most : DET_{e\times e})' := \lambda N^{e\times e\to t}\lambda P^{e\times e\to t}.most(\langle x,y\rangle^{e\times e}, N(x,y), P(x,y))$$

of the plural determiners are needed. (cf. remark 1 below).

(iii) We need new rules to construct and translate compound expressions of these categories.
The reciprocal pronoun is used to build intransitive verbs with pair subjects from transitive verbs with individual arguments, and transitive verbs with pair objects from ditransitive verbs:

$$(\text{S reci})_1 \ \frac{\delta : TV_{e,e}}{\delta \text{ each other} : IV_{e\times e}} \quad (\text{S reci})_2 \ \frac{\delta : DTV_{e,e,e}}{\delta \text{ to each other} : TV_{e,e\times e}}.$$

With these rules we can get, for example,

$$know\ each\ other : IV_{e\times e} \text{ from } know : TV_{e,e}, \text{ and}$$
$$introduce\ to\ each\ other : TV_{e,e\times e} \text{ from } introduce\ to : DTV_{e,e,e}.$$

Since transitive verbs $\delta : TV = IV/T = (t/e)/T$ and ditransitive verbs need a subject of category e but objects of category $T = t/(t/e)$, we have to "type-raise" x^e to $x^{T'} := \lambda P^{e\to t}.P(x)$ for the objects in the translation:

$$(\text{T reci})_1 \ \frac{\delta' : TV'_{e,e}}{\lambda\langle x,y\rangle^{e\times e}(\delta'(x,y^{T'}) \wedge \delta'(y,x^{T'}) \wedge x \neq y) : IV'_{e\times e}}$$

$$(\text{T reci})_2 \ \frac{\delta' : DTV'_{e,e,e}}{\begin{array}{r}\lambda R^{(e\times e\to t)\to t}\lambda z^e.R(\lambda\langle x,y\rangle^{e\times e}(\delta'(z,x^{T'},y^{T'}) \wedge \\ \delta'(z,y^{T'},x^{T'}) \wedge x \neq y)) : TV'_{e,e\times e}\end{array}}$$

For example, if $\delta : TV$ is a constant and $\delta_* : e \to (e \to t)$ the assumed extensional relation with

$$\delta' = \lambda P^{T'}.\lambda x^e.P(\lambda y^e.\delta_*(x,y)) : TV',$$

then

$$\delta'(x^e, y^{T'}) = \delta'(y^{T'})(x) = y^{T'}(\lambda y^e.\delta_*(x,y)) = \delta_*(x,y)$$

and hence

$$(\delta \text{ each other})' = \lambda\langle x,y\rangle^{e\times e}(\delta_*(x,y) \wedge \delta_*(y,x) \wedge x \neq y) : IV'_{e\times e}.$$

Thus, on the level of individuals, (T reci)$_1$ corresponds to the transition from the δ_* to its symmetric irreflexive kernel, according to

$$\frac{\delta_* : e \to (e \to t)}{\lambda\langle x,y\rangle^{e\times e}(\delta_*(x,y) \wedge \delta_*(y,x) \wedge x \neq y) : e \times e \to t = IV_{e\times e}{}'}.$$

The same holds for (T reci)$_2$ and the symmetric irreflexive kernel of a ternary relation δ_* with respect to the object arguments, except that the object of is of category $T_{e\times e}$.

Moreover, rules like (S 12) to construct boolean combinations of IVs and TVs, have to be extended to the new categories $IV_{e\times e}$, $TV_{e,e}$, $TV_{e,e\times e}$ etc., so that we can build, for example,

They$^{T_{e\times e}}$ (neither (are married)$^{IV_{e\times e}}$ nor (know each other)$^{IV_{e\times e}}$).

For singular terms, reflexive pronouns and the transition to the reflexive kernel of a binary relation can be added in a similar way:

$$(\text{S refl})_{1,n}\ \frac{\delta : TV}{\delta \text{ himself}_n : IV} \qquad (\text{T refl})_{1,n}\ \frac{\delta' : TV'}{\lambda x_n^e.\delta'(x_n, x_n^{T'}) : IV'}$$

$$(\text{S refl})_{2,n}\ \frac{\delta : DTV}{\delta \text{ to himself}_n : TV} \qquad (\text{T reci})_{2,n}\ \frac{\delta' : DTV'}{\lambda P^{(e\to t)\to t}\lambda x_n^e.P(\lambda y^e.\delta'(x_n, y^{T'}, x_n^{T'})) : TV'}$$

Suitable adjustments to (S 14)$_n$ are needed: when a term $\alpha : T$ is inserted for he_n at the subject position of a reflexive predicate, $himself_n$ has to be adapted to the gender and number of α.

(iv) We need rules to construct and translate plural noun phrases, where $pl \in \{e, e\times e, e^*\}$:

$$(\text{S }1)_{a,e}\ \frac{\alpha : T, \qquad \beta : T}{\alpha \textit{ and } \beta : T_e} \qquad (\text{T }1)_{a,e}\ \frac{\alpha' : T, \qquad \beta' : T}{\lambda P^{e\to t}(\alpha'(P) \wedge \beta'(P))}$$

$$(\text{S }1)_{a,e\times e}\ \frac{\alpha : T, \qquad \beta : T}{\alpha \textit{ and } \beta : T_{e\times e}} \qquad (\text{T }1)_{a,e\times e}\ \frac{\alpha' : T', \qquad \beta' : T'}{\lambda R^{e\times e\to t}.\alpha'(\lambda x^e.\beta'(\lambda y^e.R(x,y)))}$$

$$(\text{S }1)_{a,e^*}\ \frac{\alpha : e, \qquad \beta : e}{\alpha \textit{ and } \beta : T_{e^*}} \qquad (\text{T }1)_{a,e^*}\ \frac{\alpha' : e, \qquad \beta' : e}{\lambda P^{e^*\to t}.P([\alpha', \beta'])}.$$

$$(\text{S } 2)_{pl} \ \frac{\delta : DET_{pl}, \qquad \xi : CN}{\delta \ \xi^{pl} : T_{pl}} \qquad (\text{T } 2)_{pl} \ \frac{\delta' : (CN' \to T'_{pl}), \quad \xi' : CN'}{\delta'(\xi') : T'_{pl}}$$

$$(\text{S } 13)_{a,pl} \ \frac{\alpha : T_{pl}, \qquad \beta : T_{pl}}{\alpha \ and \ \beta : T_{pl}} \qquad (\text{T } 13)_{a,pl} \ \frac{\alpha' : T'_{pl}, \qquad \beta' : T'_{pl}}{\lambda P^{pl \to t}(\alpha'(P) \wedge \beta'(P))}$$

$$(\text{S } 13)_{b,pl} \ \frac{\alpha : T_{pl}, \qquad \beta : T_{pl}}{\alpha \ or \ \beta : T_{pl}} \qquad (\text{T } 13)_{b,pl} \ \frac{\alpha' : T'_{pl}, \qquad \beta' : T'_{pl}}{\lambda P^{pl \to t}(\alpha'(P) \vee \beta'(P))}$$

(v) We need new rules to construct and translate basic sentences with plural subject or object. Montague's rules have to be refined accordingly, for example

$$(\text{S } 4)_{pl} \ \frac{\alpha : T_{pl} \qquad \delta : IV_{pl}}{\alpha \ \delta : t} \qquad (\text{T } 4)_{pl} \ \frac{\alpha' : T'_{pl} \qquad \delta' : IV'_{pl}}{\alpha'(\delta') : t}.$$

Likewise for (S 5), (S 3) and (S 14). For the latter two, we need plural pronouns $they_{n,m} : e \times e$ that translate to pairs $\langle x_n, x_m \rangle$ of individual variables.

Remark 1. In $(\text{S } 2)_{pl}$ and the translation of determiners, we used separate categories $DET_{pl} = T_{pl}/CN$, $pl \in \{e, e \times e, e^*\}$, of determiners, but –to ease the parsing a bit– a single category for common nouns. More systematically, one could use $DET_{pl} := T_{pl}/CN_{pl}$ generally (not just for symmetric relational nouns) together with coercion rules:

$$(\text{S CN})_e \ \frac{\xi : CN}{\xi^{pl} : CN_e} \qquad (\text{T CN})_e \ \frac{\xi' : CN'}{\xi' : CN_e{}'}$$

$$(\text{S CN})_{e^*} \ \frac{\xi : CN}{\xi^{pl} : CN_{e^*}} \qquad (\text{T CN})_{e^*} \ \frac{\xi' : CN'}{\lambda x^{e^*}.\forall y^e (y \in x \to \xi'(y)) : CN_{e^*}{}'}$$

$$(\text{S CN})_{e \times e} \ \frac{\xi : CN}{\xi^{pl} : CN_{e \times e}} \qquad (\text{T CN})_{e \times e} \ \frac{\xi' : CN'}{\lambda \langle x, y \rangle^{e \times e}(\xi'(x) \wedge \xi'(y) \wedge x \neq y) : CN_{e \times e}{}'}$$

Notice, however, that except for the definite article, we did not add plural determiners for groups, as quantification over groups seems to afford a different form, as in $(many\ (groups\ of\ CN_{e^*})) : T_{e^*}$. We thus avoid the need to search through a large power set of individuals and restrict ourselves to groups that are explicitly named or can be obtained from a definable property of its members by comprehension.

Example 1. Using rule $(\text{T } 1)_{a,e \times e}$, ((John and Mary) love each other) translates to (love each other)′(john′, mary′). By the same rule, ((John and every student) love each other) (cf. [9], p.270) translates to

$$all(x, \text{student}'(x), (\text{love each other})'(\text{john}', x)).$$

Notice that $(\text{T } 1)_{a,e \times e}$ gives different meanings to (every man and some woman)$^{T_{e \times e}}$ and (some woman and every man)$^{T_{e \times e}}$, as it gives the left conjunct wide scope – not what a homomorphism would do!

Example 2. A plural noun phrase $np : T_{e\times e}$ with reciprocal reading takes a predicate $vp : IV_{e\times e}$ of individual pairs as argument; the determiners $DET_{e\times e}$ distribute the predicate vp to individual pairs in the restriction set: (most children like each other) translates to

$$most(\langle x, y\rangle, \mathsf{child}'(x) \wedge \mathsf{child}'(y) \wedge x \neq y, (\mathsf{like\ each\ other})'(x, y)).$$

Likwise, ordinary predicates $p : IV_e$ are distributed to individual components of complex noun phrases of category T_e or T. Binary quantifiers are necessary for interrogatives: *Which children are siblings?* has to be answered by a set of pairs, not by a set of individuals.

Example 3. Rules (S 13)$_{pl}$ permit us to have homogeneous plural noun phrase coordinations, where both coordinates get the same *pl*-reading. By (T 13)$_{pl}$, the meaning is a pointwise combination of those of the coordinates, so that *many men or most women fight each other* translates to *many men fight each other or most women fight each other.* We did not include inhomogeneous noun phrase coordinations, such as (np^{T_e} *and* $np^{T_{e\times e}}$) or ($np^{T_{e\times e}}$ *or* np^{T}), although perhaps those of the form (np^{T_e} *and/or* np^{T}) have a clear meaning.

3 Type Reconstruction

The meaning(s) of an expression α of the above fragment PTQ$^+$ are the typed λ-terms α' that can be assigned to α according to the translation rules. However, as in programming, it is helpful to compute the meanings in two phases. In the first phase, translate the parse tree into an untyped λ-term, and in the second phase, reconstruct the possible typings of the untyped term from the (known) types of its constants. The untyped terms are given by

$$\begin{array}{rll} t, s ::= & c & \text{(constants)} \\ \mid & x & \text{(variables)} \\ \mid & (t \cdot s) & \text{(applications)} \\ \mid & \lambda x.t & \text{(abstractions)} \\ \mid & f(t_1, \ldots, t_n) & \text{(algebraic terms)} \\ \mid & \langle s, t\rangle & \text{(pairs)} \\ \mid & \lambda\langle x, y\rangle.t & \text{(abstraction over pairs)} \\ \mid & [t_1, \ldots, t_n] & \text{(lists)} \end{array}$$

Pairs and lists are included for simplicity of notation, but could be simulated with suitable constants (and reduction rules). Algebraic terms are included as a special case, since the interpretation of their function constants f will depend on a given first-order structure (the "database"). Formulas are included via "logical" constants $\neg$, $\wedge$, $\vee$, $\rightarrow$, $\leftrightarrow$ for the boolean connectives and generalized quantifiers *ex, all, most*, i.e. $\neg\varphi := (\neg\cdot\varphi)$ and $most(x, \varphi, \psi) := ((most \cdot \lambda x\varphi) \cdot \lambda x\psi)$ etc. are formulas. Moreover, a cardinality quantifier *card* is used to express $|\{x \mid \varphi\}| \leq n$ by formulas $card(x, \varphi, n)$ where n is a constant. In addition to formula-building quantifiers, we also use two interrogative quantifiers *wh* and *whn* to build the questions $wh(x, \varphi, \psi)$ asking for the set (resp. list) of objects x which satisfy φ and ψ, and the question $whn(x, \varphi, \psi)$ asking for the number of elements in the set returned by $wh(x, \varphi, \psi)$.

The intented interpretations of formulas and questions are type structures based on a finite universe of individuals. Although definable higher-order functions are needed, we do not wish to perform searching and quantification over those, but stick to "database"-queries with first-order quantification.

3.1 Simple Types and Subtyping Rules

Among our base types, we need *nat*, *t* (Bool), *e* (entities). Among the complex types we need functions $(\sigma \to \tau)$, pairs $(\sigma \times \tau)$ at least as argument types of functions, and homogeneous lists σ^*.

We use a finite set of base types and a partial ordering among them, such that *nat* and *t* neither have nor are subtypes, and *e* is the top element among the remaining base types. A *subtype context* is a finite list of subtyping assumptions $\sigma \leq \tau$, extending the partial ordering among base types. To derive subtyping statements, we use the following rules for structural subtyping:

$$(\leq_{basic}) \frac{}{\sigma \leq \tau, \Delta \vdash \sigma \leq \tau} \quad (\leq_{basic}) \frac{\Delta \vdash \sigma \leq \tau}{\sigma' \leq \tau', \Delta \vdash \sigma \leq \tau} \quad (\leq_*) \frac{\Delta \vdash \sigma \leq \tau}{\Delta \vdash \sigma^* \leq \tau^*}$$

$$(\leq_\times) \frac{\Delta \vdash \sigma \leq \sigma' \qquad \Delta \vdash \tau \leq \tau'}{\Delta \vdash (\sigma \times \tau) \leq (\sigma' \times \tau')} \qquad (\leq_\to) \frac{\Delta \vdash \sigma \leq \sigma' \qquad \Delta \vdash \tau \leq \tau'}{\Delta \vdash (\sigma' \to \tau) \leq (\sigma \to \tau')}$$

We assume a fixed subtyping context Δ in the typing rules below.

3.2 Typing Rules

A *type context* Γ is a list of *typing assumptions* $x : \tau$ with variable x and type τ. A type σ *is a type of the term s in the context* Γ, if $\Gamma \vdash s : \sigma$ according to the following rules; this is only possible if the context contains typing assumptions for all free variables of s. To type a term which binds a variable x, the context is temporarily extended by a typing assumption $x : \sigma$ and the body of the term is typed in the extended context. A context is searched from left to right to find an assumption for a free variable x, and only the first assumption of the form $x : \sigma$ is used. Thus, all free occurrences of x in the scope of its binding get the same type, and assumptions for x from bindings of wider scope are hidden in the rest of the context.

For simplicity of description, a context may also contain typing assumptions $c : \sigma$ for constants c. A constant may be (finitely) *overloaded*, i.e. there may be several assumptions for the same constant, and the typing rules below allow to find all of them. The two list constructors $[]$ and $[\cdot \mid \cdot]$ are built into the language with special typing rules, so that we have homogeneous lists of any type.

$$\text{(Var)}\ \frac{}{x:\sigma,\Gamma\vdash x:\sigma}\qquad \text{(Var)}\ \frac{x\not\equiv y,\quad \Gamma\vdash x:\sigma}{y:\tau,\Gamma\vdash x:\sigma}$$

$$\text{(Const)}\ \frac{}{c:\sigma,\Gamma\vdash c:\sigma}\qquad \text{(Const)}\ \frac{\Gamma\vdash c:\sigma}{d:\tau,\Gamma\vdash c:\sigma}\qquad \text{(Nat)}\ \frac{n\in\mathbb{N}}{\Gamma\vdash n:nat}$$

$$\text{(App)}\ \frac{\Gamma\vdash f:\rho\to\sigma,\quad \Gamma\vdash t:\tau,\quad \Delta\vdash\tau\leq\rho}{\Gamma\vdash(f\cdot t):\sigma}\qquad \text{(Abs)}\ \frac{x:\rho,\Gamma\vdash t:\tau}{\Gamma\vdash\lambda x\,t:(\rho\to\tau)}$$

$$(f)\ \frac{\Gamma\vdash f:\sigma_1\times\ldots\times\sigma_n\to\sigma,\quad \Gamma\vdash t_i:\tau_i,\quad \Delta\vdash\tau_i\leq\sigma_i\quad (1\leq i\leq n)}{\Gamma\vdash f(t_1,\ldots,t_n):\sigma}$$

$$\text{(nil)}\ \frac{}{\Gamma\vdash[\,]:\sigma^*}\qquad \text{(cons)}\ \frac{\Gamma\vdash r:\sigma^*\qquad \Gamma\vdash h:\sigma}{\Gamma\vdash[h|r]:\sigma^*}$$

$$(\wedge)\ \frac{\Gamma\vdash\varphi:t,\quad \Gamma\vdash\psi:t}{\Gamma\vdash(\varphi\wedge\psi):t}\qquad (\vee)\ \frac{\Gamma\vdash\varphi:t,\quad \Gamma\vdash\psi:t}{\Gamma\vdash(\varphi\vee\psi):t}\qquad (\neg)\ \frac{\Gamma\vdash\varphi:t}{\Gamma\vdash\neg\varphi:t}$$

$$(\to)\ \frac{\Gamma\vdash\varphi:t,\qquad \Gamma\vdash\psi:t}{\Gamma\vdash(\varphi\to\psi):t}\qquad (\leftrightarrow)\ \frac{\Gamma\vdash\varphi:t,\qquad \Gamma\vdash\psi:t}{\Gamma\vdash(\varphi\leftrightarrow\psi):t}$$

$$(\exists)\ \frac{x:\sigma,\Gamma\vdash\varphi:t,\qquad x:\sigma,\Gamma\vdash\psi:t}{\Gamma\vdash ex(x:\sigma,\varphi,\psi):t}\qquad (\forall)\ \frac{x:\sigma,\Gamma\vdash\varphi:t,\quad x:\sigma,\Gamma\vdash\psi:t}{\Gamma\vdash all(x:\sigma,\varphi,\psi):t}$$

$$(most)\ \frac{x:\sigma,\Gamma\vdash\varphi:t,\quad x:\sigma,\Gamma\vdash\psi:t}{\Gamma\vdash most(x:\sigma,\varphi,\psi):t}\qquad (card)\ \frac{x:\sigma,\Gamma\vdash\varphi:t,\quad \Gamma\vdash n:nat}{\Gamma\vdash card(x:\sigma,\varphi,n):t}$$

$$(wh)\ \frac{x:\sigma,\Gamma\vdash\varphi:t,\quad x:\sigma,\Gamma\vdash\psi:t}{\Gamma\vdash wh(x:\sigma,\varphi,\psi):\sigma^*}\qquad (whn)\ \frac{x:\sigma,\Gamma\vdash\varphi:t,\quad x:\sigma,\Gamma\vdash\psi:t}{\Gamma\vdash whn(x:\sigma,\varphi,\psi):nat}$$

Since we want to use structured variables $\langle x,y\rangle$ for pairs (and avoid decomposition functions), variants of the abstraction and quantification rules are needed as well, such as

$$(\text{Abs}_\times)\ \frac{x:\sigma,y:\tau,\Gamma\vdash r:\rho}{\Gamma\vdash\lambda\langle x,y\rangle.r:\sigma\times\tau\to\rho}.$$

Verbs with different categories, in particular with arguments that may be read both collectively and distributively, are captured by the typing rule (f), where f is the predicate constant of the verb.

Remark 2. In the quantifier rules (Q), for $Q\in\{ex, all, most, wh, whn\}$, it may be more natural to let the type of the quantified "search" variable be determined by the restriction predicate φ, and let the scope ψ have a less restrictive type:

$$(Q,\leq)\ \frac{x:\sigma,\Gamma\vdash\varphi:t,\quad x:\tau,\Gamma\vdash\psi:t,\quad \Delta\vdash\sigma\leq\tau}{\Gamma\vdash Q(x:\sigma,\varphi,\psi):t}.$$

3.3 Type Reconstruction

Hindley's[7] type reconstruction algorithm for simply typed λ-calculus assigns a type variable to each subterm of its untyped input term and an equational constraint $\alpha = (\beta \to \gamma)$ for each application $(f^\alpha \cdot t^\beta)^\gamma$, and then solves the constraints by unification. If a solution exists, there is a most general one, the principal type scheme of the input term. The algorithm is often presented as a function $type(\Gamma, t)$ which takes a type context Γ and an untyped term t and either fails or returns a pair (U, τ) where $U : TypeVar \to Type$ is the most general substitution such that $U\Gamma \vdash t : \tau$.

We adapt this algorithm in two respects: (i) each non-logical constant may have finitely many types, all of which must be closed (i.e. without type variables), and (ii) a subtype relation based on the structure of type expressions and on a partial order among base types is used when typing application terms (and generalized quantifiers, perhaps). By (i), we can have verb constants with different types for the same argument, and by (ii), functions can be applied to arguments whose type is a subtype of the function's argument type. Obviously terms no longer have a principal type scheme, but it seems clear that each term has a finite number of type schemes whose instances are the types of the term.

Thus, in our setting, *type reconstruction* means that given a typing context Γ and an untyped term t, a list of pairs (U, τ) is constructed where $U : TypeVar \to Type$ and τ is a type such that $U\Gamma \vdash t : \tau$ according to the typing rules. We assume that type assumptions $c : \sigma$ for constants c have closed types σ, and for each free variable x of t, there is a single type assumption $x : \sigma$ in Γ, where σ need not be closed. In the following selection of defining clauses for *type*, we write $type(\Gamma, e) \rightsquigarrow (U, \tau)$ for $(U, \tau) \in type(\Gamma, e)$, and use α, β for "fresh" type variables:

$$
\begin{aligned}
type([x:\sigma,\Gamma],x) &\rightsquigarrow (Id,\sigma)\\
type([y:\sigma,\Gamma],x) &\rightsquigarrow \text{let } (S,\tau)\leftarrow type(\Gamma,x) \text{ in } (S,\tau) \quad (\text{for } y\not\equiv x)\\
type(\Gamma,\lambda x.t) &\rightsquigarrow \text{let } (S,\tau)\leftarrow type([x:\alpha,\Gamma],t) \text{ in } (S,(S\alpha\to\tau))\\
type(\Gamma,\lambda\langle x,y\rangle.t) &\rightsquigarrow \text{let } (S,\tau)\leftarrow type([x:\alpha,y:\beta,\Gamma],t) \text{ in } (S,(S\alpha\times S\beta\to\tau))\\
type(\Gamma,(t\cdot s)) &\rightsquigarrow \text{let } \ (S,\sigma)\leftarrow type(\Gamma,s)\\
&\qquad\qquad (T,\tau)\leftarrow type(S\Gamma,t)\\
&\qquad\qquad U = subtype(TS\Gamma,\tau,(T\sigma\to\alpha))\\
&\qquad \text{in } \ (UTS,U\alpha)\\
type(\Gamma,Q(x,\varphi,\psi)) &\rightsquigarrow \text{let } \ (S,(\sigma\to\tau))\leftarrow type(\Gamma,\lambda x\varphi),\\
&\qquad\qquad (S',(\sigma'\to\tau'))\leftarrow type(S\Gamma,\lambda x\psi)\\
&\qquad\qquad U = subtype(S'S\Gamma,(\sigma'\to\tau'),S'(\sigma\to\tau))\\
&\qquad\qquad V = unify(U(\sigma'\to\tau'),(\alpha\to t)) \quad (\text{ where } t = \text{ bool })\\
&\qquad \text{in } \ (VUS'S,V\alpha^*)
\end{aligned}
$$

$$
\begin{aligned}
\mathit{type}(\Gamma, f(t_1,\ldots,t_n)) \rightsquigarrow \text{let } & (S_1,\tau_1) \leftarrow \mathit{type}(\Gamma,t_1),\ldots,\\
& (S_n,\tau_n) \leftarrow \mathit{type}(S_{n-1}\cdots S_1\Gamma,t_n)\\
& (\mathit{Id},\sigma_1\times\cdots\times\sigma_n\to\sigma) \leftarrow \mathit{type}(\Gamma,f)\\
& U = \mathit{subtype}(S_n\cdots S_1\Gamma,(\sigma_1\times\cdots\times\sigma_n\to\sigma),\\
& \qquad\qquad (\tau_1\times\cdots\times\tau_n\to\alpha))\\
\text{in } & (US_n\cdots S_1, U\alpha)
\end{aligned}
$$

In type-checking applications $(f\cdot t)$ and algebraic terms $f(t_1,\ldots,t_n)$, the purpose of *subtype* is to weaken the type of an argument to the corresponding argument type of the function.

$$
\begin{aligned}
\mathit{subtype}(\Gamma,\sigma\to\tau,\sigma'\to\tau') &:= \text{let } S = \mathit{subtype}(\Gamma,\sigma',\sigma)\\
&\qquad\quad T = \mathit{subtype}(S\Gamma,S\tau,S\tau')\\
&\quad\ \text{in } TS\\
\mathit{subtype}(\Gamma,\sigma\times\tau,\sigma'\times\tau') &:= \text{let } S = \mathit{subtype}(\Gamma,\sigma,\sigma')\\
&\qquad\quad T = \mathit{subtype}(S\Gamma,S\tau,S\tau')\\
&\quad\ \text{in } TS\\
\mathit{subtype}(\Gamma,\sigma^*,\tau^*) &:= \text{let } S = \mathit{subtype}(\Gamma,\sigma,\tau) \text{ in } S\\
\mathit{subtype}(\Gamma,\sigma,\tau) &:= \mathit{Id}, \qquad \text{if } \sigma \text{ and } \tau \text{ are base types and } \Delta\vdash\sigma\leq\tau\\
\mathit{subtype}(\Gamma,\sigma,\tau) &:= \mathit{unify}(\sigma,\tau), \qquad \text{else}.
\end{aligned}
$$

Note that *subtype* is functional and instantiates variable arguments through unification. Note also that overloading is restricted to constants; free and abstracted variables must be used with the same type at each occurrence.

Example 4. Let Δ contain $m \leq h, h \leq e,\ s \leq e$, and $\Gamma = \{v : h\times s\to t, v : s\times m\to t, p : m\to t, d : \alpha\}$ where α is a type variable. Then the question "which p's stand in relation v to d?", $wh(x,p(x),v(x,d))$, is typed in the context Γ as follows: $\mathit{type}(\Gamma,\lambda x.p(x)) = \{(\mathit{Id},m\to t)\}$, $\mathit{type}(\Gamma,\lambda x.v(x,d)) = \{([\alpha/s],h\to t),([\alpha/m],s\to t)\}$, $\mathit{subtype}([\alpha/s]\Gamma,(h\to t),(m\to t)) = \mathit{Id}$ and $\mathit{subtype}([\alpha/m]\Gamma,(s\to t),(m\to t))$ fails. Hence, $\mathit{type}(\Gamma,wh(x,p(x),v(x,d))) = \{([\alpha/s],m^*)\}$.

3.4 Application

We demonstrate the effect of type reconstruction on an example involving both reciprocal and reflexive pronouns. Our implementation differs slightly from the description above: it does not fully implement the quantifying-in rule (S 14), but produces a traditional parse tree with noun phrases in place, then adds untyped λ-terms (with various relative scopes for the noun phrases), and finally types these λ-terms. The reciprocal only transforms a pair-predicate to its symmetric kernel; the switch from a transitve verb to a pair-predicate is done on the fly when combining the verb with its objects.

Example 5. The interrogative plural determiner *which* in *Which barbers who shave themselves shave each other?* has two readings, leading to two readings of the subject noun phrase, one expecting a property of individuals, and the other expecting a property of pairs of individuals.[5]. Only the second reading, of type `(m*m->t)->list(m*m)`, can be applied to the meaning of the (not explicitly represented) verb phrase *shave each other*, whose type is `m*m->t`.

```
?- debug(typisieren), parses.
|: welche Barbiere, die sich rasieren, rasieren einander.
Baum:
    s([qu], [praes, ind, vz])
    + qu((X, Y):m*m,
        barbier(X)&rasieren(X, X)& (barbier(Y)&rasieren(Y, Y)),
        rasieren(X, Y)&rasieren(Y, X)):list(m*m)
      np([qu, 3, mask], [pl, nom])
      + lam(X: (m->t), qu(Y:m, barbier(Y)&rasieren(Y, Y), X*Y)):
                                                ((m->t)->list(m))
      + lam(X: (m*m->t), qu((Y, Z):m*m,
            barbier(Y)&rasieren(Y, Y)&barbier(Z)&rasieren(Z, Z),
              X* (Y, Z))): ((m*m->t)->list(m*m))
        det([qu], [mask, pl, nom]) welche
        + lam(X: (Y->t), lam(Z: (Y->t), qu(A1:Y, X*A1, Z*A1))):
                                         ((Y->t)-> (Y->t)->list(Y))
        + lam(X: (Y->t), lam(Z: (Y*Y->t),
                         qu((A1, B1):Y*Y, X*A1&X*B1, Z*(A1, B1))))
                                   : ((Y->t)-> (Y*Y->t)->list(Y*Y))
        n([mask], [pl, nom]) 'Barbiere'
        + lam(X:m, barbier(X)): (m->t)
        s([rel(mask, pl)], [praes, ind, vl])
        + lam(X:m, rasieren(X, X)): (m->t)
          np([rel(mask, pl), 3, mask], [pl, nom])
          + lam(X: (Y->Z), lam(A1:Y, X*A1)): ((Y->Z)->Y->Z)
            pron([rel], [mask, pl, nom]) die
            + lam(X: (Y->Z), lam(A1:Y, X*A1)): ((Y->Z)->Y->Z)
          np([refl, 3, mask], [sg, akk])
          + lam(X: (Y->Z), lam(A1:Y, X*A1)): ((Y->Z)->Y->Z)
            pron([refl], [mask, sg, akk]) sich
            + lam(X: (Y->Z), lam(A1:Y, X*A1)): ((Y->Z)->Y->Z)
          v([nom, akk], [3, pl, praes, ind]) rasieren
          + lam(X:m, lam(Y:m, rasieren(X, Y))): (m->m->t)
    v([nom, akk], [3, pl, praes, ind]) rasieren
```

[5] Meaning terms are marked with + and written underneath the nodes of the syntax tree. m is the type of humans, a subtype of e.

```
    + lam(X:m, lam(Y:m, rasieren(X, Y))): (m->m->t)
    np([rezi, 3, mask], [pl, akk])
   + lam(X: (Y*Y->t), lam((Z, A1):Y*Y, X* (Z, A1)&X* (A1, Z))):
                                              ((Y*Y->t)->Y*Y->t)
       pron([rezi], [mask, pl, akk]) einander
   + lam(X: (Y*Y->t), lam((Z, A1):Y*Y, X* (Z, A1)&X* (A1, Z))):
                                            ((Y*Y->t)->Y*Y->t)
```

4 Conclusion and Open Questions

In a framework like Montague's where noun phrases are functionals applying to properties provided by verb phrases, the existence of verbs of different logical types implies that noun phrases may have several meanings, differing in their argument type (at least). In particular, this occurs when verb arguments may be individuals, pairs, or groups, or when the domain of individuals is structured by subtypes. Adapting type inference algorithms for programming languages, we can resolve some of the ambiguities of (in particular: plural) noun phrases.

Given that we use a modest modification of Hindley's algorithm, it seems plausible that the set of typings of a term can be described by finitely many type schemes, and that these can be computed efficiently from the input term. However, we havn't yet tried to prove this. On the practical side, there is a question of when and how to report type errors: if function and argument each can have several types, should one suppress all type error messages as long as one compatible combination of types remains, and if not, how should one report errors relative to a particular choice of function and argument type?

From the linguistic point of view, the typing system presented here is not quite satisfying in that we assume that abstractions are monomorphic.. But a single plural noun phrases may be "used" with several types in its scope, as in

$$\text{(John and Mary)}^{?\to t}\ \text{(like Bill)}^{e\to t}\ \text{but (don't like each other)}^{e\times e\to t},$$

so that we cannot generally select one of its possible types, here $(e \to t) \to t$ and $(e \times e \to t) \to t$. One might try to use intersection types [2] to type the subject here; but we have no well-typed verb phrase in the first place.

5 Related Work

In Link's[10] influential algebraic semantics of plural noun phrases, individuals are atoms and groups are suprema of sets of atoms in a complete lattice. Unless atoms and non-atoms are seen as types, this is rather incompatible with our way of distinguishing individuals, pairs and groups by their type. Kamp and Reyle[8] let plural noun phrases be ambiguous between distributive and collective readings, where the former provides the predicate with an atomic discourse referent, the latter with a non-atomic one. Using underspecified discourse representation structures, Frank and Reyle[6] formulate a semantics principle for

HPSG where likewise the "argument type" of the verb is determined only when the corresponding noun phrase is disambiguated. Chaves[1] modifies this account by letting plural noun phrases be collective (introduce non-atomic discourse referents) generally and by locating plural disambiguation in the lexical entries of verbs: they may require atomic or non-atomic noun phrase arguments or admit both via underspecification. A special plural resolution constraint resolves the underspecification and relates the discourse referent of the noun phrase with the argument type of the verb. Our overloading of verbs is technically simpler; however, we generate all readings of a noun phrase and let type reconstruction filter out the ones compatible with the verb's types. The constraint-based approach aims at a more compact description of the set of all readings.

Plural noun phrases with the pairwise reading occur in mathematical texts, where symmetric predicates are frequent. Cramer and Schröder[3] present a plural disambiguation algorithm for such cases, using discourse representation structures. We think that our type reconstruction algorithm ought to be able to correctly select the pairwise reading in most of their examples.

The functional programming community has studied type reconstruction for various extensions of the system of simple types. Mitchell [11] showed that type reconstruction for simple types with a subsumption rule

$$\frac{\Gamma \vdash t : \sigma, \quad \sigma \leq \tau}{\Gamma \vdash t : \tau},$$

where $\leq$ is a partial order on the set of types, reduces to solving a set of inequalities over the set of types, and if the partial order is generated by its restriction to base types, it reduces to solving a set of inequalities over the set of base types, which can be done in NEXPTIME by nondeterministic choice. Tiuryn and Wand [15] give a DEXPTIME algorithm for the more general problem of type reconstruction for recursive types and a partial order generated by a subtype relation on base types; however, constants must not be overloaded. Smith[14] gives an extension of the Hindley/Milner type system (for λ-terms with local declarations) that covers both subtyping and overloading. Its principal type schemes have unversal type quantifiers bounded by a set of inequality constraints to restrict type instantiations; since the solvability of constraints is undecidable in general, restrictions on overloading (and subtyping) have to be imposed to make it (efficiently) decidable. Since we have no local declarations, it seems best to relate our system to Mitchell's, by replacing an overloaded constant by different non-overloaded constants. But our way of solving subtype constraints through unification is somewhat ad-hoc and not equivalent to satisfiability.

Acknowledgement. I thank Shuqian Wu for pushing me to write down this material and a referee for the hint to [14].

References

1. Chaves, R.P.: DRT and underspecification of plural ambiguities. In: Bunt, H., Geertzen, J., Thijse, E. (eds.) Proceedings of the 6th IWCS, pp. 78–89 (2005)

2. Coppo, M., Giannini, P.: Principal types and unification for simple intersection type systems. Information and Computation 122(1), 70–96 (1995)
3. Cramer, M., Schröder, B.: Interpreting plurals in the Naproche CNL (2010), http://staff.um.edu.mt/mros1/cnl2010/TALKS/Plurals_in_Naproche.pdf
4. Dalrymple, M., Kanazawa, M., Kim, Y., Mchombo, S., Peters, S.: Reciprocal expressions and the concept of reciprocity. Linguistics and Philosophy 21(2), 159–210 (1998)
5. Franconi, E.: A treatment of plurals and plural quantifications based on a theory of collections. Minds and Machines, special issue on Knowledge Representation for Natural Language Processing 3(4), 453–474 (1993)
6. Frank, A., Reyle, U.: Principle based semantics for HPSG. In: Proceedings of the 6th Meeting of the Association for Computational Linguistics, European Chapter, pp. 9–16 (1995)
7. Hindley, R.: The principal type-scheme of an object in combinatory logic. Transactions of the American Mathematical Society 146, 29–60 (1969)
8. Kamp, H., Reyle, U.: From Discourse to Logic. Kluwer (1983)
9. Keenan, E.L., Faltz, L.M.: Boolean Semantics for Natural Language. D. Reidel, Dordrecht (1985)
10. Link, G.: Algebraic Semantics for Language and Philosophy. CSLI Publications (1989)
11. Mitchell, J.: Type infernce with simple subtypes. Journal of Functional Programming 1, 245–285 (1991)
12. Montague, R.: The proper treatment of quantification in ordinary english. In: Thomason, R. (ed.) Formal Philosophy. Selected Papers of Richard Montague. Yale Univ. Press (1974)
13. Sabato, S., Winter, Y.: From semantic restrictions to reciprocal meanings. In: Rogers, J. (ed.) Proceedings FG-MoL 2005. 10th Conference on Formal Grammar and 9th Meeting on Mathematics of Language. CSLI online publications (2005)
14. Smith, G.S.: Principal type schemes for functional programs with overloading and subtyping. Science of Computer Programming 23, 197–226 (1994)
15. Tiuryn, J., Wand, M.: Type reconstruction with recursive types and atomic subtyping. In: Gaudel, M.-C., Jouannaud, J.-P. (eds.) CAAP 1993, FASE 1993, and TAPSOFT 1993. LNCS, vol. 668, pp. 686–701. Springer, Heidelberg (1993)

Weak Familiarity and Anaphoric Accessibility in Dynamic Semantics*

Scott Martin

Department of Linguistics
Ohio State University
Columbus, Ohio 43210 USA
http://www.ling.ohio-state.edu/~scott/

Abstract. The accessibility constraints imposed on anaphora by dynamic theories of discourse are too strong because they rule out many perfectly felicitous cases. Several attempts have been made by previous authors to rectify this situation using various tactics. This paper proposes a more viable approach that involves replacing Heim's notion of familiarity with a generalized variant due to Roberts. This approach is formalized in hyperintensional dynamic semantics, and a fragment is laid out that successfully deals with some problematic examples.

Keywords: Anaphora, accessibility, familiarity, dynamic semantics, discourse.

1 Overview

Dynamic theories such as discourse representation theory (DRT: Kamp 1981, Kamp and Reyle 1993), file change semantics (FCS: Heim 1982, 1983), and dynamic Montague grammar (DMG: Groenendijk and Stokhof 1991) are able to successfully treat 'donkey anaphora' because they appropriately constrain cross-clausal anaphoric links. Unfortunately, for certain classes of examples, these constraints are too strong. A number of attempts have been made to appropriate loosen these constraints in different frameworks using widely varying tactics, including scope extensions, so-called 'E-type' pronouns, and presupposition accommodation.

I argue below that these previous attempts miss an empirical generalization due to Roberts (2003) that many cases of seemingly inaccessible anaphora can be described by a weak variant of Heim's **familiarity**. I then show how Roberts' weak version of familiarity can be incorporated into a formal model of discourse following Martin and Pollard (in press, to appear). A fragment shows how the extended theory can deal with some recalcitrant counterexamples to Heim's familiarity-based theory. I also examine the possibility of further extending this theory with Roberts' **informational uniqueness** and give a discussion of its interaction with certain pragmatic effects.

* Thanks to Carl Pollard for comments on an earlier draft, and to Craige Roberts for discussion of the relevant data. Of course, any errors are my own.

P. de Groote and M.-J. Nederhof (Eds.): Formal Grammar 2010/2011, LNCS 7395, pp. 287–306, 2012.

The rest of this paper is organized as follows. Section 2 describes the problem of anaphora occurring across inaccessible domains with motivating examples, and then lays out some other attempts to deal with it. In section 3, I discuss Heim's notion of (strong) familiarity and contrast it with Roberts' generalization of it to weak familiarity. An overview of Martin and Pollard's hyperintensional dynamic semantics (HDS) is given in section 4, along with some proposed extensions for modeling weak familiarity. This extended framework is then applied to some examples of anaphora across inaccessible domains in section 5, and a discussion of some apparent cases of overgeneration is provided. Section 6 summarizes and indicates some avenues for possible future work.

2 The Problem of Anaphora across Inaccessible Domains

One of the central triumphs of dynamic semantic theories in the tradition of DRT, FCS, and DMG is that they make pronominal anaphora possible only under certain conditions. This notion of **anaphoric accessibility** explains the difference in felicity between the examples in (A) and (B).

(A) If Pedro owns $\left\{\begin{array}{c}\text{a}\\ \text{\#every}\end{array}\right\}$ donkey$_i$ he beats it$_i$.
(Kamp 1981, examples 1, 17)

(B) 1. Everybody found a cat$_i$ and kept it$_i$.
2. #It$_i$ ran away.
(Heim 1983, example 5)

In these examples, the quantifying expression *every* limits the anaphoric accessibility of discourse referents (DRs) introduced within its scope. The quantifier *no* exhibits similar behavior:

(C) 1. $\left\{\begin{array}{c}\text{A}\\ \text{\#No}\end{array}\right\}$ donkey$_i$ brays.
2. Its$_i$ name is Chiquita.

In (C), as in (A) and (B), a quantifying expression constrains the scope of DRs occurring in its scope in a way that the indefinite does not. Although it is encoded differently in each, DRT, FCS, and DMG tell very similar stories to explain these facts. These dynamic accounts of anaphoric accessibility rest on the same basic idea that indefinites introduce DRs and quantifiers limit the scope of DRs. Indefinites themselves do not place bounds on DR scope because they are treated either as non-quantifying (and thus as scopeless) or as extending their scope across discourse (unless they are outscoped by a 'true' quantifier). Many authors have subsequently adopted the essential details of this treatment of anaphoric accessibility (Chierchia 1992, 1995; van der Sandt 1992; Muskens 1994, 1996; Geurts 1999; Beaver 2001; de Groote 2006, 2008, among others).

However, the treatment of anaphoric accessibility found in dynamic semantics is not without problems, as the following 'bathroom' example[1] shows.

(D) Either there's no bathroom$_i$ in this house or it$_i$'s in a funny place. (Roberts 1989)

Examples like (D) seem to pose a direct counterexample to anaphoric inaccessibility: a DR introduced in the scope of a quantifier (here, *no*) is clearly accessible from pronouns that occur outside of the quantifier's scope. Unless somehow elaborated, a dynamic theory in the tradition of DRT/FCS/DMG would incorrectly predict that *bathroom* cannot serve as an antecedent for *it*.

This problem isn't simply limited to disjunctions or intrasentential anaphora, as (E) shows.

(E) 1. Every farmer owns a donkey.
2. Pedro is a farmer, and his donkey is brown.

The discourse in (E) is unproblematic. But the anaphoric accessibility constraints in dynamic theories would predict that the anaphora associated with *his donkey* is not resolvable. Yet we seem to have no problem understanding that Pedro's donkey ownership is a result of his being a farmer and the fact that, as previously mentioned, all farmers have a donkey. Various attempts have been made to square the idea of anaphoric accessibility with problematic examples like (D) and (E). I examine some of these attempts in the next section.

2.1 Some Attempts to Rectify the Problem

Groenendijk and Stokhof (1991) entertain the possibility of accounting for certain cases of anaphora in inaccessible domains by allowing some dynamic quantifiers and connectives to extend their scope further than the accessibility constraints dictate. The resulting extension of DMG accounts for the anaphora in examples involving disjunction such as (D), but it also gives rise to other undesirable predictions. For instance, the scope-extension variant of their theory is unable to rule out cases where anaphora is truly inaccessible, such as the following.

(F) 1. Every farmer owns a donkey$_i$.
2. # Pedro is a farmer, and he beats it$_i$.

(G) 1. Every farmer$_i$ owns a donkey.
2. # The farmer$_i$'s name is 'Pedro.'

For both (F) and (G), the proposed extension to DMG would allow the pronoun in the second utterance to have as its antecedent the indicated DR in the first utterance.

[1] Example (D) is attributed to Barbara Partee both by Roberts and by Chierchia (1995, p. 8), who gives a slight variant of it. A similar class of examples is discussed by Evans (1977).

In Chierchia (1995), an ambiguity is posited for pronouns between the "dynamically bound" case (in which the accessibility constraints are followed) and the 'E-type' case of e.g. Cooper (1979), in which anaphora across inaccessible domains is allowable in certain cases. Chierchia successfully applies his theory both to donkey anaphora and to the bathroom sentence (D), and although he does not treat parallel examples, a straightforward account of (E) in Chierchia's theory using an E-type pronoun for *his* is not beyond imagination. However, even leaving aside the arguments advanced by Roberts (2004) against the viability of the E-type approach in general, it would be desirable if a single mechanism could account for discourse anaphora without needing an ambiguity between dynamically bound and E-type pronouns. Below, I argue that such a unified treatment of pronominal anaphora in discourse is possible.

Lastly, an approach to bathroom sentences like (D) is laid out in Geurts (1999), which in turn is an extension of the presuppositional DRT of van der Sandt (1992). In this theory, the anaphora in (D) is treated as an instance of presupposition accommodation: an antecedent for the pronoun *it* is added to the right disjunct in order to allow felicitous interpretation. I would argue that construing such examples as involving accommodation is somewhat odd, since felicitous interpretation actually seems to require an overt (non-accommodated) antecedent. To illustrate this, consider the following example, which contains only the right disjunct of (D).

(H) #It's in a funny place.

The use of the pronoun in (H) gives rise to infelicity because no antecedent can be found. Yet Geurts seems to predict that an antecedent would simply be accommodated in a way similar to the accommodation that his theory predicts for (D). Similarly, Geurts predicts that an antecedent for *his donkey* in (E) must be accommodated into the global discourse context in order for it to be felicitous. Yet the fact that Pedro is a farmer, coupled with the fact that every farmer owns a donkey, seems to be what allows the anaphora in *his donkey* to be resolved. I present an account below in which the seemingly accessible NPs *a bathroom* in (D) and *a donkey* in (E) are crucial to permitting the observed anaphoric links.

3 Strong and Weak Familiarity

For Heim (1982), a semantic representation containing a definite NP (e.g., a pronoun) requires for its felicity that the definite NP be **familiar** in the discourse context. Following Roberts (2003), I refer to Heim's notion of familiarity as the **strong** variant, for reasons that are clarified below. The details of Heim's formalization of strong familiarity are given in Defintion 1.

Definition 1 (Strong Familiarity). Let i be the index of a definite NP d in a semantic representation r. Then the DR i is **strongly familiar** in a discourse context c iff

1. The DR i is among the active DRs in c, and
2. If d has descriptive content, then c entails that i has the relevant descriptive content.

Heim's familiarity has the effect of requiring pronouns and other definites to have an adequate, previously established antecedent in the discourse context. The dynamic meanings of quantifiers, conditionals, etc. are then set up in a way that guarantees the anaphoric accessibility conditions discussed in section 2, above.

Taking (B) as an example, Heim's theory correctly predicts the felicity of the first occurrence of *it* and the infelicity of the second occurrence. The first *it* is meets the familiarity condition because an antecedent DR, introduced by *every*, is accessible. The second, however, is infelicitous because it occurs outside the scope of *every*, where no antecedent DR is available.

The problem of accounting for anaphora in inaccessible domains arises for Heim's theory as a direct result of her formulation of familiarity and the accessibility conditions on anaphora. For example, the familiarity condition requires that *his donkey* in (E) have an accessible DR in the discourse context that has the property of being a donkey. But since the donkey-DR introduced by *a donkey* in the first sentence of (E) has its accessibility limited by the scope of *every*, the occurrence of *his donkey* in the second utterance does not meet the familiarity condition. This is the reason Heim's theory incorrectly predicts infelicity for (E). The bathroom example (D) represents an analogous situation: the quantifier *no* limits the scope of the bathroom-DR introduced in the first conjunct, which results in the pronoun *it* failing to satisfy the condition for familiarity imposed on definites. Here again, Heim's theory predicts infelicity for a perfectly felicitous discourse.

Roberts (2003) reworks Heim's familiarity condition on definites into a more general notion of "weak" familiarity. In Definition 2, I give a simplified version of Roberts' formalization of this idea.

Definition 2 (Weak Familiarity). Let i be the index of a definite NP d in a semantic representation r. Then the DR i is **weakly familiar** iff c entails the existence of an entity bearing the descriptive content of d (if any).

As this definition shows, for a definite to be weakly familiar in a certain discourse context, the context does not necessarily have to contain an active DR with the relevant descriptive content, if any. Weak familiarity only requires that the discourse context entails that an entity bearing the relevant description exists.

For example, supplanting Heim's strong familiarity with Roberts' weak familiarity renders examples like (E) felicitous. The definite *his donkey* meets the weak familiarity condition because the discourse context entails that a donkey exists that is owned by Pedro. Example (D) is also felicitous under weak familiarity for a similar reason. Although the would-be antecedent *a bathroom* is inaccessible, its use in the first disjunct results in a discourse context that entails the existence of a bathroom. This entailment allows the pronoun *it* in the second disjunct to satisfy the weak familiarity condition.

As Roberts notes, the strong version of the familiarity condition, coupled with Heim's definitions for e.g. quantifiers, is essentially just anaphoric accessibility. The reason weak familiarity is called 'weak' is that it subsumes strong familiarity: a definite's being strongly familiar entails that it is weakly familiar, but not the other way around. In the next section, I implement Roberts' more general weak familiarity into an essentially Heim-like formal theory of discourse.

4 A Formalization in Hyperintensional Dynamic Semantics

To formalize weak familiarity, I extend the hyperintensional dynamic semantics (HDS) of Martin and Pollard (in press, to appear), which implements a version of Heim's strong familiarity condition for definites. HDS is a theory of discourse built on the hyperintensional (static) semantics of Pollard (2008a, 2008b) that additionally extends the Montagovian dynamics of de Groote (2006, 2008). It is expressed in a classical higher-order logic (HOL) in the tradition of Church (1940), Henkin (1950), and Montague (1973) that is augmented with some of the extensions proposed by Lambek and Scott (1986), as I describe below. The next four sections are mostly review of HDS. Below, in section 4.5, I propose extensions to HDS for dealing with weak familiarity.

As usual, pairing is denoted by $\langle\ ,\ \rangle$. For f a function with argument x, application is written $(f\ x)$ rather than the usual $f(x)$. Application associates to the left, so that $(f\ x\ y)$ becomes shorthand for $((f\ x)\ y)$. Variables that are λ-abstracted over are written as subscripts on the lambda, following Lambek and Scott. Successive λ-abstractions are usually simplified by collapsing the abstracted variables together onto a single lambda, so that $\lambda_{xyz}.M$ is written instead of $\lambda_x\lambda_y\lambda_z.M$. I sometimes use the . symbol to abbreviate parentheses in the usual way, with e.g. $\lambda_x.M\ N$ shorthand for $\lambda_x(M\ N)$. Lastly, parentheses denoting application are sometimes omitted altogether when no confusion can arise.

4.1 Types and Constants

The basic types e of entities and t of truth values are inherited from the underlying logic, as are the usual type constructors U (unit), $\times$ (product), and $\rightarrow$ (exponential). HDS follows Lambek and Scott (1986) in adopting the following extensions to HOL:

- The type natural number type ω, which is linearly ordered by $<$ and equipped with the successor function $\mathsf{suc} : \omega \rightarrow \omega$.
- Lambda-definable subtypes: for any type A, if φ is a formula with $x : A$ free, then $\{x \in A \mid \varphi\}$ denotes the subtype consisting of those inhabitants of A for which φ is true.

A partial function from A to B is written $A \rightharpoonup B$, i.e., as a function from a certain subtype of A to B. I also use dependent coproduct types parameterized by ω, so that $\coprod_{n\in\omega} T_n$ denotes the dependent coproduct type whose cofactors are all the types T_n, for n a natural number. I sometimes drop the subscript denoting the natural number parameter when the parameter is clear from context.

Discourses generally involve a set of DRs. Accordingly, I introduce a set of subsets of ω:

$$\omega_n =_{\text{def}} \{i \in \omega \mid i < n\}$$

Since natural numbers are used to represent DRs, the type ω_n is intuitively the first n DRs.

The type $\mathrm{a_n}$ of n-**anchors** are mappings from the first n DRs to entities, analogous to Heim's assignments.

$$\mathrm{a}_n =_{\text{def}} \omega_n \to \mathrm{e}$$

The constant functions $\bullet_n : \mathrm{a}_n \to \mathrm{e} \to \mathrm{a}_{(\mathbf{suc}\,n)}$ extend an anchor to map the 'next' DR to a specified entity, subject to the following axioms:

$$\vdash \forall_{n\in\omega}\forall_{a\in\mathrm{a}_n}\forall_{x\in\mathrm{e}}.(a \bullet_n x)\, n = x$$
$$\vdash \forall_{n\in\omega}\forall_{a\in\mathrm{a}_n}\forall_{x\in\mathrm{e}}\forall_{m:\omega_n}.(a \bullet_n x)\, m = (a\, m)$$

These axioms together ensure that for an n-anchor a, the extended anchor $(a\bullet_n x)$ maps n to x, and that none of the original mappings in a are altered.

Relative salience for the DRs in an n-anchor is encoded by an n-resolution r_n, axiomatized as the subtype of binary relations on ω_n that are preorders (this property is denoted by preo_n):

$$\mathrm{r}_n =_{\text{def}} \{r \in \omega_n \to \omega_n \to \mathrm{t} \mid (\mathsf{preo}_n\, r)\}$$

Analogously to anchors, an n-resolution can be extended to cover the 'next' DR using $\star_n : \mathrm{r}_n \to \mathrm{r}_{(\mathbf{suc}\,n)}$. For an n-resolution r, $(\star_n\, r)$ is the resolution just like r except that n is added and axiomatized to be only as salient as itself (and unrelated to any $m < n$).

I adopt the basic type p of propositions from Pollard's (2008b) static semantics. This type, which is preordered by the entailment relation $\mathsf{entails} : \mathrm{p} \to \mathrm{p} \to \mathrm{t}$, is used to model the **common ground (CG)** following Stalnaker (1978). Certain natural language entailments are central to the analysis of anaphora I propose below. The hyperintensional entailment axioms pertaining to the (translations of the) English 'logic words' that impact the analysis I propose in section 5 are given in Equations 1 through 5.

$$\vdash \forall_{p\in\mathrm{p}}.p \;\mathsf{entails}\; p \tag{1}$$
$$\vdash \forall_{p,q,r\in\mathrm{p}}.(p \;\mathsf{entails}\; q) \to ((q \;\mathsf{entails}\; r) \to (p \;\mathsf{entails}\; r)) \tag{2}$$
$$\vdash \forall_{p,q\in\mathrm{p}}.(p \;\mathsf{and}\; q) \;\mathsf{entails}\; p \tag{3}$$
$$\vdash \forall_{p,q\in\mathrm{p}}.(p \;\mathsf{and}\; q) \;\mathsf{entails}\; q \tag{4}$$
$$\vdash \forall_{p\in\mathrm{p}}.(\mathsf{not}\;(\mathsf{not}\; p)) \;\mathsf{entails}\; p \tag{5}$$

The first two of these simply state that the entailment relation on p forms a preorder (reflexive, transitive relation). Equations 3 and 4 require that a conjunction of two propositions entails either conjunct, and Equation 5 axiomatizes double negation elimination. See Pollard (2008b, (42)–(44)) for a complete axiomatization of $\mathsf{entails}$.

Discourse contexts are defined as tuples of an anchor, resolution, and a CG, inspired both by Heim and by Lewis (1979).

$$\mathrm{c}_n =_{\mathrm{def}} \mathrm{a}_n \times \mathrm{r}_n \times \mathrm{p}$$

$$\mathrm{c} =_{\mathrm{def}} \coprod_{n \in \omega} \mathrm{c}_n$$

For each $n \in \omega$, a discourse context of type c_n is one that 'knows about' the first n DRs. The type c is simply the type of n-contexts of any arity.

Several functions are useful in HDS for managing discourse contexts. The projection functions for the three components of a context are mnemonically abbreviated as $\mathbf{a} : \mathrm{c} \to \mathrm{a}$ (for *anchor*), $\mathbf{r} : \mathrm{c} \to \mathrm{r}$ (for *resolution*) and $\mathbf{p} : \mathrm{c} \to \mathrm{p}$ (for *proposition*). As a shorthand, I further abbreviate $(\mathbf{a}cn)$, the entity anchoring the DR n in the context c, as follows.

$$\vdash \forall_{m \in \omega} \forall_{c \in \mathrm{c}_m} \forall_{n \in \omega_m} . [n]_c = (\mathbf{a}\, c\, n)$$

As long as no confusion is possible, I usually drop the subscript c and write simply $[n]$. The 'next' DR for an n-context is always the natural number n, retrievable by next_n:

$$\vdash \forall_{n \in \omega} \forall_{c \in \mathrm{c}_n} . (\mathsf{next}_n\, c) = n$$

The constants $::_n$ and $+_n$ are used to extend the anchor/resolution and CG of a context, respectively:

$$\vdash \forall_{n \in \omega} . ::_n = \lambda_{cx} \langle (\mathbf{a}\, c) \bullet_n x, \star_n (\mathbf{r}\, c), (\mathbf{p}\, c) \rangle$$
$$\vdash \forall_{n \in \omega} . +_n = \lambda_{cp} \langle (\mathbf{a}\, c), (\mathbf{r}\, c), (\mathbf{p}\, c)\ \mathsf{and}\ p \rangle$$

These axioms ensure that $::_n$ maps a specified entity to the 'next' DR and adds it to the resolution, while $+_n$ adds a specified proposition to the CG.

Lastly, the definedness check $\downarrow : (A \rightharpoonup B) \to A \to \mathrm{t}$ (written infix) tests whether a given partial function is defined for a given argument.

$$\vdash \downarrow = \lambda_{fx} . \mathsf{dom}\ f\ x$$

Where for a given partial function $f : A \rightharpoonup B$, $(\mathsf{dom}\ f)$ is the characteristic function of the subset of A that is the domain of f.

4.2 Context-Dependent Propositions, Updates, and Dynamic Propositions

Context-Dependent Propositions (CDPs), type k, are partial functions from contexts to propositions.

$$\mathrm{k} =_{\mathrm{def}} \mathrm{c} \rightharpoonup \mathrm{p}$$

The partiality of this type reflects the fact that an utterance is sensitive to the discourse context in which it is situated: not every context is suitable to yield an interpretation for a given utterance, only those where conditions like familiarity are met. The **empty CDP** $\top =_{\mathrm{def}} \lambda_c.\mathsf{true}$ 'throws away' whatever context it is passed, returning the contentless proposition true (a necessary truth).

Updates, of type u, map CDPs to CDPs:

$$\mathrm{u} =_{\mathrm{def}} \mathrm{k} \rightarrow \mathrm{k}$$

The type u is used to model the dynamic meanings of declarative sentences.

Dynamic properties are the dynamicized analogs of static properties, where static properties is defined as follows:

$$\mathrm{R}_0 =_{\mathrm{def}} \mathrm{p}$$
$$\mathrm{R}_{(\mathbf{suc}\ n)} =_{\mathrm{def}} \mathrm{e} \rightarrow \mathrm{R}_n$$

Note that in particular, nullary properties are equated with propositions, and the arity of a static proposition is simply the number of arguments of type e it takes. The type hierarchy for dynamic properties is obtained from the one for static properties by replacing the base type p with the type u of updates, and replacing the argument type e with the type ω of DRs:

$$\mathrm{d}_0 =_{\mathrm{def}} \mathrm{u}$$
$$\mathrm{d}_{(\mathbf{suc}\ n)} =_{\mathrm{def}} \omega \rightarrow \mathrm{d}_n$$

Again, note that nullary dynamic properties are just updates. Since d_1 is used most frequently, I write d to abbreviate the type d_1.

The **dynamicizer** functions $\mathbf{dyn}_n$ map a static property of arity n to its dynamic counterpart:

$$\mathbf{dyn}_0 =_{\mathrm{def}} \lambda_{pkc}.p \mathsf{\ and\ } (k\ (c + p)) : \mathrm{R}_0 \rightarrow \mathrm{d}_0$$
$$\forall_{n:\omega}.\mathbf{dyn}_{(\mathbf{suc}\ n)} =_{\mathrm{def}} \lambda_{Rm}.(\mathbf{dyn}_n\ (R\ [m])) : \mathrm{R}_{(\mathbf{suc}\ n)} \rightarrow \mathrm{d}_{(\mathbf{suc}\ n)}$$

(Here, and is Pollard's (2008b) propositional conjunction.) I write static propositions in lowercase sans-serif (e.g. donkey) and their dynamic counterparts in smallcaps (e.g., DONKEY). Some examples of dynamicization:

$$\textsc{rain} =_{\mathrm{def}} (\mathbf{dyn}_0\ \mathsf{rain}) = \lambda_{kc}.\mathsf{rain\ and\ } (k\ (c + \mathsf{rain}))$$
$$\textsc{donkey} =_{\mathrm{def}} (\mathbf{dyn}_1\ \mathsf{donkey}) = \lambda_{nkc}.(\mathsf{donkey}\ [n]) \mathsf{\ and\ } (k\ (c + (\mathsf{donkey}\ [n])))$$
$$\textsc{own} =_{\mathrm{def}} (\mathbf{dyn}_2\ \mathsf{own}) = \lambda_{mnkc}.(\mathsf{own}\ [m]\ [n]) \mathsf{\ and\ } (k\ (c + (\mathsf{own}\ [m]\ [n])))$$

These examples show the central feature of dynamic properties: the static proffered content is added to the discourse context that is used for evaluating subsequent updates.

Reducing a dynamic proposition to its static counterpart is handled by the **staticizer** function **stat** : u → k, which is defined as follows:

$$\mathbf{stat} =_{\text{def}} \lambda_u . u \top$$

The partiality of **stat** reflects the fact that a dynamic proposition can only be reduced to a static proposition in contexts that satisfy its presuppositions. To demonstrate, consider (for a hypothetical DR n) the staticizer applied to the dynamic proposition (DONKEY n):

$$\begin{aligned} n : \omega \vdash (\mathbf{stat}\ (\text{DONKEY}\ n)) &= (\lambda_{kc}.(\mathsf{donkey}\ [n])\ \mathsf{and}\ (k\ (c + (\mathsf{donkey}\ [n]))) \top) \\ &= \lambda_c.((\mathsf{donkey}\ [n])\ \mathsf{and}\ (\top\ (c + (\mathsf{donkey}\ [n])))) \\ &= \lambda_c.(\mathsf{donkey}\ [n])\ \mathsf{and}\ \mathsf{true} \\ &\equiv \lambda_c.\mathsf{donkey}\ [n] \end{aligned}$$

where ≡ denotes equivalence of CDPs.

4.3 Connectives and Quantifiers

The dynamic conjunction AND : u → u → u essentially amounts to composition of updates, as it is for Muskens (1994, 1996):

$$\text{AND} =_{\text{def}} \lambda_{uvk}.u\ (v\ k) \tag{6}$$

The effect of dynamic conjunction is that the modifications to the discourse context made by the first conjunct are available to the second conjunct. For example (again with a hypothetical DR n), the conjunction (DONKEY n) AND (BRAY n) : u is treated as follows:

$$\begin{aligned} & n : \omega \vdash (\text{DONKEY}\ n)\ \text{AND}\ (\text{BRAY}\ n) \\ &= \lambda_{kc}.(\text{DONKEY}\ n)\ ((\text{BRAY}\ n)\ k)\ c \\ &= \lambda_{kc}.(\lambda_{kc}(\mathsf{donkey}\ [n]\ \mathsf{and}\ k\ (c + \mathsf{donkey}\ [n]))\ \lambda_c(\mathsf{bray}\ [n]\ \mathsf{and}\ k\ (c + \mathsf{bray}\ [n])))\ c \\ &= \lambda_{kc}.(\mathsf{donkey}\ [n])\ \mathsf{and}\ (\mathsf{bray}\ [n])\ \mathsf{and}\ k\ (c + \mathsf{donkey}\ [n] + \mathsf{bray}\ [n]) \end{aligned}$$

(Here, DONKEY = ($\mathbf{dyn}_1$ donkey) and BRAY = ($\mathbf{dyn}_1$ bray).)

The dynamic existential quantifier EXISTS : d → u introduces the 'next' DR:

$$\text{EXISTS} =_{\text{def}} \lambda_{Dkc}.\mathsf{exists}\ \lambda_x.D\ (\mathsf{next}\ c)\ k\ (c :: x) \tag{7}$$

As Equation 7 shows, the dynamic existential introduces a new DR mapped to an entity that is existentially bound at the propositional level. This new DR is added to the discourse context that is used by subsequent updates.

Dynamic negation limits the accessibility of DRs introduced within its scope, while negating the proffered content of its complement but propagating any presuppositions.

$$\text{NOT} =_{\text{def}} \lambda_{uk}\lambda_{c \mid (u\ k)\downarrow c}.\mathbf{dyn}_0\ (\mathsf{not}\ (\mathbf{stat}\ u\ c))\ k\ c \tag{8}$$

The partiality condition $(u\ k) \downarrow c$ on the variable c is designed to require that any presuppositions of the complement of NOT become presuppositions of the dynamic negation. This is best illustrated with an example, as follows for (DONKEY n).

$$\begin{aligned} n : \omega \vdash\ & (\text{NOT}\ (\text{DONKEY}\ n)) \\ &= \lambda_k\lambda_{c \mid ((\text{DONKEY}\ n)\ k)\downarrow c}.\mathbf{dyn}_0\ (\mathsf{not}\ (\mathbf{stat}\ c\ (\text{DONKEY}\ n)))\ k\ c \\ &= \lambda_k\lambda_{c \mid ((\text{DONKEY}\ n)\ k)\downarrow c}.(\mathsf{not}\ (\mathsf{donkey}\ [n]))\ \mathsf{and}\ (k\ (c + (\mathsf{not}\ (\mathsf{donkey}\ [n]))))) \end{aligned}$$

Here, the (static) proffered content of (DONKEY n) is negated and this negation is added to the CG of the discourse context passed to the incoming update. As the condition on the variable c shows, (NOT (DONKEY n)) also requires of the incoming update that the DR n can be retrieved from the discourse context used to interpret it. Note that if the complement of NOT introduced any DRs, these new DRs would be unavailable to subsequent updates, as desired.

I also extend HDS with a dynamic disjunction, which will be used below to analyze a bathroom example like (D).

$$\text{OR} =_{\text{def}} \lambda_{uv}.\text{NOT}\ ((\text{NOT}\ u)\ \text{AND}\ (\text{NOT}\ v)) \tag{9}$$

This definition is analogous to the treatment of dynamic disjunction by Groenendijk and Stokhof (1991).

4.4 Dynamic Generalized Determiners

To model the English discourse meanings, several dynamic generalized determiners (all of type $\mathrm{d} \to \mathrm{d} \to \mathrm{u}$) are needed. First, the dynamic indefinite article A:

$$\text{A} =_{\text{def}} \lambda_{DE}.\text{EXISTS}\ \lambda_n.(D\ n)\ \text{AND}\ (E\ n) \tag{10}$$

Similarly to the usual treatment of the generalized indefinite determiner in static semantics, the dynamic indefinite introduces a new DR and passes it to two conjoined dynamic properties. There is no need to state a novelty condition for indefinites, as Heim (1982) does, because the newly-introduced DR will always be as yet unused (see Equation 7, above).

I use the dynamic negation NOT and the definition of A in Equation 10 to build the dynamic generalized quantifier NO, which models the meaning of the generalized determiner *no*.

$$\text{NO} =_{\text{def}} \lambda_{DE}.\text{NOT}\ (\text{A}\ D\ E) \tag{11}$$

Along with AND and EXISTS, dynamic negation is also used to build the dynamic universal EVERY : $\mathrm{d} \to \mathrm{d} \to \mathrm{u}$.

$$\text{EVERY} =_{\text{def}} \lambda_{DE}.\text{NOT}\ (\text{EXISTS}\ \lambda_n.(D\ n)\ \text{AND}\ (\text{NOT}\ (E\ n))) \qquad (12)$$

This definition ensures that any DR that has the property specified in the restrictor D also has the property in the restrictor E. (Note that this definition of the dynamic universal yields only the so-called 'strong' readings for donkey sentences, but describing how the 'weak' readings arise is well beyond the scope of this paper. See e.g. Rooth (1987), Chierchia (1992), and Kanazawa (1994) for discussion.)

4.5 Extensions for Modeling Weak Familiarity

The dynamic generalized quantifier meaning $\text{IT}_\text{s} : \text{d} \rightarrow \text{u}$ uses the **def** operator to select the uniquely most salient nonhuman DR in the discourse context:

$$\text{IT}_\text{s} =_{\text{def}} \lambda_{Dk}\lambda_{c \mid (\mathbf{def}\ \text{NONHUMAN})\downarrow c}.D\ (\mathbf{def}\ \text{NONHUMAN}\ c)\ k\ c \qquad (13)$$

The difference between the definition of IT used here is that a partiality condition is used on the variable c to explicitly require that the context contain a DR with the property NONHUMAN. Since this is the strong version of Heim's familiarity condition (see Definition 1), I also add the subscript s. The ω-parameterized definiteness operator $\mathbf{def}_n : \text{d} \rightarrow \text{c} \rightharpoonup \omega_\text{n}$ is defined as follows to yield the most salient DR in the discourse context with a given dynamic property:

$$\mathbf{def}_n =_{\text{def}} \lambda_{Dc}.\bigsqcup_{(\mathbf{r}\ c)} \lambda_{i \in \omega_n}.(\mathbf{p}\ c)\ \mathsf{entails}\ (\mathbf{stat}\ (D\ i)\ c) \qquad (14)$$

where $\bigsqcup_{(\mathbf{r}\ c)}$ denotes the unique least upper bound operation on the resolution preorder of the context c. Note that $\mathbf{def}_n$ is partial, since for any given dynamic property D and context c, there may be no DR that is uniquely most salient among the DRs with the property D according to c's resolution.

To model weak familiarity for the pronoun *it*, I add a separate definition for *it* that is built on top of the strongly familiar version in Equation 13.

$$\text{IT}_\text{w} =_{\text{def}} \lambda_{Dk}\lambda_{c \mid \varphi}.\mathsf{exists}\ \lambda_x.(\mathsf{nonhuman}\ x)\ \mathsf{and}\ \text{IT}_\text{s}\ D\ k\ (c :: x + \mathsf{nonhuman}\ x) \qquad (15)$$

Here the condition φ on the context variable c is as follows:

$$\varphi = (\neg\ ((\text{IT}_\text{s}\ D\ k) \downarrow c)) \wedge ((\mathbf{p}\ c)\ \mathsf{entails}\ (\mathsf{exists}\ \lambda_x.\mathsf{nonhuman}\ x))$$

In this weak version of *it*, the condition φ that describes which contexts it is defined for is broken into a conjunction.

The first conjunct $(\neg\ ((\text{IT}_\text{s}\ D\ k) \downarrow c))$ ensures that the strongly familiar IT_s is *not* defined. This is done in order to force the strong familiarity version to be used whenever an overt discourse referent is actually present in the context, rather than merely being entailed. This clause is important since IT_w has the potential to introduce DRs. Without it, the weak familiarity *it* in Equation 15 could introduce DRs into a context when a suitable antecedent already existed.

The second conjunct expresses Roberts' (2003) notion of weak familiarity as given in Definition 2: the CG of the discourse context must entail that a nonhuman entity exists. The body of the abstract of IT_w just invokes the strong version with a modified context that contains a newly introduced nonhuman DR. So the fundamental difference between the strong and weak versions of *it* are that one references a DR present in the context, and another introduces a new DR based on certain existential entailments of the CG.

I extend HDS to handle anaphora by possessive determiners by giving strong and weak versions of the pronoun *his.* The strong familiarity version of the dynamic generalized determiner HIS_s resembles the strong version of *it* in IT_s in Equation 13.

$$\text{HIS}_\text{s} =_{\text{def}} \lambda_{DEk}\lambda_{c\,|\,\varphi}.E\,(\mathbf{def}\;\lambda_n((D\;n)\;\text{AND}\;(\text{POSS}\;n\;(\mathbf{def}\;\text{MALE}\;c)))\;c)\;k\;c \quad (16)$$

(Here, $\text{MALE} = (\mathbf{dyn}_1\;\mathsf{male})$ and $\text{POSS} = (\mathbf{dyn}_2\;\mathsf{poss})$, where $\mathsf{poss} : \text{R}_2$ is the two-place static relation of possession). For Equation 16, the condition on the context variable c is represented by

$$\varphi = ((\mathbf{def}\;D) \downarrow c) \wedge ((\mathbf{def}\;\text{MALE}) \downarrow c)$$

As the partiality condition φ shows, HIS_s is only defined for contexts where both a male DR and a DR with the property D are overtly accessible. This strong version of *his* takes two dynamic properties as arguments to return an update. It then applies the second dynamic property to the most salient DR with the property D that is possessed by the most salient male DR.

As for *it*, the weak familiarity version of *his* is defined in terms of the strong version HIS_s.

$$\begin{aligned}&\text{HIS}_\text{w} =_{\text{def}} \lambda_{DEk}\lambda_{c\,|\,\varphi}.\mathsf{exists}\;\lambda_x.\\&\quad((D\;(\mathsf{next}\;c)\;\text{AND}\;(\text{POSS}\;(\mathsf{next}\;c)\;[\mathbf{def}\;\text{MALE}\;c]))\;k\;(c :: x))\;\mathsf{and}\\&\quad\text{HIS}_\text{s}\;D\;E\;k\;(c :: x + (\mathbf{stat}\;(D\;(\mathsf{next}\;c)\;\text{AND}\;(\text{POSS}\;(\mathsf{next}\;c)\;[\mathbf{def}\;\text{MALE}\;c]))\;c :: x))\end{aligned}$$

In the case of HIS_w, the definedness condition φ on c is

$$\begin{aligned}&\varphi = (\neg\,((\text{HIS}_\text{s}\;D\;E\;k) \downarrow c))\\&\quad\wedge\;(\mathbf{p}\;c)\;\mathsf{entails}\;\mathsf{exists}\;\lambda_x.\mathbf{stat}\;(D\;(\mathsf{next}\;c)\;\text{AND}\;\text{POSS}\;(\mathsf{next}\;c)\;[\mathbf{def}\;\text{MALE}\;c])\;c :: x\end{aligned}$$

This version requires that the strong version of *his* is undefined in the discourse context it is passed. In particular, this implies that there is no uniquely most salient DR overtly represented in the context that bears the property D. It further requires that the CG entails the existence of an entity possessed by the uniquely most salient male DR, and that the possessed entity additionally has the property D. Similarly to the weak version of *it*, HIS_w invokes the strong *his* with a modified context that is extended with a DR bearing the weakly entailed property.

5 A Small Fragment Demonstrating Weak Familiarity

The weak familiarity version of *it* is best illustrated with an example.

(I) Either no donkey is walking around, or it's braying.

The example discourse in (I) is a simplification of bathroom examples of the kind in (D). But the principle is the same: no DR is accessible to serve as the anaphoric antecedent of the pronoun *it*. Noting that DONKEY = ($\mathbf{dyn}_1$ donkey), WALK = ($\mathbf{dyn}_1$ walk), and BRAY = ($\mathbf{dyn}_1$ bray), the dynamic meaning of (I) is as follows.

$$\begin{aligned}&\vdash (\text{NO DONKEY WALK})\ \text{OR}\ (\text{IT}_\text{w}\ \text{BRAY})\\&= (\text{NOT}\ (\text{A DONKEY WALK}))\ \text{OR}\ (\text{IT}_\text{w}\ \text{BRAY})\\&= \text{NOT}\ (\text{NOT}\ (\text{NOT}\ (\text{A DONKEY WALK})))\ \text{AND}\ (\text{NOT}\ (\text{IT}_\text{w}\ \text{BRAY}))\end{aligned}$$

Note that the left conjunct of the argument to the widest-scope negation is the dynamic double negation of *a donkey walks*:

$$\begin{aligned}&\vdash \text{NOT}\ (\text{NOT}\ (\text{EXISTS}\ \lambda_n.(\text{DONKEY}\ n)\ \text{AND}\ (\text{WALK}\ n)))\\&\equiv \lambda_{kc}(\mathsf{not}\ (\mathsf{not}\ (\mathsf{exists}\ \lambda_x((\mathsf{donkey}\ x)\ \mathsf{and}\ (\mathsf{walk}\ x))))) \ \mathsf{and}\ (k\ (c+\varpi))\end{aligned}$$

Here, the proposition contributed to the CG by the first conjunct is represented as

$$\varpi = \mathsf{not}\ (\mathsf{not}\ (\mathsf{exists}\ \lambda_x.(\mathsf{donkey}\ x)\ \mathsf{and}\ (\mathsf{walk}\ x)))$$

This proposition, along with the axiomatization of entailment for and and not in Equations 3, 4 and 5, together mean that the CG of the discourse context passed to the right disjunct entails the proposition $\mathsf{exists}\,\lambda_x.(\mathsf{donkey}\,x)\ \mathsf{and}\ (\mathsf{walk}\,x)$. This entailment therefore satisfies the requirement of the weak familiarity version of *it* that the CG must entail the existence of a nonhuman (with the assumption that any discourse context we would ever practically consider contains only nonhuman donkeys).

In view of this, (IT_w BRAY) in the right disjunct reduces as follows, where the conditions on the context are suppressed for readability since they are satisfied.

$$\begin{aligned}\vdash (\text{IT}_\text{w}\ \text{BRAY}) &= \lambda_{kc}.\mathsf{exists}\ \lambda_x.(\mathsf{nonhuman}\ x)\ \mathsf{and}\ (\text{IT}_\text{s}\ \text{BRAY}\ k\ \kappa)\\&= \lambda_{kc}.\mathsf{exists}\ \lambda_x.(\mathsf{nonhuman}\ x)\ \mathsf{and}\ (\text{BRAY}\ (\mathbf{def}\ \text{NONHUMAN}\ \kappa)\ k\ \kappa)\end{aligned}$$

Here $\kappa = c + \varpi :: x + (\mathsf{nonhuman}\ x)$ is the updated context produced by IT_w in the second conjunct, which in turn contains the proposition ϖ contributed by the first conjunct. Clearly, the conditions placed on the discourse context by IT_s are satisfied since the CG contains the information that the newly-introduced DR is nonhuman.

To demonstrate that this weak familiarity treatment extends to other definites besides pronouns, consider the following example, a simplification of (E).

(J) 1. Every man owns a donkey.
2. One man beats his donkey.

In (J), as in (E), the antecedent for *his donkey* is not overtly present in the discourse context, but is only inferable from entailments introduced by the first utterance.

Equation 17 shows an HDS analysis of the discourse in (J) that uses the weak familiarity variant of *his*.

$$\vdash \text{EVERY MAN } \lambda_j.\text{A DONKEY } \lambda_i.\text{OWN } i\ j \text{ AND A MAN } \lambda_j.\text{HIS}_\text{w} \text{ DONKEY } \lambda_i.\text{BEAT } i\ j \tag{17}$$

Starting with the analysis of the first utterance (J1) shows the entailment it introduces.

$$\begin{aligned}
&\vdash \text{EVERY MAN } \lambda_j.\text{A DONKEY } \lambda_i.\text{OWN } i\ j\\
&= \text{NOT (EXISTS } \lambda_n.(\text{MAN } n)\\
&\quad \text{AND (NOT (EXISTS } \lambda_m.(\text{DONKEY } m) \text{ AND OWN } m\ n))\\
&\equiv \lambda_{kc}(\mathsf{not}\ (\mathsf{exists}\ \lambda_x((\mathsf{man}\ x)\\
&\quad \mathsf{and}\ (\mathsf{not}\ (\mathsf{exists}\ \lambda_y((\mathsf{donkey}\ y)\ \mathsf{and}\ (\mathsf{own}\ y\ x))))))\ \mathsf{and}\ (k\ (c + \varpi))
\end{aligned}$$

Here, MAN = $(\mathbf{dyn}_1\ \mathsf{man})$ and the variable ϖ represents the modifications to the discourse context made by the utterance in (J1):

$$\varpi = \mathsf{not}\ (\mathsf{exists}\ \lambda_x.(\mathsf{man}\ x)\ \mathsf{and}\ (\mathsf{not}\ (\mathsf{exists}\ \lambda_y.(\mathsf{donkey}\ y)\ \mathsf{and}\ (\mathsf{own}\ y\ x))))$$

This modified context, which is passed to the second utterance, is crucial because it contains an entailment that for each man, there exists some donkey that man owns. It is this entailment which allows the use of the weak familiarity version HIS_w. Importantly, though the weak familiarity *his* is defined in the second utterance of (J), the strong version is not. This is because the discourse context $c + \varpi$ passed to (J2) does not contain a DR with the property of being a donkey owned by the uniquely most salient male. However, the existence of such an individual is entailed by the CG.

The analysis of (J2) is repeated in Equation 18.

$$\vdash \text{A MAN } \lambda_j.\text{HIS}_\text{w} \text{ DONKEY } \lambda_i.\text{BEAT } i\ j : \mathrm{u} \tag{18}$$

To show how the weak version of *his* allows the desired anaphoric reference, I start by reducing a subterm:

$$\begin{aligned}
&\vdash \lambda_j.\text{HIS}_\text{w} \text{ DONKEY } \lambda_i.\text{BEAT } i\ j\\
&= \lambda_{jkc}.\mathsf{exists}\ \lambda_y.(\mathsf{donkey}\ y)\ \mathsf{and}\ (\mathsf{poss}\ y\ [\mathbf{def}\ \text{MALE}\ c])\\
&\quad \mathsf{and}\ ((\text{HIS}_\text{s} \text{ DONKEY } \lambda_i.\text{BEAT } i\ j)\ k\ \kappa)\\
&= \lambda_{jkc}.\mathsf{exists}\ \lambda_y.(\mathsf{donkey}\ y)\ \mathsf{and}\ (\mathsf{poss}\ y\ [\mathbf{def}\ \text{MALE}\ c])\\
&\quad \mathsf{and}\ ((\text{BEAT } (\mathbf{def}\ \lambda_n(\text{DONKEY } n \text{ AND POSS } n\ (\mathbf{def}\ \text{MALE}\ \kappa))\ \kappa)\ j)\ k\ \kappa)
\end{aligned}$$

where $\kappa = c :: y +$ (donkey y) and (poss y [**def** MALE c]) represents the context as modified by HIS_w DONKEY, and the constraints placed on c by HIS_w are suppressed since they are satisfied. This reduction shows how the weak version of *his* interacts with the strong version: the DR j is required by HIS_s to beat the most salient donkey possessed by the most salient male, and HIS_w provides a context extended with an entity y that has exactly that property.

The reduction of the full term in Equation 18 is then as follows:

$$\vdash \text{A MAN } \lambda_j.\text{HIS}_\text{w} \text{ DONKEY } \lambda_i.\text{BEAT } i\, j$$
$$= \text{EXISTS } \lambda_n.(\text{MAN } n) \text{ AND } (\text{HIS}_\text{w} \text{ DONKEY } \lambda_i.\text{BEAT } i\, n)$$
$$= \lambda_{kc}.\mathsf{exists}\ \lambda_x.(\mathsf{man}\ x)\ \mathsf{and}\ \mathsf{exists}\ \lambda_y.(\mathsf{donkey}\ y)\ \mathsf{and}\ (\mathsf{poss}\ y\ x)\ \mathsf{and}\ (\mathsf{beat}\ y\ x)$$
$$\mathsf{and}\ (k\ (c + \varpi :: x + (\mathsf{man}\ x) :: y + (\mathsf{donkey}\ y)\ \mathsf{and}\ (\mathsf{poss}\ y\ x) + (\mathsf{beat}\ y\ x)))$$

Here, the proposition ϖ is the contribution to the CG made by the first utterance (as shown in the analysis of (J1), above) that permits the use of the weakly familiar version of *his*. Note that the first argument MAN to the dynamic indefinite A allows **def** in the second argument to select the most salient male DR in κ.

5.1 Overgeneration and Pragmatic Effects

Carl Pollard (personal communication) points out that the approach to weak familiarity I describe here seems to overgenerate. He gives (K) as an example.

(K) 1. Not every donkey brays.
 2. # It's brown.

This discourse is clearly odd, because the pronoun seems to lack an anaphoric antecedent. Yet the theory I have presented thus far licenses (K) because an entailment is present that permits the weak familiarity version of *it* to be used in analyzing (K2). To see why, note that the following analysis of (K1) is permitted in HDS with the extensions I propose:

$$\vdash \text{NOT (EVERY DONKEY BRAY)}$$
$$= \text{NOT (NOT (EXISTS } \lambda_n.(\text{DONKEY } n) \text{ AND (NOT (BRAY } n))))$$
$$\equiv \lambda_{kc}.(\mathsf{not}\ (\mathsf{not}\ (\mathsf{exists}\ \lambda_x.(\mathsf{donkey}\ x)\ \mathsf{and}\ (\mathsf{not}\ (\mathsf{bray}\ x))))) \ \mathsf{and}\ (k\ (c + \varpi))$$

Here, the updates made to the context are represented by

$$\varpi = \mathsf{not}\ (\mathsf{not}\ (\mathsf{exists}\ \lambda_x.(\mathsf{donkey}\ x)\ \mathsf{and}\ (\mathsf{not}\ (\mathsf{bray}\ x))))$$

Similarly as for the analysis of (I), above, this means that the resulting CG entails the proposition $\mathsf{exists}\,\lambda_x.(\mathsf{donkey}\,x)\ \mathsf{and}\ (\mathsf{not}\,(\mathsf{bray}\,x))$. It is this entailment that incorrectly allows the conditions imposed by IT_w to be met for (K).

By way of illuminating this seeming overgeneration, consider the difference between the bathroom example (D), repeated here, and the discourses in (L).

(D) Either there's no bathroom$_i$ in this house, or it$_i$'s in a funny place.

(L) 1. Either there is no seat$_i$ in this theater that isn't taken, or ?it$_i$'s in the front row.
2. Either there are no seats in this theater that aren't taken, or #it's in the front row.

The discourses in (D) and (L) are only mild variants of one another, yet (D) is perfectly felicitous, (L1) is somewhat odd, and (L2) is infelicitous. A similar class of examples is due to Barbara Partee:

(M) 1. I lost ten marbles and found only nine of them.
2. $\left\{\begin{array}{r}\text{The missing marble}\\ \text{?It}\end{array}\right\}$ is probably under the sofa.

In (M), the missing marble can be anaphorically referenced by a sufficiently descriptive definite NP. But the descriptively impoverished *it* does not seem to suffice.

In attempting to explain away the apparent overgeneration in (K) in light of the difference in judgments reflected in these discourses, an appeal could be made to the **informational uniqueness** of Roberts (2003). Such a move would involve arguing that (K) is infelicitous because weak familiarity alone is not enough, and that definite NPs also presuppose that their antecedents are unique among the DRs in the context that are contextually entailed to have the relevant descriptive content. Since, in the discourses in (K) and (L), it is impossible to tell whether the existential entailment only applies to a single weakly familiar DR, attempting to anaphorically reference the weakly entailed DR with a uniqueness-presupposing pronoun like *it* results in a presupposition failure. Example (M) is similar, except that there are multiple possible antecedents for *it* that are overtly (and not merely weakly) familiar. So in (K), there could be multiple non-braying donkeys; in (L), more than one seat could be available; and in (M), *it* is insufficient to pick out the marble that is probably under the sofa.

For cases like (M), in which overtly familiar DRs are present, HDS correctly requires that a candidate antecedent be informationally unique (see the axiomatization of **def** in Equation 14). Ascribing the infelicity in (K) and (L) to informational uniqueness in an analogous way seems promising, but it leaves open one obvious question: what about the original bathroom example (D)? It does not seem reasonable to assume for any house that either it does not have a bathroom or it has a unique bathroom that is in a funny place. The house could easily have multiple bathrooms, all situated in odd locales. Yet, as mentioned above, the discourse in (D) is completely felicitous. In fact, it would seem strange in the extreme to follow up (D) with the question *Which bathroom are you referring to?*, possibly because (D) does not seem to be about a specific bathroom, just one that might be locatable.

I would argue that such apparent counterexamples to the informational uniqueness requirement are due to pragmatic effects. In the case of (D), a kind of pragmatically conditioned informational uniqueness is likely responsible for the felicity of the use of *it*. It is straightforward to imagine a discourse context for (D) in which the interlocutors are not so much interested in whether the house in question has a unique bathroom, but whether there is one that is usually designated for guests to use that can be located. Such a pragmatic explanation would be unavailable for examples like (L), because none of the (possibly multiple) available seats is in any sense expected by convention. Likewise, for (K), there is no designated non-braying donkey that can be picked out from all of the possible non-brayers.

However, I stop short of building Roberts' informational uniqueness into the lexical meaning of the weakly familiar versions of *it* and *his*. It seems preferable for the semantics to generate readings for felicitous discourses like (D), even if it means licensing some infelicitous examples like (L). My argument for this is simply that it is the job of the semantic theory to generate readings, and that pragmatic effects are beyond its scope. Since examples like (D), in which a pronoun is used even when there is no informational uniqueness, may well be at least as common as the examples like (K) where the lack of informational uniqueness is problematic, it does not seem appropriate to forcefully exclude one class of examples or another.

6 Conclusions and Remaining Issues

The extension to hyperintensional dynamic semantics I present in this paper represents the first attempt I am aware of to implement Roberts' (2003) weak familiarity in a dynamic framework. The resulting formal model lays out a fragment that deals with problematic examples of anaphora across inaccessible domains in a way that only mildly extends Heim's (1982) familiarity condition on definites. Rather than resort to tactics like scope extension, E-type pronouns, or presupposition accommodation, this account allows all definites to be construed by two similar mechanisms: anaphoric links are licensed by entailments of the common ground, and an overt DR is only required to be present in certain cases. The apparent cases of overgeneration of this approach seem less like true overgeneration and more like instances of pragmatic effects.

One formal issue that remains is that the dynamic meanings posited for *it* and *his* seem very similar. Each has two cases, one of which requires an overtly accessible DR in the discourse context with a certain property, the other merely requires the existence of an entity with that property. Since both function so similarly, it seems desirable to find a way to unify and simplify their definitions that clarifies this deep similarity between them. Another topic for future work is to explain the apparent similarity between certain aspects of the approach described here and the tactic for modeling proper names via presupposition accommodation given by de Groote and Lebedeva (2010).

Finally, the account I give here should be expanded to deal with problematic examples of the kind pointed out by Groenendijk and Stokhof (1991, (46)).

(N) Every player chooses a pawn. He puts it on square one.

In cases like these, there is neither an overly accessible DR available to serve as the anaphoric antecedent of *he*, nor is the existence of an antecedent entailed by the CG. It seems that weak familiarity, as formulated here, cannot capture this instance of anaphora across an inaccessible domain any more than strong familiarity can.

References

Beaver, D.I.: Presupposition and Assertion in Dynamic Semantics. CSLI Publications (2001)

Chierchia, G.: Anaphora and dynamic binding. Linguistics and Philosophy 15, 111–183 (1992)

Chierchia, G.: The Dynamics of Meaning: Anaphora, Presupposition, and the Theory of Grammar. University of Chicago Press (1995)

Church, A.: A formulation of the simple theory of types. Journal of Symbolic Logic 5, 56–68 (1940)

Cooper, R.: The interpretation of pronouns. Syntax and Semantics 10, 61–92 (1979)

Evans, G.: Pronouns, quantifiers and relative clauses. Canadian Journal of Philosophy 7, 467–536 (1977)

Geurts, B.: Presuppositions and Pronouns. Current Research in the Semantics/Pragmatics Interface, vol. 3. Elsevier (1999)

Groenendijk, J., Stokhof, M.: Dynamic Montague grammar. In: Stokhof, M., Groenendijk, J., Beaver, D. (eds.) DYANA Report R2.2.A: Quantification and Anaphora I. Centre for Cognitive Science, University of Edinburgh (1991)

de Groote, P.: Towards a Montagovian account of dynamics. In: Proceedings of Semantics and Linguistic Theory, vol. 16 (2006)

de Groote, P.: Typing binding and anaphora: Dynamic contexts as $\lambda\mu$-terms. Presented at the ESSLLI Workshop on Symmetric Calculi and Ludics for Semantic Interpretation (2008)

de Groote, P., Lebedeva, E.: Presupposition accommodation as exception handling. In: Proceedings of SIGDIAL 2010: the 11th Annual Meeting of the Special Interest Group on Discourse and Dialogue (2010)

Heim, I.: The Semantics of Definite and Indefinite Noun Phrases. Ph.D. thesis. University of Massachusetts, Amherst (1982)

Heim, I.: File change semantics and the familiarity theory of definiteness. In: Meaning, Use and the Interpretation of Language. Walter de Gruyter, Berlin (1983)

Henkin, L.: Completeness in the theory of types. Journal of Symbolic Logic 15, 81–91 (1950)

Kamp, H.: A theory of truth and semantic representation. In: Groenendijk, J., Janssen, T., Stokhof, M. (eds.) Formal Methods in the Study of Language. Mathematisch Centrum, Amsterdam (1981)

Kamp, H., Reyle, U.: From Discourse to Logic. Kluwer Academic Publishers, Dordrecht (1993)

Kanazawa, M.: Weak vs. strong readings of donkey sentences and monotonicity inference in a dynamic setting. Linguistics and Philosophy 17(2), 109–158 (1994)

Lambek, J., Scott, P.: Introduction to Higher-Order Categorical Logic. Cambridge University Press (1986)

Lewis, D.: Scorekeeping in a language game. Journal of Philosophical Logic 8, 339–359 (1979)
Martin, S., Pollard, C.: Hyperintensional Dynamic Semantics: Analyzing Definiteness with Enriched Contexts. In: de Groote, P., Nederhof, M.-J. (eds.) Formal Grammar 2010/2011. LNCS, vol. 7395, pp. 114–129. Springer, Heidelberg (2012)
Martin, S., Pollard, C.: A higher-order theory of presupposition. Studia Logica Special Issue on Logic and Natural Language (to appear)
Montague, R.: The proper treatment of quantification in ordinary English. In: Hintikka, K., Moravcsik, J., Suppes, P. (eds.) Approaches to Natural Language, D. Reidel, Dordrecht (1973)
Muskens, R.: Categorial grammar and discourse representation theory. In: Proceedings of COLING (1994)
Muskens, R.: Combining Montague semantics and discourse representation theory. Linguistics and Philosophy 19, 143–186 (1996)
Pollard, C.: Hyperintensional Questions. In: Hodges, W., de Queiroz, R. (eds.) WoLLIC 2008. LNCS (LNAI), vol. 5110, pp. 272–285. Springer, Heidelberg (2008)
Pollard, C.: Hyperintensions. Journal of Logic and Computation 18(2), 257–282 (2008)
Roberts, C.: Modal subordination and pronominal anaphora in discourse. Linguistics and Philosophy 12, 683–721 (1989)
Roberts, C.: Uniqueness in definite noun phrases. Linguistics and Philosophy 26(3), 287–350 (2003)
Roberts, C.: Pronouns as definites. In: Reimer, M., Bezuidenhout, A. (eds.) Descriptions and Beyond. Oxford University Press (2004)
Rooth, M.: Noun phrase interpretation in Montague grammar, file change semantics, and situation semantics. In: Gärdenfors, P. (ed.) Generalized Quantifiers. Reidel, Dordrecht (1987)
van der Sandt, R.A.: Presupposition projection as anaphora resolution. Journal of Semantics 9, 333–377 (1992)
Stalnaker, R.: Assertion. Syntax and Semantics 9: Pragmatics, 315–332 (1978)

Author Index